COHERENCE AND ENERGY TRANSFER IN GLASSES

NATO CONFERENCE SERIES

I Ecology
II Systems Science
III Human Factors
IV Marine Sciences
V Air–Sea Interactions
VI Materials Science

VI MATERIALS SCIENCE

Volume 1 Molecular Metals
Edited by William E. Hatfield

Volume 2 Materials for Advanced Batteries
Edited by D. W. Murphy, J. Broadhead, and B. C. H. Steele

Volume 3 Adhesion in Cellulosic and Wood-Based Composites
Edited by John F. Oliver

Volume 4 Adhesion Problems in the Recycling of Concrete
Edited by Pieter C. Kreijger

Volume 5 Atomistics of Fracture
Edited by R. M. Latanision and J. R. Pickens

Volume 6 Electronic Structure and Properties of Hydrogen in Metals
Edited by P. Jena and C. B. Satterthwaite

Volume 7 Soot in Combustion Systems and Its Toxic Properties
Edited by J. Lahaye and G. Prado

Volume 8 Surface Modification and Alloying by Laser, Ion, and Electron Beams
Edited by J. M. Poate, G. Foti, and D. C. Jacobson

Volume 9 Coherence and Energy Transfer in Glasses
Edited by Paul A. Fleury and Brage Golding

COHERENCE AND ENERGY TRANSFER IN GLASSES

Edited by

Paul A. Fleury

and

Brage Golding

AT & T Bell Laboratories
Murray Hill, New Jersey

Published in cooperation with NATO Scientific Affairs Division

PLENUM PRESS · NEW YORK AND LONDON

Library of Congress Cataloging in Publication Data

NATO Workshop on Coherence and Energy Transfer in Glasses (1982: Clare College of Cambridge University) Coherence and energy transfer in glasses.

(NATO conference series. VI, Materials science; v. 9)

"Proceedings of the NATO Workshop on Coherence and Energy Transfer in Glasses, held at Clare College of Cambridge University, during September 1982, in Cambridge, England"—T.p. verso.

"Published in cooperation with NATO Scientific Affairs Division."

Includes bibliographical references and index.

1. Glass—Optical properties—Congresses. 2. Amorphous semiconductors—Congresses. 3. Energy transfer—Congresses. I. Fleury, Paul A. II. Golding, Brage. III. North Atlantic Treaty Organization. Scientific Affairs Division. IV. Title. V. Series.

TA450.N38 1982 620.1′4495 84-2160

ISBN 0-306-41638-7

Proceedings of the NATO Workshop on Coherence and Energy Transfer in Glasses, held at Clare College of Cambridge University, during September 1982, in Cambridge, England

A Division of Plenum Publishing Corporation
233 Spring Street, New York, N.Y. 10013

Printed in the United States of America

PREFACE

In recent years the physics of disordered systems has been one of the most active and fruitful areas of research in condensed matter science. In contrast to the considerable attention paid by conferences, schools and workshops to the static and structural aspects of glasses, there has been no forum devoted primarily to the dynamic and energetic aspects of amorphous solids. The NATO Workshop on Coherence and Energy Transfer in Glasses was organized to address this deficiency. The intent was to bring together in an intense and interactive environment, experts in several rather disparate subfields relating to the dynamics and energetics of disordered systems. This volume represents the Proceedings of that Workshop, which took place in September 1982 at Clare College of Cambridge University. Forty-three scientists from eight NATO countries participated. These included representatives from universities and industrial laboratories, as well as government research institutions. The meeting was organized into eight formal sessions and one informal session devoted entirely to unstructured discussion. Each formal session featured two comprehensive lectures. An additional 60 to 90 minutes was devoted in each session to discussions and contributions related to the lectures. Since only about 60% of the session time was devoted to formal presentations, the discussions formed an equally important part of the workshop. The chairmen and discussion leaders - as well as the workshop participants themselves - brought forth lively and illuminating discussions for each session. We have attempted to capture the essence and highlights of these in this volume, along with the lectures themselves.

The opening session was intended to establish common language and to review the basic amorphography and linear response of disordered systems. Professor A. C. Wright gave a masterful and comprehensive presentation on experimental techniques and related models for the structural aspects of amorphous materials. Both conventional oxide glasses and amorphous semiconductors were reviewed. Professor E. A. Davis next surveyed the experimental situation on frequency dependent linear response in amorphous materials ranging from transport and dielectric measurements through higher frequency microwave and infrared phenomena. His delineation of the different signatures in temperature and frequency dependence of various microscopic phenomena contributing to the responses formed an excellent basis for many of the workshop's subsequent sessions.

The second session was devoted to the low energy excitations uniquely associated with the amorphous state as probed by low temperature experiments. Thermal and acoustic phenomena in glasses below 1K were reviewed by Dr. M. von Schickfus. Coherent phenomena, including phonon, photon and electric field echoes in disordered materials were reviewed by Professor J. Joffrin. Dr. Golding led a comprehensive complementary discussion on related low temperature coherent phenomena. The presentations introduced the participants to many of the ideas of energy transfer in disordered systems and in particular to the tunneling systems which were to receive increased emphasis in later sessions.

Sessions 3 and 4 dealt with optical energy transfer in glasses and other solids. In Session 3, rare earth doped silicate glasses and related materials were reviewed from the point of view of a theorist by Professor D. L. Huber and by Professor W. Yen as an experimentalist. The ways in which optically active impurities in amorphous materials may be probed and may be influenced by their environment formed the main themes of these presentations. Energy propagation from one defect or impurity ion to another in amorphous media is currently a subject of some controversy. A number of competing theoretical models were compared with one another by Huber and with experimental observations by Yen. The temperature dependence of laser excited fluorescence spectra in a number of such systems remains an unresolved puzzle.

Session 4 was devoted to coherent optical processes and built nicely upon the earlier sessions. Dr. R. L. Brewer described his elegant experiments on optical free induction decay in rare earth doped crystals. His experiments definitively demonstrated the rich analogs between the optical spectra of probe ions in solids and the more familiar coherent spin dynamic studies traditionally done in magnetic systems. Observations of optical phenomena with typical frequencies in the kilohertz range demonstrate the remarkable power of coherent optical techniques which permit measurement accuracies approaching one part in 10^{12}. Although Dr. Brewer's talk involved crystals rather than glasses, in the discussion it became clear that a wealth of information and new phenomena await exploration in glassy systems using coherent optical techniques.

Some preliminary steps in that direction were described by Dr. MacFarlane who gave the second talk in the fourth session. MacFarlane's experiments on spectral hole burning in rare earth doped glasses revealed that homogeneous line widths of the impurity ions in the disordered environment could be obtained in the presence of severe inhomogeneous broadening. These early studies have already revealed a number of interesting puzzles and apparent differences regarding the manner in which energy relaxes following optical excitation of different rare earth ions and indeed different states of the same ion within the amorphous environment. Many formal analogies between these optical studies and the microwave studies at very low temperatures presented in Session 2 were elucidated in the discussion. Several opportunities for more experiments and the need for more comprehensive theories on energy transfer of both deliberately placed impurities as well as native defects in amorphous solids emerged.

Session 5 emphasized the theme of optical information transfer, storage, and processing in glasses. Professor S. D. Smith described experiments on optical bistability and optical logic devices, detailed the conditions and criteria for various optical switching processes and described simple devices which have been already operated. Here again little effort thus far has been devoted to optimizing such phenomena in amorphous or glassy materials. But the possibility that amorphous materials (particularly chalcogenide glasses) might offer advantages both in fabrication and efficiency over the crystalline materials studied to date formed the central topic of the discussion.

The second major application in optical information technology for glasses is their use as lightguide media. Dr. R. H. Stolen reviewed several nonlinear optical propagation effects in optical fibers including stimulated Raman scattering, four-wave mixing, self-phase modulation and the recently observed propagation of solitons. The subsequent discussion centered largely around solitons in terms of their potential application in real transmission systems and the differences between fiber solitons and those associated with optical and microwave self-induced transparency.

The workshop focus then shifted toward semiconducting glasses. Their optical properties were reviewed by Professor P. C. Taylor, and the transport phenomena associated with both intrinsic and extrinsic defects were reviewed by Professor A. Owen. These lectures also made contact with other important traditional focus of interest in amorphous solids - localization effects. The specific natures of certain defects in both the chalcogenide glasses and in tetrahedrally-coordinated semiconductors were examined in some detail by both speakers. Taylor emphasized the emerging importance of time-resolved optical fluorescence and luminescence experiments whereby the migration of energy among local environments could be followed in real time, the time scale being adjustable over several orders of magnitude by changing the temperature. In the discussion the relationship between the tunneling systems studied at low temperatures in chalcogenide glasses and the photo-induced defects observed previously by both optical and spin resonance techniques was clearly brought out.

Session 7 was devoted to defects in amorphous semiconductors beginning with Dr. J. C. Knight's lecture on hydrogenated amorphous silicon. He emphasized the dependence of observed properties upon preparative techniques and on incorporated defects, noting that defects have at least as profound an influence on the properties of amorphous silicon as they do in the case of crystalline silicon. Professor B. C. Cavenett then described the use of optically-detected magnetic resonance in the study of defects in amorphous materials. In addition to permitting selective pumping and selective observation of particular defect states of particular symmetries, this technique and its forthcoming time dependent modification should begin to give us insights on defects in glasses of a quality similar to those we now enjoy in crystalline materials.

The final formal session of the workshop was devoted to localization phenomena. Professor D. Weaire provided an elegant description of electron localization in one, two, and three dimensions. He compared several theoretical models employed to describe these, including scaling theories, noting that only in three dimensions do the latest theories predict a true Anderson transition between localized and delocalized electronic states. The language and concepts of critical phenomena flowed throughout this talk, with Professor Weaire suggesting that it makes sense to begin to talk about scaling phenomena and exponents for the disappearance of conductivity.

The final formal lecture was presented by Professor H. Gibbs. His elegant review described unsuccessful attempts to observe Anderson localization in the optically-accessible excitons in ruby, theoretical reasons for this failure, and the necessary attributes a system must have in order to exhibit the Anderson transition. Possible candidates for successful experiments were described.

Following the final lecture Dr. P. A. Fleury gave a brief conference summary in which some of the central points made throughout the week were reviewed. Attention was called to analogies between the microwave low temperature experiments on the one hand and the optical experiments on the other; to the potential for optical logic devices based in amorphous materials; to the intriguing similarities and yet important differences among solitons in resonant coherent systems on the one hand and in fiber-like extended media on the other. The observation that defects in amorphous solids must be given the same kind of care and attention traditionally accorded the crystalline materials (particularly in their influence upon both optical and electronic energy transfer and transport) was emphasized.

In closing, he proposed a broader view of disorder was proposed which drew some analogies between the evolution of chaos in the time regime, and the evolution of disorder in the space regime. For example, it was noted that the evolution of turbulence in a fluid can, in certain models, proceed via the nonlinear interaction of a very few incommensurate periodicities. This leads to the speculation that it might be possible to model spatial disorder at least in one or two dimensions by a compounding of incommensurate transitions, perhaps as few as three, beginning from the ordered crystalline state. The role of fluctuations in strongly nonequilibrium systems are encountered in the transition to turbulence in the time regime, offers some intriguing possibilities for exploration in the spatial regime.

The atmosphere and interactive modes achieved in this meeting were every bit as important as the formal presentations themselves. They were in no small measure aided by the ambience provided by the conference setting, Clare College, and by the close proximity in which the conferees lived, ate and worked. A special note of thanks is extended to Dr. W. A. Phillips of Cambridge, who in addition to helping organize the program, leading discussions and contributing ideas, did much to identify and secure the Clare College site and to oversee the local arrangements.

The workshop participants, drawn from at least four major subfields of glasses and disordered materials, from the first moment engaged in vigorous crosstalk and interactions. Indeed, every participant at some point in the conference remarked to the chairmen favorably upon the degree of interaction, level of discussion and general esprit de corps which developed throughout the week. In that regard the meeting was quite successful. The whole clearly exceeded the sum of the parts.

An especially memorable note was lent to the meeting by Sir Neville Mott who addressed the conference dinner on Thursday evening. His remarks, generally in a light vein, charged the conferees with redoubling their efforts to put the traditional long-standing competition between theorists and experimentalists into a renewed state of vigor.

In sum the workshop appeared to be quite successful. The chemistry among the participants was evident in all the discussions and interactions. The quality of the lectures was uniformly high as was that of the contributed remarks. The manuscripts and transcriptions of the discussions form the contents of this volume, which we hope will serve not only to document the proceedings, but to help nucleate a more interactive attack on the fascinating problem associated with Coherence and Energy Transfer in Glasses.

P. A. Fleury
B. Golding
Bell Laboratories
Murray Hill, NJ 07974 USA

CONTENTS

BASIC AMORPHOGRAPHY

Adrian C. Wright

J. J. Thomson Physical Laboratory
Whiteknights
Reading, RG6 2AF, U.K.

An account is presented of the quantification of amorphous network structures. Experimental techniques for the investigation of such structures are discussed and it is demonstrated that modern instrumentation is capable of providing highly accurate structural data. The greatest barrier to progress lies in the generation of structural models which can reproduce these data within their known experimental uncertainties.

I. INTRODUCTION

This introductory chapter will consider the quantification and experimental investigation of the structure of amorphous solids. As with crystalline materials, the major structural probes are X-ray and neutron diffraction. There is, however, an important difference between diffraction studies of crystalline solids and the corresponding investigation of amorphous materials. For a crystal the structure may be specified in terms of a unit cell and translational symmetry, the latter leading to characteristic sharp Bragg peaks in the diffraction pattern. Relatively few parameters are required to specify the atomic positions within the unit cell and, given a

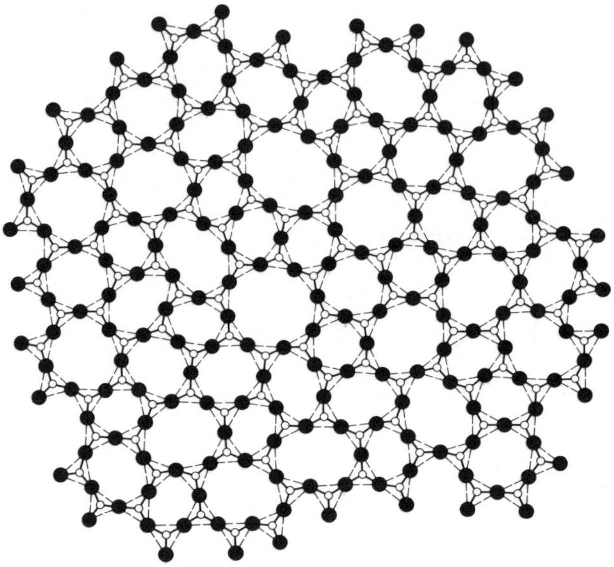

Fig. 1. Two dimensional A_2X_3 random network. ○A and ●X.

simple single crystal sample and good diffraction data over a reasonable region of reciprocal space, it is possible to determine structure absolutely. The same is not true for amorphous materials. The lack of periodicity, together with the fact that they are normally isotropic on a macroscopic scale, means the maximum that can be obtained from a diffraction experiment is a one-dimensional correlation function, from which the regeneration of the underlying three-dimensional structure can never be unique. It is for this reason that modelling plays such an important role in structural studies of amorphous solids and why the choice between possible models involves not only diffraction data but also results from other techniques such as EXAFS, vibrational and magnetic resonance spectroscopy.

The materials of interest to the present workshop include both conventional melt-quenched* glasses and other amorphous solids formed by techniques such as vapour deposition, sputtering etc. Most, however, have structures dominated by a network in which the bonding is predominantly covalent, and have been traditionally described in terms of the random network theory, which was introduced for oxide glasses by Zachariasen[1] and has subsequently been extended to many other classes of covalent amorphous

* The terms "vitreous" and "glass" will be reserved for materials covered by the standard A.S.T.M. definition of a glass as "an inorganic product of fusion which has cooled to a rigid condition without crystallizing".

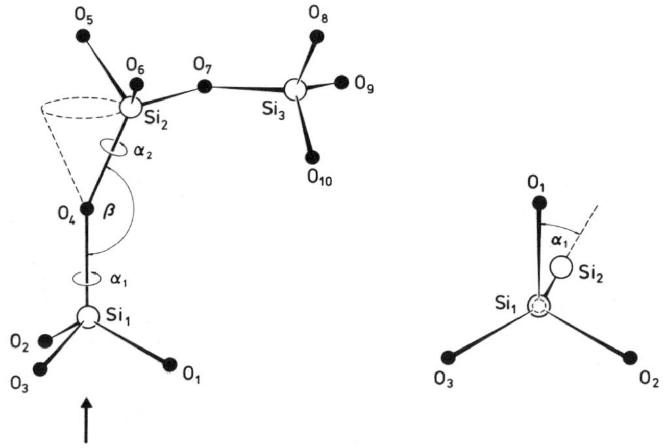

Fig. 2. Definition of the torsion angles α_1 and α_2 and the bond angle β for vitreous SiO_2.

solid. Alternative structural models have also been proposed and these are discussed in a later Section.

As originally formulated for oxide glasses, the random network hypothesis arose from a consideration of the corresponding crystalline materials. The atoms in a glass are linked by forces essentially the same as those in the crystal and, if the free energies of the glass and crystal are to be comparable, it is necessary that their structural units are similar. As in the crystal, extended three dimensional networks are formed and the atoms vibrate about definite equilibrium positions. The principal difference is that the glass lacks periodicity, symmetry and long range order as indicated schematically for a two dimensional glass-former A_2X_3 in Figure 1. With 4-connected units in three dimensions this figure would represent a possible structure for the AX_2 glass-formers such as SiO_2. The structural units in the glass are as regular as those in the crystal and, in a three dimensional structure, disorder is introduced via a distribution of the torsion angles α_1 and α_2 and the bond angle β as defined for vitreous SiO_2 in Figure 2.

The extension of Zachariasen's ideas to the amorphous group IV semiconductors was initially due to Polk[2] who constructed a mechanical model containing 440 atoms bonded together to form a perfect 4-connected random network (cf. Figure 3 which shows 3-connected units in two dimensions). Subsequently similar models have been built for other elemental amorphous solids and an even-membered ring model[3,4] to simulate the structure of the tetrahedrally bonded amorphous III-V semiconductors. It is unfortunate that the word "continuous" was appended to the term "random network" for just those materials where there is the greatest evidence of

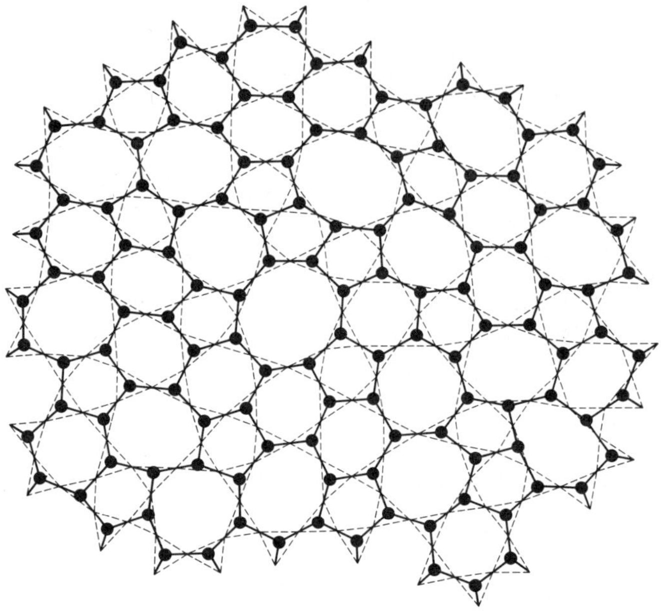

Fig. 3. Two dimensional 3-connected elemental random network.

discontinuities in the network in the form of broken bonds and/or terminal monovalent impurity atoms. Nevertheless, a perfect random network containing no internal broken bonds is a useful baseline from which to discuss their true structure.

Chalcogenide amorphous solids are in general structurally more complicated than their oxide counterparts in that even two element systems such as Ge–S exist over a wide range of stoichiometry and contain bonds between like atoms. Models for the numbers of each type of bond present range from the random covalent model[5,6] to complete chemical ordering. The former places no restriction on the bonds which may be formed, the number of each type of bond being determined solely from the covalency and concentration of each atomic species. A chemically ordered system, on the other hand, contains the maximum number of bonds between unlike atoms. In the vitreous state, the relevant bond strengths are a useful guide to the relative number of each type of bond likely to be formed, but in vapour deposited materials the balance may be significantly affected by the molecules present in the vapour phase (cf. Section VII E). Most elements in chalcogenide systems obey the 8-N rule although there are exceptions such as boron in vitreous B_2S_3 and perhaps pentavalent phosphorous.

The variation of structure with preparation conditions means that amorphous solids exhibit continuous polymorphism as opposed to the discrete polymorphism of crystalline materials. Similarly in most systems crystalline polymorphs only occur at discrete stoichiometries whereas amorphous solids are formed over a continuous range of stoichiometry. The polymorphism and stoichiometry of the crystalline state are constrained by the need for a unit cell containing a fixed number of atoms in well defined positions. Conversely any model of the amorphous state must be able to accommodate continuous changes in both structure and stoichiometry. The ability to do this in a simple straightforward way is one of the major successes of the random network theory.

A continuous random network is to the amorphous state what crystal structure is to crystalline solids - a perfect defect-free system. Any real material will of course contain a range of intrinsic and extrinsic defects, which will be considered in later chapters. As with treatises on crystallography, however, the present chapter on amorphography will concentrate on idealized structures, since by their nature diffraction methods are insensitive to small numbers of defects.

II. THE REAL SPACE CORRELATION FUNCTION

Although the concept of a random network is a very simple one, the absence of a unit cell means the resulting structure cannot be completely defined without specifying the co-ordinates of all the atoms present, which is clearly impossible for any real material. The isotropic nature of an amorphous solid leads naturally to a structural description in terms of a one-dimensional real space correlation function, the various forms of which are illustrated for a monatomic material in Figure 4. The radial density $\rho(r)$ describes on average the number density a distance r from an arbitrarily chosen origin atom and is a gross average over all the atoms in the sample taken at centre. ρ^0 is the average number density.

The number of neighbours between r and r + dr is given by

$$g(r)dr = 4\pi r^2 \rho(r)dr \tag{1}$$

in which g(r) is known as the radial distribution function. The correlation function

$$t(r) = 4\pi r \rho(r) \tag{2}$$

is best used in the analysis of diffraction data since it is in this function that experimental broadening is both symmetric and r-independent. The

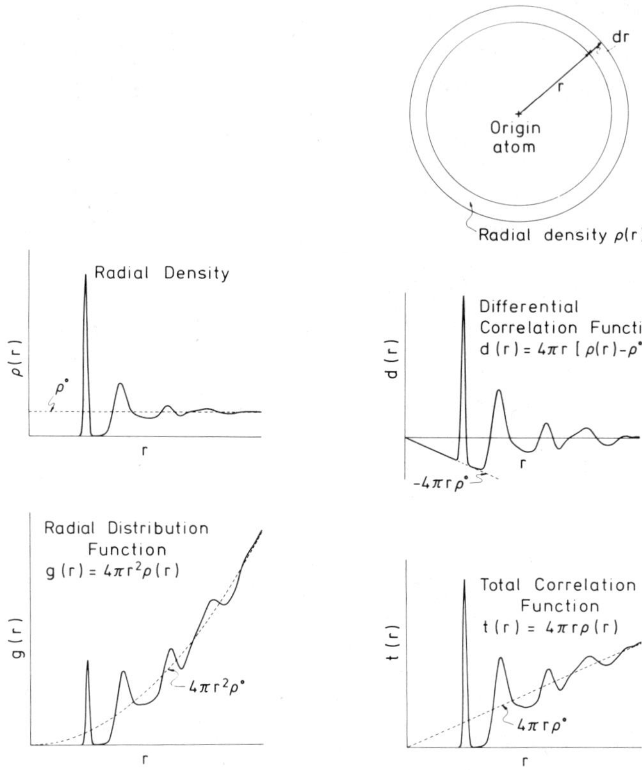

Fig. 4. Real space correlation functions for a monatomic material.

differential correlation function d(r) is obtained on subtracting the average density contribution $4\pi r\rho^{o}$ from t(r)

$$d(r) = 4\pi r\,[\rho(r) - \rho^{o}] \tag{3}$$

For a sample containing n elements the structure may be described in terms of n(n + 1)/2 independent component or partial correlation functions of the form

$$t_{jk}(r) = 4\pi r\rho_{jk}(r) \tag{4}$$

which describe the distribution of element k atoms about the jth atom in the sample composition unit. Any single X-ray or neutron diffraction experiment measures a linear combination of these components as detailed in the next Section.

III. X-RAY AND NEUTRON DIFFRACTION

A. OUTLINE THEORY

The corrected X-ray or neutron diffraction pattern takes the general form[7,8]

$$I(Q) = I^S(Q) + i(Q) \tag{5}$$

where Q is the scattering vector of magnitude $(4\pi/\lambda)\sin\theta$, λ being the incident wavelength and 2θ the scattering angle. $I^S(Q)$ is the self (independent) and $i(Q)$ the distinct (interference) scattering. The structural information contained within $i(Q)$ may be extracted by means of a Fourier transformation of the interference function $Qi(Q)$ as outlined for neutrons in Figure 5.

$$T(r) = T^o(r) + \frac{2}{\pi}\int_0^\infty Qi(Q)\, M(Q) \sin rQ\, dQ \tag{6}$$

$M(Q)$ is a modification function to allow for the fact that data can only be obtained for Q less than or equal to some maximum value Q_{max} and is zero for $Q>Q_{max}$. $T^o(r)$ is given by

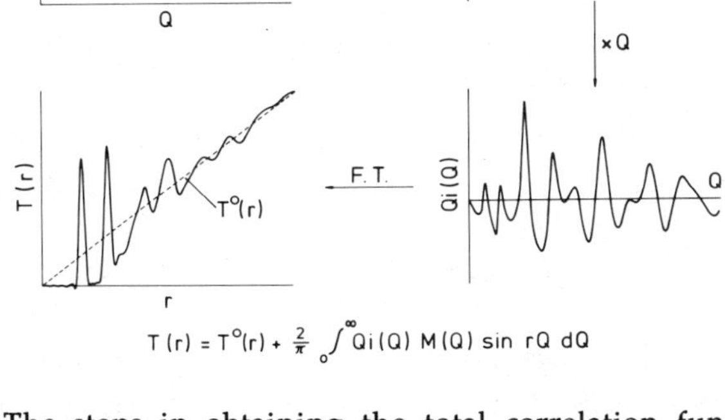

Fig. 5. The steps in obtaining the total correlation function T(r) from corrected neutron diffraction pattern I(Q). The data are for vitreous SiO_2.

$$T^o(r) = 4\pi r \rho^o \left[\sum_j \mathbf{b}_j\right]^2 \tag{7}$$

for neutrons and a similar expression for X-rays except that the atomic number Z replaces the neutron scattering length **b**. The j summation is taken over the atoms in one unit of composition.

In principle Qi(Q) and T(r) contain the same information expressed in a different form, although in practice the use of a modification function slightly reduces the information content of T(r). Both are one dimensional representations of a three dimensional structure and a gross average over the whole irradiated volume. The finite upper limit in Q is reflected in T(r) by the fact that each component $t_{jk}(r)$ is convolved with an appropriate peak function $P'_{jk}(r)$ such that

$$T(r) = \sum_j \sum_k t'_{jk}(r) \tag{8}$$

$$t'_{jk}(r) = \int_o^\infty t_{jk}(u) \left[P'_{jk}(r-u) - P'_{jk}(r+u)\right] du$$

The k summation is over atom types (elements), u is a dummy convolution variable and the prime indicates N or X for neutrons or X-rays respectively. The peak functions

$$P^N_{jk}(r) = \frac{\mathbf{b}_j \mathbf{b}_k}{\pi} \int_o^\infty M(Q) \cos rQ \, dQ \tag{9}$$

$$P^X_{jk}(r) = \frac{1}{\pi} \int_o^\infty \frac{f_j(Q) f_k(Q)}{f_e^2(Q)} M(Q) \cos rQ \, dQ$$

define the experimental resolution in real space, $f_j(Q)$ and $f_k(Q)$ being atomic scattering factors and $f_e(Q)$ the average form factor per electron for the sample in question. For neutrons the scattering lengths **b** are independent of Q and the factor $\mathbf{b}_j\mathbf{b}_k$ is a simple scaling factor, whereas for X-rays it is conventional to divide Qi(Q) by the "sharpening" function $f_e^2(Q)$ before Fourier transformation to yield the so-called atomic correlation function. Since the Q dependence of f(Q) is different for each element, the shape of $P^X_{jk}(r)$ changes for each independent component. A typical reduced neutron peak function ($\mathbf{b}_j = \mathbf{b}_k = 1$) is shown in Figure 6 appropriate to the Lorch[9] modification function

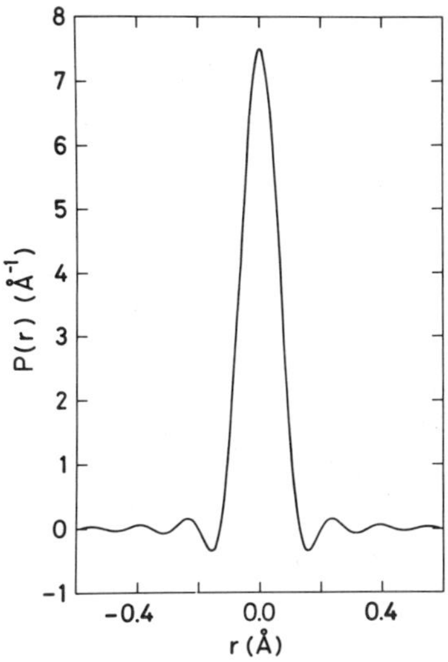

Fig. 6. Reduced neutron peak function $\left(b_j = b_k = 1;\ Q_{max} = 40\ \text{Å}^{-1}\right)$.

$$M(Q) = \sin(\pi Q/Q_{max})/(\pi Q/Q_{max}) \tag{10}$$

and $Q_{max} = 40\,\text{Å}^{-1}$. Note that the width of the central maximum is inversely proportional to $Q_{max}(\Delta r_{1/2} = 5.437/Q_{max})$ so that to obtain good real space resolution it is necessary to make measurements to high values of Q.

B. EXPERIMENTAL TECHNIQUES

As indicated in Eq. (5) an X-ray or neutron diffraction experiment involves a measurement of the scattered intensity as a function of the scattering vector Q. In order to achieve the necessary variation in Q it is possible to scan 2θ at a fixed incident wavelength (conventional technique) or to make measurements as a function of λ at constant scattering angle (dispersion or time-of-flight technique).

The two techniques for neutrons are compared in Figure 7. In the conventional steady state reactor twin-axis experiment the flux $\phi(\lambda)$ of thermal neutrons extracted from the moderator/reflector of a steady-state reactor has a Maxwellian distribution of velocities and is time independent. A beam of wavelength λ, selected by the monochromator crystal, is incident on the sample and scattered into the detector through a variable angle 2θ.

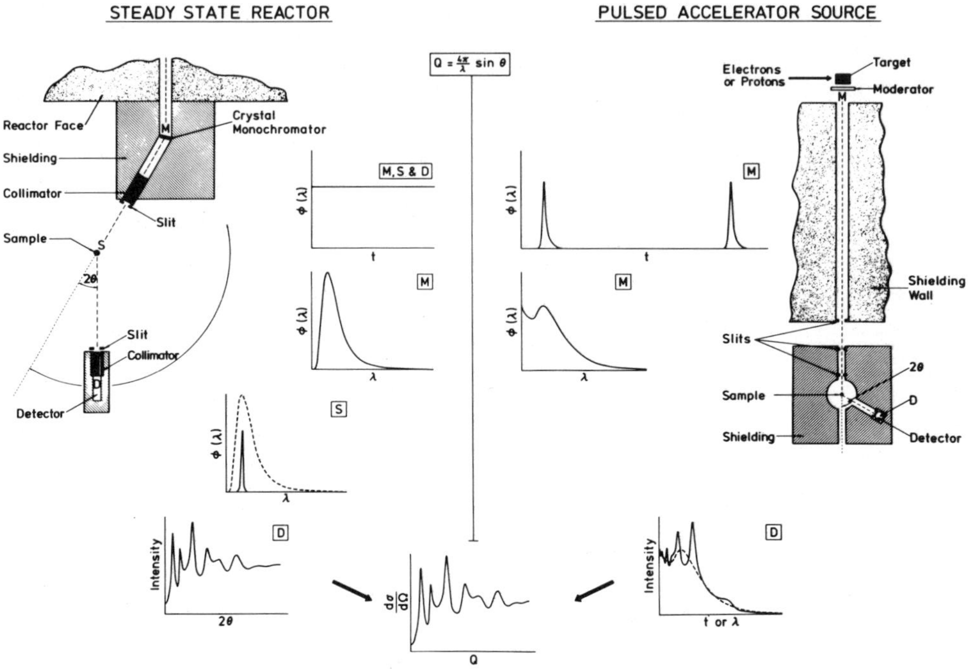

Fig. 7. A comparison of conventional and time-of-flight neutron diffraction techniques.

The variable λ time-of-flight technique is usually employed with a pulsed accelerator source. Electrons or protons strike a heavy metal target producing pulses of fast neutrons which are partially moderated before being incident on the sample and scattered into a detector situated at a fixed angle 2θ. The detector records the scattered intensity as a function of time-of-flight for the distance from the moderator via the sample to the detector. For any neutron the time-of-flight is simply related to λ through the velocity and De Broglie's relationship and the diffraction pattern I(Q) may be extracted by dividing the measured intensity by the incident neutron spectrum shape. The great advantage of a pulsed accelerator source over a steady-state reactor is that the former is under-moderated which gives rise to a strong epithermal component in $\phi(\lambda)$. These short wavelength neutrons allow data to be obtained to much higher values of Q resulting in a corresponding increase in real space resolution.

Accurate quantitative X-ray diffraction measurements on amorphous solids usually employ a modified powder counter diffractometer. The situation for X-rays is complicated by the presence of incoherent Compton scattering which at high values of Q can completely swamp the required coherent contribution. Since neither the wavelength distribution nor the integrated intensity of the Compton scattering from a given sample can be satisfactorily calculated it is necessary to remove this contribution by means of a diffracted beam monochromator. Many authors have used a single monochromator in the diffracted beam but the difficulty with this arrangement is that white background radiation from the X-ray tube can be Compton scattered into the monochromator envelope. A much better arrangement is that due to Warren and Mavel,[10] shown in Figure 8, which employs a monochromator in both the incident and diffracted beams. A conventional curved crystal monochromator is used in the incident beam (Ag K_{α_1}: $\lambda = 0.5594\,\text{Å}$), but that in the diffracted beam comprises a foil with an absorption edge at a wavelength slightly longer than the characteristic line of the X-ray tube (Ru K edge: $\lambda = 0.5605\,\text{Å}$) such that the coherent intensity will excite fluorescence whereas the Compton scattering will not. The fluorescent radiation is then recorded by the detector.

The dispersion technique is also possible with X-rays using a semiconductor detector. The advantage of this method is that for complex sample environments (e.g. cryostats, furnaces, high pressure cells etc.) only two or three small windows are needed, a fixed angle apart. Similarly, all values of Q are examined simultaneously making the technique ideally suited for following phase changes or studying kinetic effects. Outside such special applications, however, X-ray variable wavelength experiments are unlikely to

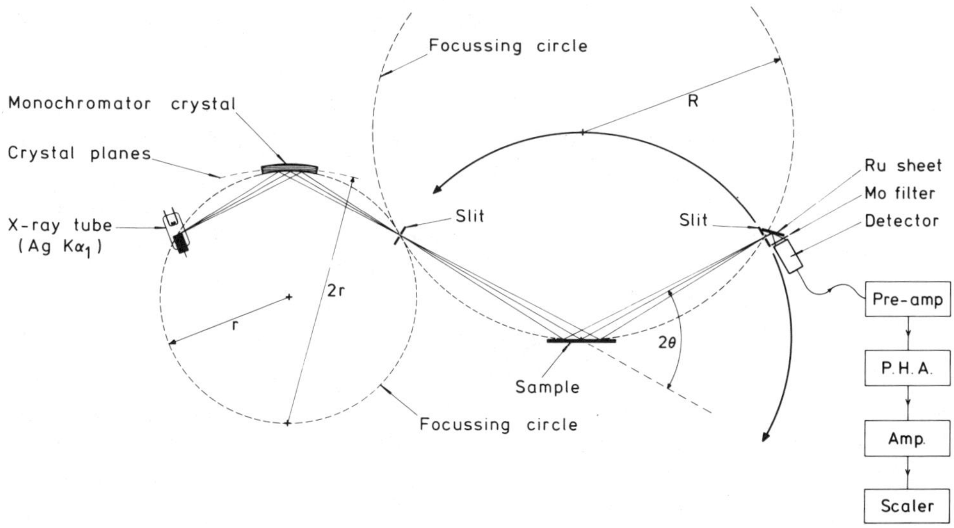

Fig. 8. Fluorescent foil X-ray diffraction technique.[10]

prove important for work on amorphous materials owing to the difficulty of accurately determining the incident spectrum and making corrections for absorption and Compton scattering.

Any measurement involves a certain number of corrections to the raw data but the good experiment minimizes these corrections or puts them in a form in which they are easily handled. In addition to the experimental background, corrections are normally included for absorption, self-shielding and multiple scattering. Extra corrections in the case of X-rays have to be made for polarization and residual Compton scattering not removed by the diffracted beam monochromator, while for neutrons it is necessary to take account of distortions due to departures from the static approximation (Placzek corrections).

The accuracy of experimental data can most easily be judged by the behaviour of the resulting transform at low r, below the first true peak. If T(r) is well behaved in this region then the data are of reasonable quality. Great care is needed in making this assessment, however, as some authors either plot the radial distribution function rT(r), which has the effect of reducing the amplitude of error ripples at low r, or use these false oscillations as a criterion for tidying their data before publication and do not include the original unadulterated transform. In the absence of other information such data must be treated with the utmost suspicion. Similar techniques are sometimes used to "remove" the effects of terminating the data at finite Q_{max}. Information theory, however, indicates that it is impossible to replace the unmeasured data without making some assumption about the material under investigation so that the resulting correlation function merely becomes one possible model which fits the results rather than an unbiased Fourier transform.

As indicated in Eq. (8), a single diffraction experiment on a multi-element sample yields only a weighted sum of the individual components $t'_{jk}(r)$. Various methods exist in the literature for experimentally determining the individual components of Qi(Q) and T(r) for a polyatomic system,[7,8] all of which involve a variation of f(Q) or **b** for one or more of the constituent elements. For both X-rays and neutrons it is possible to utilize the phenomenon of anomalous dispersion (the wavelength dependence of f(Q) or **b** near an absorption edge or resonance respectively) and for neutrons there are also the techniques of isotopic substitution (**b** changes for different isotopes of the same element) and magnetic diffraction. The successful use of each of these techniques is, however, limited to a relatively few favourable samples and in most cases the maximum that can be achieved is a combination of X-ray and neutron diffraction techniques for which the component weighting functions are different. (Note that the neutron scattering length **b** is not a smooth function of Z). An example of the use of a combination of X-ray and neutron data is afforded by the peak fit for vitreous SiO_2 discussed in Section V-A.

IV. EXAFS SPECTROSCOPY

Extended X-ray Absorption Fine Structure (EXAFS) measurements are rapidly gaining in popularity as a complimentary probe to diffraction studies. For a multi-element system a Fourier transformation of the EXAFS yields a linear combination of the same radial density functions $\rho_{jk}(r)$ as defined in Section II. There are, however, important differences between the two techniques.[11]

An EXAFS investigation at the edge of a given element j yields information only about the components involving that element, i.e. the sum over j is eliminated leaving only that over k and n components, rather than the $n(n + 1)/2$ obtained in a diffraction experiment. This leads to a considerable simplification of the resulting correlation function as does the fact that the peak functions for EXAFS spectroscopy vary greatly between components making peak fitting much easier (cf. Section V A). It should also be noted that the EXAFS signal is obtained directly, while to get the same information from diffraction measurements involves taking the difference between diffraction patterns, as **b** or f(Q) is varied, with a consequent degeneration in the signal-to-noise ratio.

The data range covered in a typical EXAFS experiment is equivalent to $5 \lesssim Q \lesssim 30{-}40$ Å^{-1}. The lack of data below ~ 5 Å^{-1} means that EXAFS is only sensitive to the sharp peaks at low r in the correlation function. It also causes problems in the determination of accurate co-ordination numbers. On the other hand Q_{max} is much higher than in many diffraction studies yielding better real space resolution.

Finally, EXAFS studies are much more sensitive for an element present at low concentration since the EXAFS signal comprises only components from that element, whereas the contribution from the same components to the total diffraction pattern is negligibly small.

V. THE EXTRACTION OF STRUCTURAL INFORMATION FROM EXPERIMENTAL DATA

The data available for an amorphous material may range from a single total correlation function T(r) to a complete determination of all the individual components $t'_{jk}(r)$. The objective is always the same, however, namely to extract the maximum structural information. Most amorphous solids contain one or more well defined structural units which give rise to relatively sharp peaks in the correlation function at low r. For such materials it is useful to distinguish between short range order, involving the configuration within a single structural unit and its connection to its immediate neighbours, and the intermediate range order or network topology.

Long range order is absent in amorphous materials but there may nevertheless be long range fluctuations in the average density ρ^o, due to phase separation, voids etc., which give rise to scattering at small Q and are beyond the scope of the present chapter. Even where there is no phase separation there are still long range thermodynamic fluctuations. For a glass these are linked to the isothermal compressibility of the melt at the fictive temperature and define the limit of I(Q) as $Q \rightarrow 0$[7,8].

A. SHORT RANGE ORDER

The advent of modern peak fitting techniques and the development of experimental methods for the isolation of individual component correlation functions[7,8] have greatly increased the short range order information which can be extracted from diffraction and EXAFS experiments. If it is assumed that the distribution of distances about a mean value r_{jk} is Gaussian in $t_{jk}(r)$ with a root mean square deviation $\left(\overline{u_{jk}^2}\right)^{1/2}$ then the resulting contribution to Qi(Q) is of the form[8]

$$Qi_{jk}(Q) = n_{jk}\overline{f_j(Q)}\,\overline{f_k(Q)}\ \frac{\sin r_{jk}Q}{r_{jk}}\ \exp\left(-\tfrac{1}{2}Q^2\overline{u_{jk}^2}\right) \tag{11}$$

in which n_{jk} is the number of k type atoms around atom j at this distance and $\overline{f(Q)}$ represents either f(Q) or **b** for X-rays or neutrons respectively. The corresponding real space correlation function, including the contribution from P(r), can be obtained by Fourier transforming Eq. (11) with the same value of Q_{max} and modification function as used experimentally. It is thus possible to obtain the parameters n_{jk}, r_{jk} and $\left(\overline{u_{jk}^2}\right)^{1/2}$ by performing a fit to the experimental data either in real space or to the experimental interference function at high Q, where only the sharpest real space peaks contribute.

An example of such a peak fit is shown for vitreous silica in Figure 9, the peak parameters being summarized in Table 1. In the case of silica, the neutron scattering length for oxygen is greater than that for silicon, with the result that the area under the first O-O peak is greater than that beneath the

TABLE 1

Peak Parameters for Vitreous SiO_2

j–k	n_{jk}	r_{jk}(Å)	$\left(\overline{u_{jk}^2}\right)^{1/2}$ (Å)
Si–O	3.91	1.610	0.050
O–O	6.00	2.632	0.089
Si–Si	4.00	3.080	0.100

first Si-Si contribution. The converse is true for X-rays where Si is the dominant scatterer ($f(Q) \propto Z$). Thus the fit to the neutron data of Figure 9 was made only to the first Si-O and O-O interactions; parameters for the Si-Si peak were taken from the X-ray work of Konnert and Karle.[12]

Peak fitting techniques may thus be employed to establish the identity of the underlying structural unit and to quantify such parameters as

(i) First neighbour co-ordination numbers, n_{jk}, and hence the relative contributions from the various bond types. e.g. A-A, A-X and X-X in an $A_{1-x}X_x$ system.

(ii) Bond lengths, r_{jk}, together with their r.m.s. variations, $\left(\overline{u_{jk}^2}\right)^{1/2}$. Note that, since the total diffraction technique is conventionally used to study amorphous solids, $\left(\overline{u_{jk}^2}\right)^{1/2}$ represents the variation in the instantaneous bond length and does not include effects due to long wavelength acoustic phonons where the atoms concerned vibrate together with no change in bond length.

(iii) Bond angle distributions $B(\beta)$.

(iv) Torsion (dihedral) angle distributions $A(\alpha)$.

The exponential factor in Eq. (11) means that only the sharpest peaks in T(r) contribute to the interference function at high Q. Figure 10 shows a fit to the reduced neutron interference function for amorphous Ge appropriate to the first peak in t(r).[13] (Since Ge is monatomic an X-ray or neutron diffraction measurement gives $t'(r)$ directly). The exponential factor

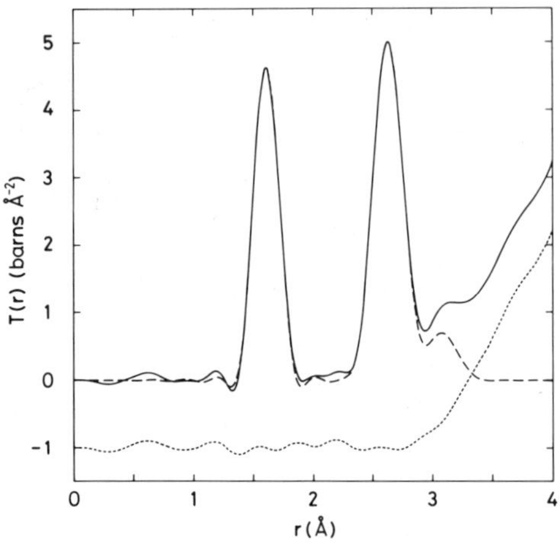

Fig. 9. Peak fit to the neutron correlation function for vitreous SiO_2. __________ Experiment, — — — — fit and .... residual.

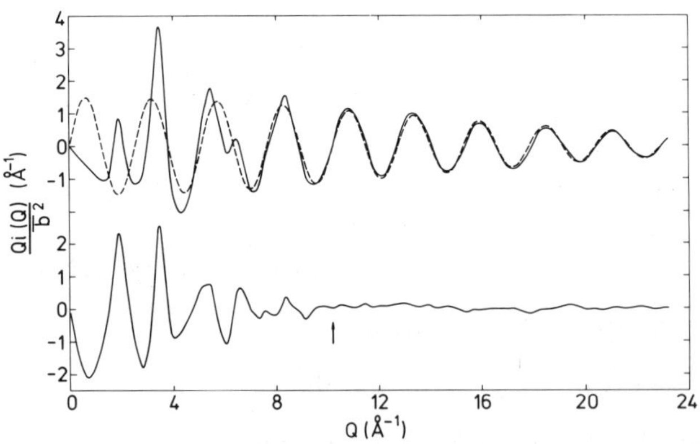

Fig. 10. The contribution of the first peak in the correlation function, t(r), to the reduced interference function, $Qi(Q)/\bar{b}^2$ for amorphous Ge. Upper curves: __________ experiment and — — — — contribution from first peak in t(r). Lower curve: residual. The arrow indicates the value of Q_{max} used in the subsequent Fourier transformation.[13]

corresponding to the second peak in t(r) has already fallen to ~0.02 by $Q = 10$ Å^{-1} and consequently, for $Q > 10$ Å^{-1}, Qi(Q) is almost solely determined by r_1. The residual interference function after subtracting the contribution from the first peak ($n_1 = 3.68$; $r_1 = 2.463$ Å and $\left(\overline{u_j^2}\right)^{1/2} = 0.074$ Å) is also shown in Figure 10 and it can be seen that the noise contribution dominates above 10 Å^{-1}. Since the residual interference function has effectively decayed to zero within the range of the experimental measurement, it is possible to Fourier transform these data without the use of a modification function. This was done with $Q_{max} = 10.2$ Å^{-1} and the appropriate correlation function is illustrated in Figure 11 together with the calculated first peak. It should however be stressed that the correlation function obtained by combining these two contributions is only one possible model fit to the experimental data in that it assumes the distribution of first neighbour distances to be Gaussian in t(r). This method of obtaining the best estimate of t(r) is entirely equivalent to the often used technique of artificially extending diffraction data to "eliminate" termination effects. Like the latter it is merely a fit to the experimental data, but does not try to conceal this fact by multiple Fourier transformation.

B. INTERMEDIATE RANGE ORDER : NETWORK TOPOLOGY

Recently there has been increasing interest in the so-called intermediate or medium range order which is closely associated with network topology. It is useful to introduce the concept of an underlying topological network[14,15] which can be decorated in various ways to represent different amorphous solids. The value of this approach is that it highlights the structural

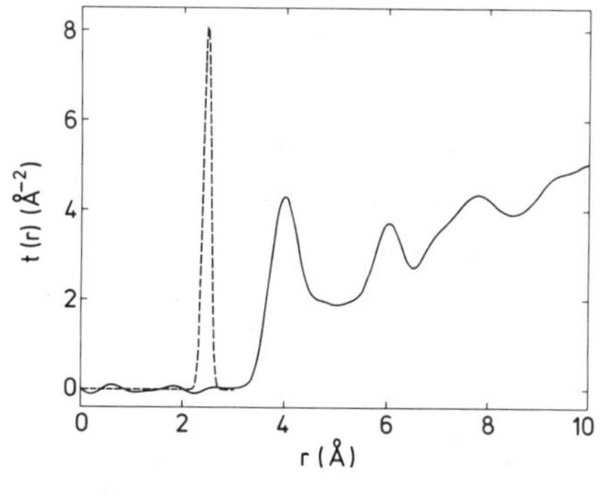

Fig. 11. The residual correlation function obtained from the residual interference function of Figure 10 (________), together with the best fitting first peak (— — — —).

relationships between the various types of amorphous solid and forms a basis for their systematic classification. It also finds practical application in the generation of new structural models from those which already exist in the literature as demonstrated in Section VI.

Figures 1 and 3 are in fact topologically equivalent, being simply different decoration modes of the same topological network shown schematically in Figure 12. Such a topological network may be completely specified either in terms of a symmetric connectivity matrix c_{pq} or a first neighbour table. As with the atomic co-ordinates, it is impossible to specify either c_{pq} or the near neighbour table for any real material and so it is necessary to work with a much simpler statistical average. The quantity usually employed is the ring statistics obtained by shortest path analysis, M(m), which have the advantage that the total number of rings per topological unit is well defined ($m(m-1)/2$ for an m-connected unit) and at large ring sizes tends to zero.[14] The shortest path ring statistics are obtained by taking every pair of vertices on each topological unit in turn and then finding the smallest sized ring, expressed as a number of topological units, in which they are both contained. The shortest path ring statistics for the topological network of Figure 12 are given

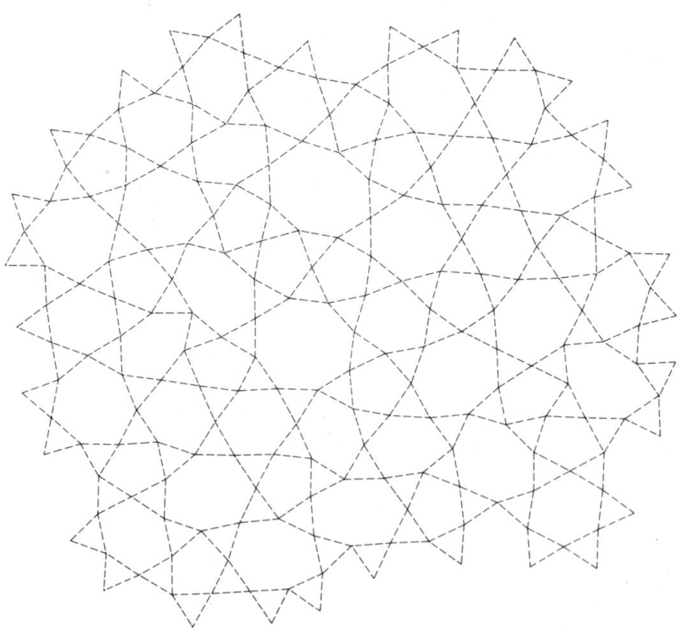

Fig. 12. Topological network appropriate to Figures 1 and 3. The topological units are represented by the dashed triangles.

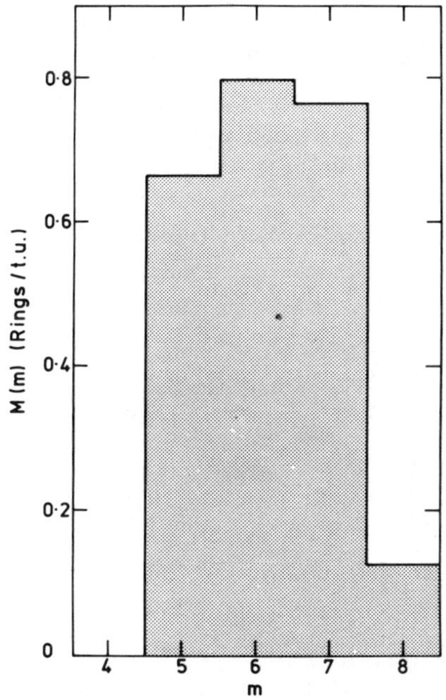

Fig. 13. Shortest path ring statistics for the topological network of Figure 12.

in Figure 13. If all the topological units in a finite sized model are taken as centre, the percentage ring closure is less than 100%, due to edge effects. The percentage closure for Figure 13 is 78%.

There is an inherent difference between the shortest path ring statistics for infinite two and three dimensional models. In two dimensions, if the shortest path ring is determined for two vertices on a given topological unit, this will also be the shortest path ring for the appropriate pair of vertices of all the other topological units involved. Hence it may be shown that the average ring sizes

$$\overline{m} = \frac{\sum_{m} M(m)}{\sum_{m} \frac{M(m)}{m}} \tag{12}$$

for 3 and 4 connected networks are 6 and 4 respectively.[16,17] In three dimensions the situation is more complicated as may be seen from Figure 14, which represents a plan view of part of a three dimensional 4-connected network. The topological units are represented by circles and their interconnections by straight lines. The shortest path ring for unit 1 is 7-membered and involves two vertices from unit 4. The same two vertices from unit 4, however, yield a shortest path 5-membered ring through units 8 and 9.

No method has yet been devised for directly measuring M(m) for an amorphous solid. Gaseous absorption studies provide information on ring statistics and as a result of such measurements Shackleford[18] has suggested that the distribution of ring sizes in a random network glass is log-normal. The greatest progress, however, has been made for small (m = 2 or 3) planar rings. A ring will tend to be planar if the sum of the equilibrium bond angles for all the atoms in the ring is greater than that dictated by the geometrical constraint that the sum of the interior angles of a planar n-sided polygon must equal $(n-2)\pi$. The most well known example of such a planar ring is the B_3O_6 boroxol group (Figure 15) found in vitreous B_2O_3 which gains additional stability in the planar configuration from aromatic π-bonding. A fit to the neutron correlation function for vitreous B_2O_3 is illustrated in Figure 16 for a fraction f = 0.6 of the boron atoms in boroxol rings.[19] Evidence for the existence of some well defined super structural unit is afforded by the peak in T(r) at 3.6 Å which is extremely sharp showing that the interatomic spacing concerned must either involve no unconstrained bond torsion angles or be insensitive to bond rotations. This distance corresponds to r_4 in Figure 15.

Raman scattering is particularly sensitive to the presence of planar rings. In chemically ordered stoichiometric oxides and chalcogenides, the ring

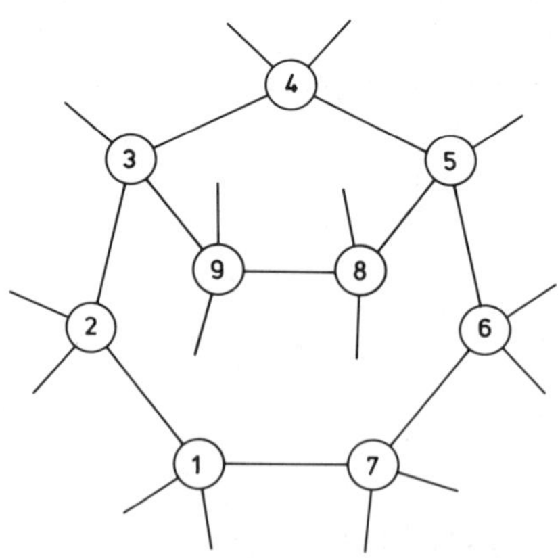

Fig. 14. Shortest path rings in three dimensions.

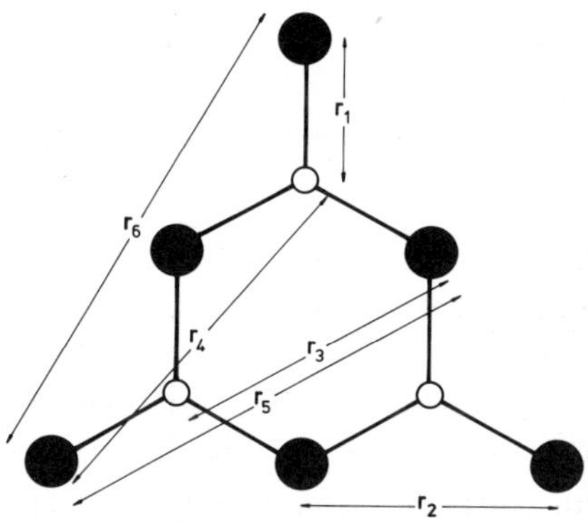

Fig. 15. B_3O_6 boroxol group. ○B and ●O.

breathing mode is dominated by motion of the divalent atoms and the frequency for a given material is mainly determined by the divalent atom bond angle.[20] Since this bond angle is well defined the breathing mode gives rise to a very sharp line in the Raman spectrum such as that found for the boroxol ring at 808 cm^{-1}. Using this technique Galeener and co-workers have identified planar rings in several simple glasses including SiO_2 (3-fold rings: $\bar{\nu} = 606$ cm^{-1})[20] and $GeSe_2$ (edge sharing $GeSe_4$ tetrahedra or 2-fold rings; $\bar{\nu} = 219$ cm^{-1}).[21] B_2S_3 is especially interesting in that the crystal (Figure 17)[22] contains no independent BS_3 triangles but only edge sharing triangles (2-membered rings) and three membered borsulphol groups. Evidence for both of these super structural units is found in the Raman spectrum of vitreous B_2S_3.[23].

VI. MODELLING TECHNIQUES

As noted in the introduction the limited nature of the experimental correlation function for an amorphous solid leads to an extensive use of modelling techniques in which an attempt is made to build a model of a "typical" region of a structure the co-ordinates of which can be completely specified and used in subsequent calculations. The difficulty in building such models is to make them large enough to predict the various structural parameters with sufficient statistical accuracy and to ensure that they are capable of effectively infinite extension. Initial models such as those of SiO_2[24–26] and amorphous Si/Ge[2–4,27] were hand built. The advantage of hand building is the feel it gives for a structure, the disadvantage is that it is tedious and difficult to vary parameters in a systematic way. The use of computer relaxation techniques, however, reduces the accuracy with which it is necessary to measure the atomic co-ordinates and also removes any anisotropy which might result from the effects of gravity on a large model. An alternative approach is to generate the model completely on a computer, the difficulty here being to produce structures which are both isotropic and homogeneous.

Structural transformation techniques may also be used to generate new models, the simplest being relaxation to a new set of equilibrium bond lengths and angles such as used by Davis et al[28] to produce models of amorphous antimony and phosphorus from an existing model of amorphous arsenic.[29] The relationship between the structures in Figures 1 and 3 has been discussed in the previous Section. Given one of them it is clearly possible to generate the other, by the addition or removal of the X atoms followed by

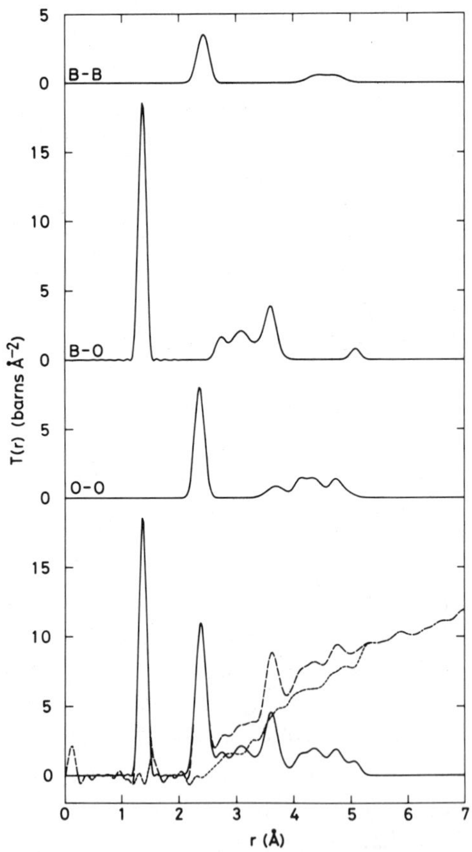

Fig. 16. Neutron correlation function for a boroxol ring model of vitreous B_2O_3 with f = 0.6.[19] __________ model, — — — — experiment and .... residual.

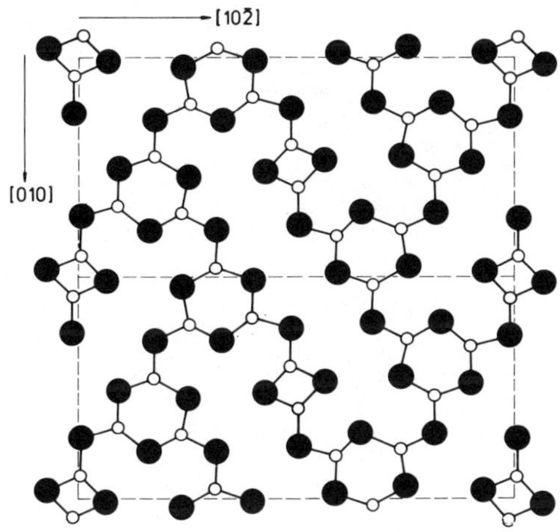

Fig. 17. Crystal structure of B_2S_3.[22] ○B and ●S.

energy relaxation. Such a transformation has been termed a decoration transformation[14,15] and is topologically invariant. Examples of decoration transformations in the literature include the As to B_2O_3 transformation of Elliott[30] and the conversion of the Evans and King[25] SiO_2 model to amorphous Si.[31,32]

Topological transformations are of two basic types. In a purely reconstructive transformation only the connectivity matrix c_{pq} is affected whereas a connectivity transformation involves a change in the connectivity of at least some of the topological units present. Two types of connectivity transformation are illustrated schematically in Figure 18, the disconnection transformation and the condensation/expansion transformation. Also shown is an example of the latter involving a topologically ordered 4:3 network. Although they are shown as complete transformations any of the above may of course be only partially carried out to yield mixed systems. Several examples of topological transformations have also been reported such as the computer restructuring of the Connell-Temkin[3,4] amorphous Si/Ge model by Beeman and Bobbs[33] (reconstructive transformation - c.f. Section VII H.) and the conversion of the Bell and Dean[26] SiO_2 model to B_2O_3-SiO_2[34] and B_2O_3.[35]

The average number density ρ^o is a very simple structural parameter, but nevertheless an extremely important one since any model must be capable of predicting the correct density. In the words of Finney,[36] "Models are wrong whose density is wrong." A difficulty in this respect is that the equilibrium bond length for a given material is frequently not known to better than 1% which is equivalent to a 3% error in the density. Uniformity of density may

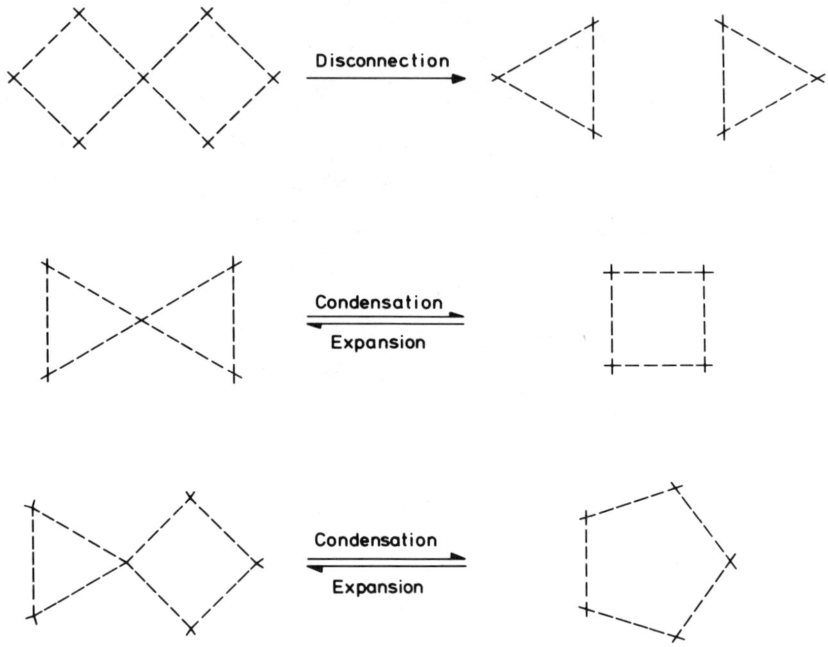

Fig. 18. Examples of connectivity transformations.

also be a problem, particularly in computer generated models, in that the average density ρ^0 is not maintained as extra units are added to an initial nucleus. A useful check[37] is to determine ρ^0 as a function of model size from the number of atoms inside a sphere of increasing diameter D, centred at the average co-ordinate $\bar{x}$, $\bar{y}$, $\bar{z}$. As D increases ρ^0 should approach a constant value and then fall off at the edge of the model. The point just before the density begins to fall off determines the largest possible spherical model for use in calculating the model component correlation functions. A spherical model is required to correct for finite model size, the model correlation functions being reduced relative to those for an infinite model by the factor[37]

$$F(r) = \left(\frac{r}{D} - 1\right)^2 \left(\frac{r}{2D} + 1\right) \qquad (r \leqslant D) \tag{13}$$

$$F(r) = 0 \qquad (r > D)$$

The functions Qi(Q) and T(r) emphasize different aspects of structure and hence it is important when comparing models with the results of diffraction studies to make such comparisons in both real and intensity space and in the former to correctly incorporate the peak functions $P'_{jk}(r)$, which as shown in Figure 6, comprise not only a central maximum at $r = 0$ but also satellite features on either side. These satellite features can significantly modify the fine structure in the correlation function. Similarly no comparison of peak widths can be made without including the effects of experimental resolution. Models which have been computer relaxed using a Keating[38] or similar potential represent an equilibrium structure with no thermal disorder (only static disorder). In such cases it is also necessary to include the effects of thermal broadening. Finally, no matter how good the agreement obtained with experiment, any given model is only one possible fit to the data and hence it is essential to consider, and hopefully eliminate, as many of the alternative models as possible.

VII. STRUCTURAL MODELS

A wide range of structural models have been reported in the literature. In many cases comparison with experiment is inadequate and few, if any, of the models so far reported reproduce the X-ray and neutron correlation functions of the material in question within the known uncertainties. The barrier to progress thus rests not so much with diffraction techniques, which at their best can produce an accurate, high resolution correlation function, but with the generation of more realistic structural models to fit such correlation functions. An idea of the problems involved can best be seen by considering some of the major classes of model in more detail.

A. RANDOM NETWORK MODELS

Random network models have been produced for a number of materials, as discussed in Section VI, a good example being those for amorphous Ge summarized in Table 2. The same models are compared with the neutron data of Etherington et al[44] in Figure 19. A difficulty with random network models is the incorporation of the correct bond angle distribution $B(\beta)$ and it can be seen that those on the right hand side of Figure 19 all contain too narrow a distribution of Ge–Ĝe–Ge angles as evidenced by the fact that the second peak in the model correlation function is too sharp. The r.m.s. bond angle deviations quoted in Table 2 for these models are all significantly less than the experimental value of 9.7°.[44] The narrow bond angle distribution is a consequence of the model topology and arises because not enough angular strain was introduced into the models when they were built. For similar reasons, all the models except that of Henderson[39] exhibit too much structure in the correlation function at high r.

TABLE 2

Random Network Models of Amorphous Ge

Model	Building Technique	No. of atoms	Reference	Diameter (Å)	Atoms used	ρ^o (atoms Å^{-3})	$\sqrt{\overline{u_1^2}}$ [a] (Å)	r_2 (Å)	$\sqrt{\overline{u_2^2}}$ [a] (Å)
Henderson	Hand (Periodic Boundary)	61	39	-	61	0.04497	0.093	3.98	0.35
Evans	Decoration Transformation	563	31,32	27.49	462	0.04247	0.065	4.00	0.28
Connell-Temkin	Hand	238	3,4	20.50	195	0.04323	0.030	4.00	0.28
Beeman	Reconstructive Transformation	238	33	20.00	187	0.04465	0.050	3.99	0.32
Polk	Computer	500	40,41	26.99	446	0.04335	0.059	4.01	0.18
Polk	Hand	519	27	28.00	492	0.04281	0.026	4.01	0.18
Steinhardt	Computer	201	42,43	20.94	190	0.03952	0.021	4.02	0.17

a) Static disorder only

b) Rings not closed

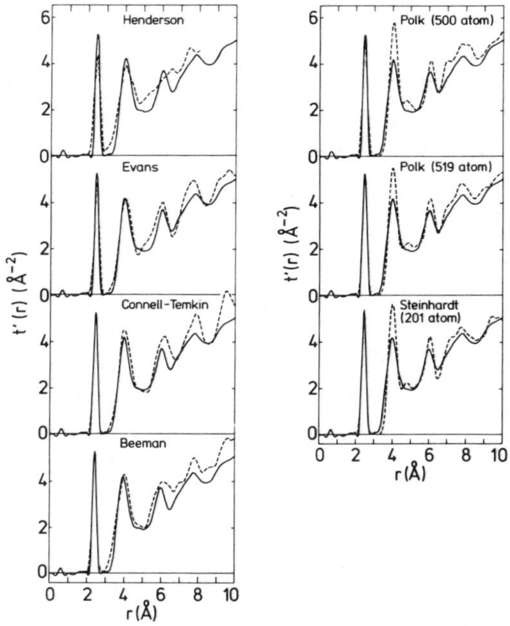

Fig. 19. Correlation functions for a series of random network models of amorphous Ge. ________ experiment $\left(Q_{max} = 23.2\ \text{Å}^{-1}\right)$ and — — — — model.[44]

The average density ρ^{o} for the model due to Steinhardt et al[42,43] is within 1% of the experimental value of 0.03975 atoms Å^{-3}. The other models have densities which are closer to that of crystalline Ge (0.04418 atoms Å^{-3}) which causes their correlation function to oscillate about a higher mean value. This is particularly apparent for $r \gtrsim 5$ Å. In this respect it should be noted that all the models comprise perfectly 4-connected random networks, containing no broken bonds within the body of the model, whereas in amorphous Ge the number of atoms surrounding each Ge atom is significantly less than 4.

B. RANDOM SPHERE PACKING MODELS

A random sphere packing is the topological inverse of a 4-connected random network and is more appropriate for materials whose structure is predominantly repulsion determined[45] (e.g. amorphous metals and ionic systems), rather than those containing well defined covalent bonding. An example of such a material is vitreous $ZnCl_2$, where it is the larger Cl^- ions which are approximately randomly close packed.[46] The Zn^{2+} ions occupy tetrahedral holes in such a way that the resulting $ZnCl_4^{2-}$ tetrahedra form a corner linked AX_2 random network. An alternative way of viewing the same structure is that it is the result of reducing the A–$\hat{X}$–A angles in a silica-like random network, allowing bond rotations, until the anions become approximately randomly close packed. The fact that vitreous $ZnCl_2$ is not purely ionic is revealed by the molecular dynamics simulations discussed below.

C. CRYSTAL BASED MODELS

As originally proposed the crystallite theory envisaged glass as an assembly of very small crystals termed crystallites, but it is now accepted that discrete crystallites do not exist in simple glasses. The modern crystallite theory[47] argues in terms of fluctuations in the degree of order, more highly ordered regions (crystallites), where the atomic arrangement approaches that in related crystalline materials, being interconnected by regions in which the degree of order is somewhat less. For multicomponent systems the crystallite theory leads to clustering and a non-uniform spatial distribution of the individual components.

Information on the structure of amorphous solids can often be obtained from a simple comparison with the corresponding crystalline polymorphs and for this reason the crystalline state is frequently a starting point for structural models. This does not necessarily imply an acceptance of the crystallite theory, merely the fact that the short range order may resemble that of the associated crystals. The great value of such models lies not in obtaining a perfect fit but in their use as a vehicle for comparing intermediate range order in the crystalline and amorphous solid states.

A convenient formalism is provided by the Quasi-Crystalline method[48] in which the component differential correlation functions for the crystal, $d_{jk}^{xtal}(r)$ are multiplied by F(r) from Eq. (13) to restrict the longer range order to interatomic spacings less than some characteristic correlation length D. The model is equivalent to embedding a crystallite of diameter D in a structureless matrix of the same average density (to eliminate small Q scattering) and then averaging over all relative orientations of the crystallite and the scattering vector Q. Thus

$$Qi(Q) = \sum_j \sum_k \overline{f_j(Q)}\ \overline{f_k(Q)} \int_o^D d_{jk}^{xtal}(r)\, F(r) \sin Qr\, dr \tag{14}$$

The model correlation function D(r) can be obtained by the usual Fourier transformation as was the case in Figure 20 which shows a Quasi-Crystalline

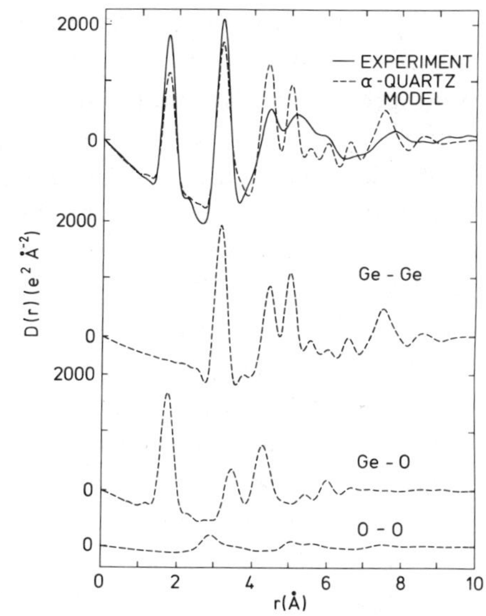

Fig. 20. Quasi-crystalline model for vitreous GeO_2 based on α-quartz GeO_2 with a correlation length of 10.5 Å. ________ model and — — — — experimental X-ray data.[49]

model for vitreous GeO_2 based on the α-quartz structure with D = 10.5 Å. The first Ge—Ge contribution for the glass is similar to that for the model, indicating a narrow distribution of Ge—O—Ge angles with a mean value close to that for the crystal (130.1°). Above r = 4 Å the structure in D(r) for the glass[49] is reduced and broadened, as a consequence of the distribution of torsion angles A(α), and is shifted to slightly higher r reflecting the lower density.

D. AMORPHOUS CLUSTER MODELS

Amorphous clusters comprise atoms in a regular but non-crystallographic configuration frequently based on pentagonal dodecahedra or fivefold symmetry. Cluster size is limited due to a slight mismatch between adjacent structural units, which leads to increasing distortion as the cluster grows. As with extreme crystallite models there exists the problem of the interconnecting regions between clusters (crystallites) which are likely to contain a significant number of defects and give rise to a large amount of small Q scattering. For this reason, amorphous cluster models are more appropriate for materials such as amorphous Ge, which exhibits small angle scattering, rather than simple melt-quenched glasses which do not.

The amorphous cluster polytetrahedral model of amorphous Ge due to Gaskell[50] was constructed using 14-atom structural units of tetrahedral shape and diamond cubic symmetry. The method of joining the tetrahedral modules was such that the resulting model is not space filling and exhibits a significant amount of distortion, particularly near its surface. In order to show the effect of strain correlation functions[15] for two spherical sections are given in Figure 21, one at the model centre containing 982 atoms and the other at the model surface containing 250 atoms. (The model comprises a total of 3094 atoms). It is obvious that $t'(r)$ for the central section exhibits far too much structure above the first peak. The agreement between the edge section and experiment is much better, but there are still serious discrepancies in the region of the second peak in $t'(r)$, particularly in respect of the high r shoulder at 4.6 Å. The correlation function for the total model would be intermediate between the two extremes shown and thus it can be concluded that the structure of amorphous Ge is much closer to the best random network models of Figure 19 than the polytetrahedral model.

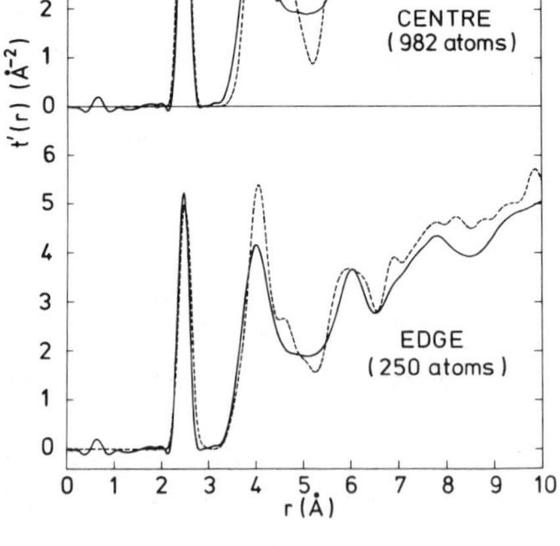

Fig. 21. Correlation functions for two spherical sections from the Gaskell[50] polytetrahedral model of amorphous Ge[15] situated at the model centre and close to its edge. ———— experiment and — — — — model.

E. MOLECULAR MODELS

The structure of an amorphous solid is a function of preparation route as shown, for example, by the fact that the atomic arrangement in glasses tends to reflect that of the melt from which they are quenched. The precursors of a vapour-deposited material are the molecular species in the vapour and thus it might be expected that the structure of these materials might show evidence of molecular fragments at least in the as-deposited state. A simple model for vapour-deposited materials is therefore a random packing of approximately spherical molecules. If there is no orientational correlation between adjacent molecules, the scattered intensity may be written[51]

$$I(Q) = \overline{F^2(Q)} + \overline{F(Q)}^2\, i_c(Q) \tag{15}$$

where F(Q) is the molecular form factor

$$F(Q) = \sum_i \overline{f_i(Q) e^{i\underset{\sim}{Q} \cdot \underset{\sim}{r}_i}} \tag{16}$$

and $i_c(Q)$ is the reduced distinct intensity for the molecular centres. The i summation is taken over the atoms in a molecule and r_i is the distance of the ith atom from the molecular centre. A convenient source of $i_c(Q)$ is the analytical expression for a Percus-Yevick hard sphere fluid which is available[52] as a function of packing density η. The packing density, effective molecular diameter, d, and the molecular density ρ_M^o are related by

$$\eta = \frac{\pi d^3 \rho_M^o}{6} \tag{17}$$

Obvious examples of vapour deposited amorphous solids include CCl_4[53] and amorphous ice.[54] Evidence for molecular fragments has also been found in more complicated materials such as arsenic sulphide,[51] but in this case the molecules have undergone at least partial polymerization. The first peaks in T(r) for vitreous ($As_{0.385}S_{0.615}$) and vapour-deposited ($As_{0.42}S_{0.58}$) arsenic sulphide are compared in Figure 22. The peak fits indicate that, while the bulk glass is chemically ordered, the vapour deposited film contains far more As—As bonds than the minimum required by stoichiometry as revealed by the high r shoulder. Moreover, the mean As—As distance (2.57 Å) exceeds the normal covalent bond length (2.49 Å) and is much nearer that (2.59 Å) found in the As_4S_4 I realgar molecule. Further evidence that realgar molecular fragments exist in vapour-deposited $As_{0.42}S_{0.58}$ is provided by

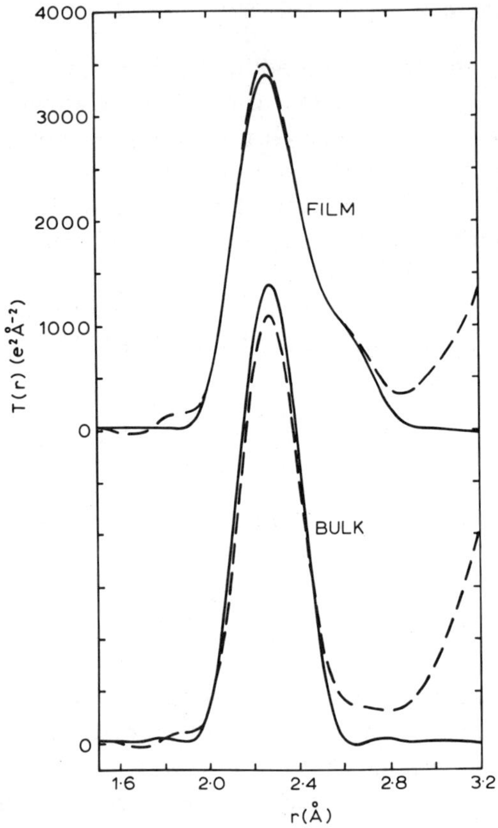

Fig. 22. A fit to the first peak in the X-ray correlation functions of bulk and thin film arsenic sulphide[51] ________ experiment and — — — — peak fit.

Figure 23, which shows random orientation molecular models for a whole series of arsenic sulphide molecules.[51] At high Q, Qi(Q) is dominated by intra-molecular contributions and in this region the model for As_4S_4 I is in best agreement with experiment. The differences between model and experiment at low Q are the result of polymerization and orientational effects arising from the fact that the As_4S_4 I molecule is far from spherical.

F. LAYER MODELS

Grigorovici[55] has associated threefold co-ordination with layer formation. Many workers have interpreted the first relatively sharp peak in the diffraction pattern for arsenic sulphide and other chalcogenide amorphous solids as confirming the presence of layers, but as may be seen from Figure 23 other explanations are equally possible. There is a wide variety of structural features which can give rise to a sharp first diffraction peak or prepeak, including layers, molecules, ionicity, chemical ordering etc. and thus it is impossible to associate the presence of such a peak with any one of them in particular. To the author's knowledge layer formation has not been

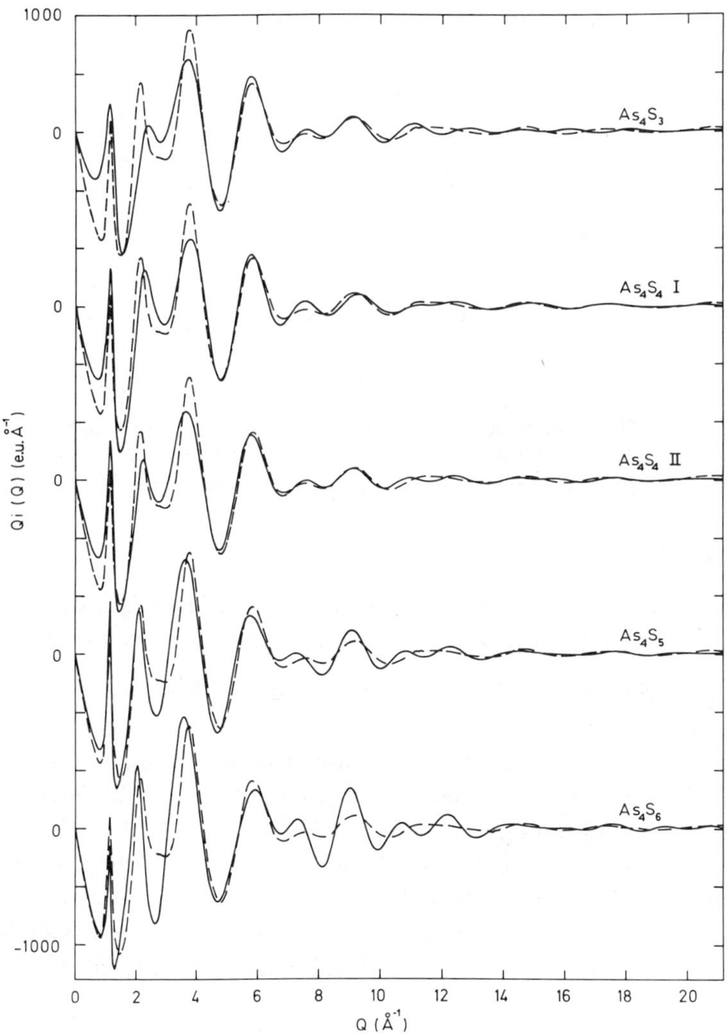

Fig. 23. X-ray interference functions for random orientation molecular models (________) of arsenic sulphide compared with data for $As_{0.42}S_{0.58}$ film (— — — —).[51]

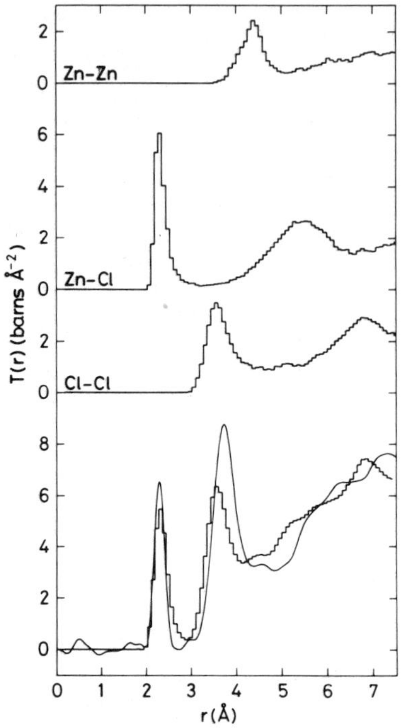

Fig. 24. The neutron correlation function for vitreous $ZnCl_2$ predicted by the Molecular Dynamics simulation of Angell and Cheeseman[57] (histogram) compared with experiment (_________).[46]

satisfactorily established in any amorphous solid and is usually suggested because the corresponding crystalline compound has a layer structure.

A recent variation on the layer theme is the "outrigger raft" model of vitreous $GeSe_2$ due to Phillips.[56] In this model a segment of the crystalline layer structure is terminated by Se—Se bonds which Phillips claims gives rise to the 219 cm^{-1} Raman line. However, as pointed out in Section V B, this line is associated with edge sharing $GeSe_4$ tetrahedra (2-membered rings) which are also present in both crystalline β—$GeSe_2$ and Phillips' model.[21] The crystal is chemically ordered and there is evidence that the same is true for the glass so that at stoichiometry it seems unnecessary to invoke the presence of Se—Se bonds which must be balanced by energetically unfavourable Ge—Ge bonds.

G. MONTE CARLO AND MOLECULAR DYNAMICS SIMULATIONS

Monte Carlo and molecular dynamics techniques have been employed to

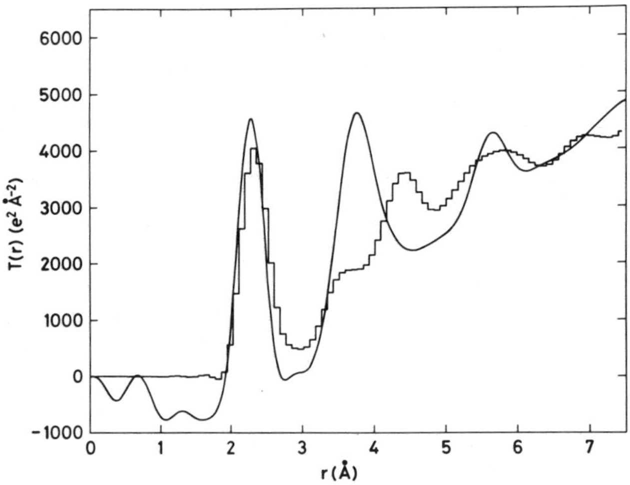

Fig. 25. The X-ray correlation function predicted for vitreous $ZnCl_2$ by the Molecular Dynamics simulation of Angell and Cheeseman[57] (histogram) compared with the X-ray data quoted by Shevchik[58] (_________).

simulate a range of simple inorganic glasses but, as with random network models, comparison with experiment is frequently inadequate in that it does not incorporate the peak functions $P'_{jk}(r)$. Given an appropriate cation:anion radius ratio there is no doubt that Monte Carlo and molecular dynamics simulations can arrive at the correct structural unit. What is not clear, however, is whether the purely ionic potentials used can accurately predict the intermediate range order.

Consider, for example, the case of vitreous $ZnCl_2$. $ZnCl_2$ exhibits a considerable degree of ionic character and would therefore appear to be an ideal candidate for computer simulation. The results of a molecular dynamics simulation of vitreous $ZnCl_2$ ($T = 300K$; $\rho^o = 1.198 \times 10^{-2}$ c.u. $Å^{-3}$)[57] are compared with experiment[46] in Figures 24 and 25. The first Zn—Cl peak is coincident with experiment and yields a co-ordination number close to 4.0. From the neutron correlation function (Figure 24), it is apparent that the Cl^- ion radius is too small causing a shift in the first Cl—Cl distance and subsequent features in T(r) to lower r than observed experimentally. The major discrepancy between the simulation and experiment, however, it much more apparent in the X-ray correlation function (Figure 25). The molecular dynamics simulation predicts a first Zn—Zn distance of ~4.4 Å whereas in order to account for the area under the second peak in both the neutron and X-ray correlation functions it is necessary to include Zn—Zn neighbours at

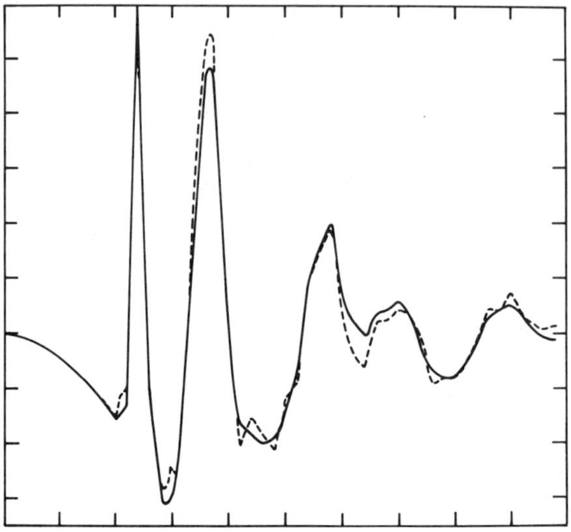

Fig. 26. Monte Carlo fit (________) to X-ray data for vitreous $As_{0.4}Se_{0.6}$ (— — — —).[59]

~3.7 Å. It can therefore be concluded that, while the molecular dynamics simulation is able to predict the correct structural unit it fails to account for the relative arrangement of the Zn^{2+} ions.

H. RANDOM DISPLACEMENT TECHNIQUES

Other related Monte Carlo techniques have been used in additional to conventional Monte Carlo simulations. These all involve random atomic displacements together with some algorithm to decide whether a particular move is to be allowed. Included in Figure 19 are data for the amorphous Ge model of Beeman and Bobbs[33] which was obtained by computer restructuring the even-membered ring Connell-Temkin[3–4] model to include odd membered rings.

Briefly the technique was to randomly displace atoms in the structure by up to $0.22\,r_1$, and then to adjust the near neighbour table entry for each atom if appropriate. The resulting structure was relaxed by minimizing bond length and bond angle distortions. In this way a sequence of models with an increasing number of odd-membered rings was generated by monitoring the ring statistics etc. of each new model and rejecting those which did not produce the desired change in structural characteristics. The correlation function shown in Figure 19 is for the model which Beeman and Bobbs state gives the closest fit to then available X-ray diffraction data.

A similar method has a criterion for accepting a given displacement that it should result in an improved fit to the experimental correlation function. Additional constraints include known co-ordination numbers and bond frequencies. The most general starting point is with the atoms in random positions, but it is also possible to use a related crystal structure or an already existing model. Good agreement with experiment is possible with this technique, as illustrated by the data for vitreous $As_{0.4}Se_{0.6}$ in Figure 26,[59] although the models concerned tend to be rather small ($\lesssim$ 100 atoms) and may contain several separate fragments.

VIII. CONCLUSION

From the foregoing paragraphs it can be seen that the current limitation in structural studies is the development of modelling techniques to the point where it is possible to accurately predict X-ray and neutron correlation functions. At their best diffraction techniques can provide high quality data, including in some cases information about the component correlation functions $t'_{jk}(r)$. Complementary information is also available from the various forms of spectroscopy (EXAFS, Raman, N.M.R. etc.) with the result that there is a wealth of structural data with which to compare models. It is essential, however, that such comparisons are carried out correctly, as discussed in Section VI, and similarly that diffraction data are compared with all the current models and not just a favoured few. Only in this way will it be possible to progress in what is undoubtedly an extremely complex field.

REFERENCES

1. W. H. Zachariasen, J. Am. Chem. Soc. **54**, 3841 (1932).
2. D. E. Polk, J. Non-Cryst. Solids **5**, 365 (1971).
3. G. A. N. Connell and R. J. Temkin, A. I. P. Conf. Proc. **20**, 192 (1974).
4. G. A. N. Connell and R. J. Temkin, Phys. Rev. **B9**, 5323 (1974).
5. F. Betts, A. Bienenstock and S. R. Ovshinsky, J. Non-Cryst. Solids **4**, 554 (1970).
6. F. Betts, A. Bienenstock, D. T. Keating and J. P. de Neufville, J. Non-Cryst. Solids **7**, 417 (1972).
7. A. C. Wright, Adv. Struct. Res. Diffr. Meth. **5**, 1 (1974).
8. A. C. Wright and A. J. Leadbetter, Phys. Chem. Glasses **17**, 122 (1976).
9. E. A. Lorch, J. Phys. C **2**, 229 (1969).
10. B. E. Warren and G. Mavel, Rev. Sci. Instr. **36** 196 (1965).
11. T. M. Hayes and A. C. Wright, "The Structure of Non-Crystalline Materials II" Eds. P. H. Gaskell, E. A. Davis and J. M. Parker (Taylor and Francis, London, 1983), in press.
12. J. H. Konnert and J. Karle, Acta Crystallogr. **A29**, 702 (1973).
13. G. Etherington, A. C. Wright, J. T. Wenzel, J. C. Dore, J. H. Clarke and R. N. Sinclair, J. Non-Cryst. Solids **48**, 265 (1982).
14. A. C. Wright, G. A. N. Connell and J. W. Allen, J. Non-Cryst. Solids **42**, 69 (1980).
15. A. C. Wright, G. Etherington, J. A. E. Desa, R. N. Sinclair, G. A. N. Connell and J. C. Mikkelsen Jr. J. Non-Cryst. Solids **49**, 63 (1982).
16. A. R. Cooper, Phys. Chem. Glasses **19**, 60 (1978).

17. A. R. Cooper, "Borate Glasses" (Mat. Sci. Res. **12**) Eds. L. D. Pye, V. D. Fréchette and N. J. Kreidl (Plenum Press, New York, 1978), 167.
18. J. F. Shackelford and B. D. Brown, J. Non-Cryst. Solids **44**, 379 (1981).
19. P. A. V. Johnson, A. C. Wright and R. N. Sinclair, J. Non-Cryst. Solids **50**, 281 (1982).
20. F. L. Galeener, J. Non-Cryst. Solids **49**, 53 (1982).
21. R. J. Nemanich, F. L. Galeener, J. C. Mikkelsen Jr., G. A. N. Connell, G. Etherington, A. C. Wright and R. N. Sinclair, Proc. 16th Int. Conf. on the Physics of Semiconductors, Montpellier, Sept. 1982.
22. H. Diercks and B. Krebs, Angew. Chem. **89**, 327 (1977).
23. A. E. Geissberger and F. L. Galeener, "The Structure of Non-Crystalline Materials II" Eds. P. H. Gaskell, E. A. Davis and J. M. Parker (Taylor and Francis, London, 1983), in press.
24. R. J. Bell and P. Dean, Nature **212**, 1354 (1966).
25. D. L. Evans and S. V. King, Nature **212**, 1353 (1966).
26. R. J. Bell and P. Dean, Phil. Mag. **25**, 1381 (1972).
27. D. E. Polk and D. S. Boudreaux, Phys. Rev. Lett. **31**, 92 (1973).
28. E. A. Davis, S. R. Elliott, G. N. Greaves and D. P. Jones, "The Structure of Non-Crystalline Materials" Ed. P. H. Gaskell (Taylor and Francis, London, 1977), 205.
29. G. N. Greaves and E. A. Davis, Phil. Mag. **29**, 1201 (1974).
30. S. R. Elliott, Phil. Mag. **B37**, 435 (1978).
31. D. L. Evans, M. P. Teter and N. F. Borrelli, A. I. P. Conf. Proc. **20**, 218 (1974).
32. D. L. Evans, M. P. Teter and N. F. Borrelli, J. Non-Cryst. Solids **17**, 245 (1975).
33. D. Beeman and B. L. Bobbs, Phys. Rev. **B12**, 1399 (1975).
34. R. J. Bell, A. Carnevale, C. R. Kurkjian and G. E. Peterson, J. Non-Cryst. Solids, **35 & 36**, 1185 (1980).
35. R. J. Bell and A. Carnevale, Phil. Mag. **B43**, 389 (1981).
36. J. L. Finney, "Diffraction Studies on Non-Crystalline Substances" Eds. I. Hargittai and W. J. Orville-Thomas Elsevier, Amsterdam, 1981, 439.
37. G. Mason, Nature **217**, 733 (1968).
38. P. N. Keating, Phys. Rev. **145**, 637 (1966).
39. D. Henderson, J. Non-Cryst. Solids **16**, 317 (1974).
40. D. S. Boudreaux, D. E. Polk and M. G. Duffy, A.I.P. Conf. Proc. **20**, 206 (1974).
41. M. G. Duffy, D. S. Boudreaux and D. E. Polk, J. Non-Cryst. Solids **15**, 435 (1974).
42. R. Alben, P. Steinhardt and D. Weaire, A. I. P. Conf. Proc. **20**, 213 (1974).
43. P. Steinhardt, R. Alben and D. Weaire, J. Non-Cryst. Solids **15**, 199 (1974).
44. G. Etherington, A. C. Wright, J. T. Wenzel, J. C. Dore, J. H. Clarke and R. N. Sinclair, J. Non-Cryst. Solids **48**, 265 (1982).
45. R. Zallen, "Fluctuation Phenomena" Eds E. W. Montroll and J. L. Lebowitz (North-Holland, Amsterdam, 1979), 177.
46. J. A. E. Desa, A. C. Wright, J. Wong and R. N. Sinclair, J. Non-Cryst. Solids **51**, 57 (1982).
47. E. A. Porai-Koshits, "The Structure of Glass" (Consultants Bureau, New York, 1958), p. 25.
48. A. J. Leadbetter and A. C. Wright, J. Non-Cryst. Solids **7**, 23 (1972).
49. A. J. Leadbetter and A. C. Wright, J. Non-Cryst. Solids **7**, 37 (1972).
50. P. H. Gaskell, Phil. Mag. **32**, 211 (1975).
51. M. F. Daniel, A. J. Leadbetter, A. C. Wright and R. N. Sinclair, J. Non-Cryst. Solids **32**, 271 (1979).
52. A. C. Wright, Disc. Faraday Soc. **50**, 111 (1970).
53. M. R. Chowdhury and J. C. Dore, J. Non-Cryst. Solids **46**, 343 (1981).
54. J. T. Wenzel, C. U. Linderstrøm-Lang and S. A. Rice, Science **187**, 428 (1975).
55. R. Grigorovici, "Electronic and Structural Properties of Amorphous Semiconductors" eds, P. G. Le Comber and J. Mort (Academic Press, London, 1973), 191.
56. J. C. Phillips, J. Non-Cryst. Solids **43**, 37 (1981).
57. C. A. Angell and P. Cheeseman, private communication (1980).
58. N. J. Shevchik, O. N. R. Rept HP-29 (ARPA-44) (1972).
59. A. L. Renninger, M. D. Rechtin and B. L. Averbach, J. Non-Cryst. Solids **16**, 1 (1974).

DISCUSSION

Glass Structure

The relationship between different descriptions of structural models was explored. Models are specified in different ways depending on the experimental results to which the models are being compared. For example, ring statistics are important in connection with electronic energy states, whereas the distribution of bond angles and lengths determine correlation functions. The use of a well defined set of known force constants might allow these distributions to be derived from ring statistics by minimizing the energy. It was felt that in four-fold, coordinated structures this was possible, although the real structure might not be in the minimum energy state corresponding to the topology.

A second point of discussion concerned the use of the word "defect". It was pointed out that the idea of a five membered ring in, for example, amorphous germanium was a defect in the sense that it does not exist in the crystal, but must be distinguished from a defect such as a dangling bond which disrupts the fully coordinated network. The former plays a limited role in determining electrical transport, although might be important in an explanation of the Hall effect, in contrast to a network breaking defect which is very important here. A line defect or disclination does not introduce gap states.

There is evidence from diffraction experiments that appreciable deviations from an ideal CRN (continuous random network) exist. For example, the area under the first peak of the radial distribution function (rdf) in a-Ge yields an average nearest-neighbor coordination number of about 3.7 instead of 4. This could be modeled, given some idea of the defect's structure, but the size of the model would undoubtedly have to be larger than those presently used due to the relatively low defect densities.

A related point is the use of the CRN for substances which are stable only as thin films (e.g., the tetrahedrals), where internal strain energies are large and appreciable atomic relaxation cannot apparently occur. It is probable that fine tuning of the models (incorporation of defects, strain) will allow fine discrepancies between theory and experiment to vanish, as pointed out in the comment below by D. Weaire.

B. Golding, Chairman
W. A. Phillips, Discussion Leader

COMMENTS

Structural Models

In discussing diffraction data for amorphous solids, some people say: it can be fitted with a wide range of models. Others say: *no* model fits the data to within experimental accuracy. How can one reconcile such statements? In the case of a-Si, the situation is as follows: the *qualitative agreement of random network models* with the experimental rdf is quite persuasive of their general validity, and seems to favor certain specific (and very similar) realizations of the random network, such as of Polk and Boudreaux and of Steinhardt *et al.* It is true that there remains some discrepancy, particularly as regards the breadth of the second nearest neighbor peak which is related to the breadth of the bond angle distribution. It can be argued that this is merely due to interatomic interactions beyond nearest neighbors not included in the theoretical model. Admittedly, this would be more convincing if it were directly demonstrated, but to do so would require rather elaborate calculations. Incorporating long-range forces into finite cluster calculations raises obvious difficulties. Up to now it has not been considered worthwhile since, if it merely closed the small remaining gap between theory and experiment, nothing new would be learned. On the other hand, we *might* get a surprise.

D. Weaire
University College, Dublin
Dublin, Ireland

Glass Structure

What are the structural constituents of glasses? In addition to the basic units of short range order mentioned by Adrian Wright, one can identify linear objects of an unusual kind.[1] They can be visualized (dotted line in Figure 1, taken from the frontispiece of Mott's and Davis' book) in continuous random networks (CRN) as imaginary, continuous lines, threading through odd-membered rings only, and closing up as loops or terminating on the surface of the material. There is one line per irreducible (shortest path) odd ring, and none per irreducible even ring. If one considers the glass as an elastic continuum, these linear objects are disclinations (rotation dislocations), characterized by their existence (oddness) rather then by their intensity. To call them "defects" may be misleading, since they are almost universal ingredients of disordered condensed matter. They are absent in CRNs containing exclusively even rings, like the Connell-Temkin model. Physically, their existence manifests itself through their core which punctures accessible space, rather then through strain energy.

Presence of these odd lines has direct physical consequences in glasses:

1. The glass at high temperatures behaves like a viscous liquid, in which odd lines can diffuse, subject to the continuity equation. Diffusion of

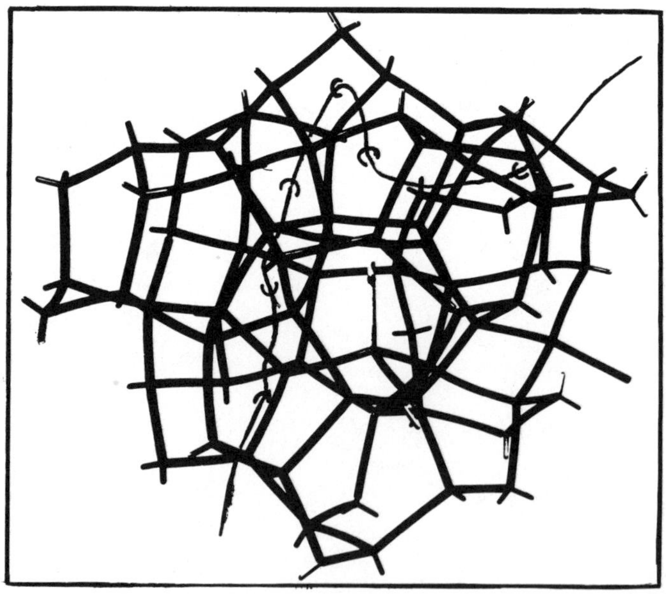

Fig. 1. Continuous random network threaded by line of odd disclinations.

the odd lines, and their interaction through elastic strain, can be shown to give a Vogel-Fulcher-WLF behavior to the viscosity, and their entropy contribution becomes negligible at and below the Fulcher temperature T (Kauzmann paradox).[2]

2. At low temperatures, one obtains not one, but two equal, lowest elastic energy configurations per odd line, between which the glass can tunnel. This is because the space into which matter is put is punctured by the core of the odd line.[3]
3. The core of the odd line also splits electronic wave packets, and this two-slit experiment set-up, with the magnetic field to vary the interfering pathlengths, may be responsible for the anomalous Hall effect in amorphous semiconductors and several amorphous metals.

Proof of existence and continuity of these odd lines

a. In CRN: consider an arbitrary surface S, homeomorphic to a sphere, and cutting the CRN at its vertices, so that S contains a 2-dimensional section of the network with edges, vertices and faces (rings). Associate with every edge i the weight $J_i = -1$, and with every ring α, the index $\phi_\alpha = \prod_{1\epsilon\alpha} J_i$. $\phi_\alpha = (+/-)1$, for even/odd ring. For *any* S, $\Pi_{\alpha\epsilon S}\phi_\alpha = \Pi_{1\epsilon S}(J_i)^2 = 1$ because every edge belongs to 2 adjacent rings, so that there is always an even number of odd rings, providing an exit for every odd line entering S.

b. In elastic continua, the local elastic configuration is single-valued. In crystals, this implies that the Burgers vector of a dislocation must belong to its space group. In glasses, the space group is trivial, yet one "odd" line defect survives because the rotation group is not simply connected. Rotation by 2π, while it returns the object to its original orientation, entangles its connections with the rest of the material. Rotation by 4π restores both object *and* connections. Hence the name: odd disclination.

N. Rivier
Imperial College
The Blackett Laboratory
London SW7 2BZ, England

REFERENCES

1. N. Rivier, Phil. Mag., A, *40*, 859 (1979); Proc. Conference on Structure of Non-Crystalline Solids, edited by P. H. Gaskell, E. A. Davis, and J. M. Parker, (Taylor and Francis, London, 1982).

2. N. Rivier and D. M. Duffy, in *Critical Phenomena*, edited by J. Della-Dora, J. Demongeot and B. Lacolle, Springer Synergetics *9*, 132 (1981).
3. N. Rivier and D. M. Duffy, J. Physique *43*, 293 (1982).

TRANSPORT, RELAXATION AND ABSORPTION IN AMORPHOUS SEMICONDUCTORS

E. A. Davis

Physics Department
University of Leicester
Leicester, LE1 7RH, U.K.

The response of non-crystalline semiconductors to alternating fields from d.c. to microwave frequencies is reviewed. Theoretical treatments appropriate to various physical models are outlined and the degree to which they succeed in explaining experimental data is discussed.

I. INTRODUCTION

Although our understanding of many of the electronic properties of amorphous semiconductors has advanced considerably over the past years, there are certain phenomena for which as yet there are no definitive explanations. The various channels for d.c. conduction are fairly well characterized and optical properties throughout the infrared, visible and ultraviolet regions of the spectrum pose no serious difficulties of interpretation. What governs the behaviour in the frequency range spanning the electrical, microwave and far-infrared regions is, however, still an open question.

In its simplest form the dielectric response is, for a great many materials, characterized by an approximately linear increase of the electrical conductivity with frequency up to about 10^{10} Hz, followed by a quadratic variation running up to the phonon spectrum. The deceptive simplicity of this behaviour parallels that of other "general phenomena" which appear to be characteristic of the amorphous state, such as the low-temperature linear term in the specific heat, the quadratic temperature variation of the thermal conductivity and the exponential fundamental optical absorption edge. For all these effects, several, sometimes phenomenological, models have been developed and it is true to say that it is not difficult to explain the *gross* behaviour in general terms. What we lack, however, are definitive microscopic atomic or electronic models to account for the *details* of the experimental results. For example, it is now known that the variation of electrical conductivity with frequency is only rarely exactly linear and, as for the specific heat, thermal conductivity and Urbach tail, the detailed behaviour is material dependent. The extent to which *extrinsic* effects, such as defects, influence these variations is not yet fully elucidated but it is clear that they do play a role.

In this paper, some of the specific models that have been proposed to explain the response of amorphous semiconductors to a.c. fields are outlined and their success or otherwise is accounting for experimental data is considered. Firstly, however, d.c. transport is reviewed.

II. D. C. CONDUCTIVITY

Under a d.c. field the current is amorphous semiconductors can be carried in extended states beyond the mobility edges or in localized states near the band edges. Alternatively, at lower temperatures and/or in the presence of a high density of states in the gap, conduction may occur by tunnelling between localized states at the Fermi level. These three channels for transport[1] are illustrated in Fig. 1a.

A. EXTENDED-STATE CONDUCTION

It is normally assumed that just beyond the mobility edges the mean free path for transport is as short as is possible — namely the average separation between atoms. The motion of carriers is then best described as diffusion, the mobility being given by

$$\mu_{ext} = eD/kT \tag{1}$$

where $D = 1/6\nu_{el}a^2$, $\nu_{el} = \hbar/ma^2$ and a is the distance in which phase coherence is lost. Thus, at room temperature, $\mu_{ext} \sim 6\ cm^2V^{-1}s^{-1}$. Direct

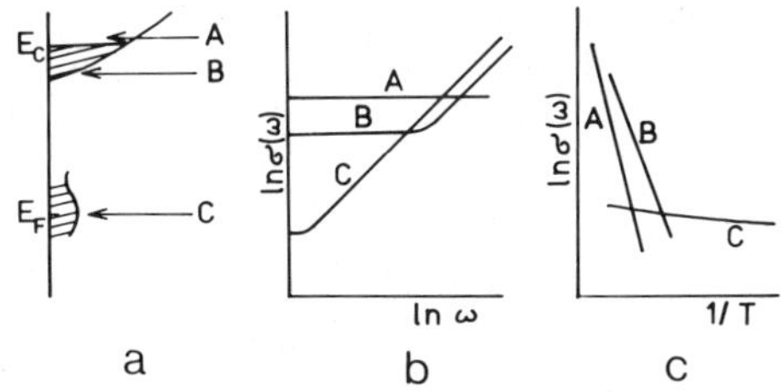

Fig. 1. a) Density of states in an amorphous semiconductor. Localized states are shaded. A, B and C refer to three separate channels for conduction. b) Schematic representation of the frequency dependence of a.c. conductivity at a fixed temperature for the three channels of conduction. c) Temperature dependence of a.c. conductivity at a fixed frequency for the three channels of conduction.

determination of μ_{ext} is difficult, but values in the range of 1-10 $cm^2V^{-1}s^{-1}$ have been *inferred* from various experiments.

The number of carriers n contributing to this mode of conduction is activated (for electrons) by $\exp[-(E_C-E_F)/kT]$ where E_C and E_F are the energies of the mobility edge and the Fermi level, respectively. The conductivity ($ne\mu_{ext}$) is not expected to be a function of frequency, at least not in accessible regions of study, because the electronic frequency ν_{el} is $\sim10^{15}$ s^{-1}.

B. LOCALIZED-STATE CONDUCTION

The mobility associated with hopping conduction between localized states is given by Eq. (1) but with the electronic frequency ν_{el} replaced by one appropriate to thermally assisted tunnelling, namely

$$\nu = \nu_o \exp(-2\alpha r) \exp(-\Delta/kT) \tag{2}$$

where ν_o is of the order of a phonon frequency, α is a tunnelling parameter equal to the inverse of the radius of the localized wavefunction, and r and Δ are the spatial and energy separations of neighbouring states. An averaging procedure over the actual r and Δ distributions is required in order to calculate the mobility. Unless conduction occurs in a narrow band of states comparable to kT then hopping is not to nearest neighbours but to more distant states at an energy Δ such that ν is maximized. In the case of conduction at the Fermi level this leads to Mott's variable-range hopping law in which $\log\sigma$ is proportional to $T^{1/4}$. For conduction in localized states at the band edge a different power law is obtained and, as for extended-state conduction, the number of carriers is exponentially dependent on temperature.

The a.c. conductivity associated with hopping conduction will be considered in § III. It will be sufficient to note here that, above a certain frequency known as the d.c. limit, the variation is *roughly* linear as illustrated schematically in Fig. 1b. The measured conductivity (d.c. or a.c.) at a given temperature will of course be the sum of the individual processes shown in Fig. 1a. Since the contributions from A and B are exponential functions of temperature (Fig. 1c), it is normally possible to separate, by experiment, the different channels for transport.

C. POLARONS

Although the experimental evidence appears to be against polaron formation for carriers in the bands of amorphous semiconductors, those in Anderson-localized states will always distort the surroundings to some extent. In the case of a *large* polaron the "distortion cloud" may extend over hundreds of atoms. If the interaction energy between the carrier and the phonons is strong in comparison to the bandwidth, a *small* polaron, with distortion confined to molecular dimensions, may form. A small polaron behaves like a free particle with enhanced mass at low temperatures but moves by thermally activated hopping at high temperatures. The following additional terms must then be included in any expressions for the transition rate

$$\exp\left(-W_H/\tfrac{1}{4}\hbar\omega_o\right) \qquad T < \tfrac{1}{2}\hbar\omega_o \tag{3a}$$

$$\exp\left(-W_H/kT\right) \qquad T > \tfrac{1}{2}\hbar\omega_o \tag{3b}$$

where ω_o is the relevant phonon frequency and W_H is equal to half the distortion (or self-trapping) energy W_p associated with the polaron.

For carriers trapped at deep, strongly localized defects, polaron formation is very likely. In the case of defects in chalcogenides the energy gained by local deformation is sufficiently large to overcome the Coulomb repulsion between two electrons on the same site and a bipolaron is formed (see § III E). The electronic states of these defects pin the Fermi level but no d.c. transport associated with them has been observed. It has been shown by Phillips[2] that, for a bipolaron, $W_H = 2W_p$ and so the low-temperature transition probability for tunnelling contains a factor $\exp(-8W_p/\hbar\omega_o)$. For W_p equal to a few tenths of an eV, this factor results in the bipolaron being essentially immobilized. At high temperatures the transition rate contains the factor $\exp(-2W_p/kT)$ which is less damaging.

The increase in relaxation time associated with deep levels in the chalcogenides is so great that one might predict that the a.c. conductivity would also vanish. This is not the case experimentally. However, it will be shown in § III E that the mechanism is probably one of *hopping over* rather than *tunnelling through* the potential barrier separating two defect sites.

III. A.C. CONDUCTIVITY

A. GENERAL

The application of an alternating electric field $E(\omega)$ of frequency ω to a sample containing polarizable units produces a polarization $P(\omega)$ given by

$$P(\omega) = \epsilon_o \chi(\omega) E(\omega)$$

where $\chi(\omega)$ is the complex susceptibility

$$\chi(\omega) = \chi'(\omega) - i\chi''(\omega) \, .$$

The imaginary component $\chi''(\omega)$ is called the dielectric loss and is related to the a.c. conductivity by

$$\sigma(\omega) = \epsilon_o \omega \, \chi''(\omega) \, . \tag{3}$$

A Debye response is one for which the polarization is characterized by a single relaxation time τ:

$$P(t) = P(0) \exp(-t/\tau) \, .$$

A Fourier transform then leads to an a.c. conductivity of the form

$$\sigma(\omega) \sim \omega^2 \tau / (1 + \omega^2 \tau^2) \, . \tag{4}$$

i.e. the a.c. conductivity varies as ω^2 for $\omega\tau < 1$ and the dielectric loss has a maximum at $\omega = 1/\tau$.

In an amorphous material it is generally assumed that there exists not one, but a spectrum of Debye-like processes and $\sigma(\omega)$ is then expressed as an integral over some distribution $n(\tau)$. If this is sufficiently broad then an a.c. conductivity varying approximately linearly with frequency is obtained for $\omega\tau < 1$.[3]

B. QUANTUM-MECHANICAL TUNNELLING

A physical process having the required broad distribution of relaxation times is that of electrons tunnelling between localized states. This mechanism was first proposed by Pollak and Geballe[4] to account for the behaviour of a.c. conductivity in heavily doped silicon. In this system, the localized states are associated with the randomly distributed impurity sites. The relaxation time (transition rate) for a pair of centres having spatial separation r and energy separation Δ was taken (see Eq. (2)) as

$$\tau^{-1} = \nu_o \exp(-2\alpha r) \exp(-\Delta/kT)$$

and the polarizability of the pair as

$$\alpha = e^2 r^2/12kT \cosh^2(\Delta/2kT) \ . \tag{5}$$

The major contribution to the a.c. conductivity for this mechanism arises for pairs of centres for which $\omega\tau \simeq 1$ and whose separation at frequency ω is given by

$$r_\omega = (1/2\alpha) \ln(\nu_o/\omega) \ . \tag{6}$$

The frequency dependence of $\sigma(\omega)$ on this model can be expressed as

$$\sigma(\omega) = A\omega \ln^4(\nu_o/\omega) \ . \tag{7}$$

Austin and Mott[5] adapted the analysis to the situation likely to be appropriate for amorphous semiconductors — namely a finite density of states $N(E_F)$ (per unit energy per unit volume) at the Fermi level E_F. They obtained the same frequency dependence as in Eq. (7), and a value for the constant A proportional to $\{N(E_F)\}^2$ T where T is the temperature.

C. FREQUENCY AND TEMPERATURE DEPENDENCES OF $\sigma(\omega)$

The variation of $\sigma(\omega)$ with frequency and temperature predicted by the tunnelling model is commonly quoted as $T^n\omega^s$ with $n = 1$ and $s \sim 0.8$. The actual value of s can be obtained from Eq. (7):

$$s = d[\ln \sigma(\omega)]/d[\ln \omega] = 1 - 4/\ln(\nu_o/\omega) \tag{8}$$

This expression is plotted in Fig. 2, from which it is clear that, for a given value of the attempt frequency ν_o, s is a decreasing function of frequency and

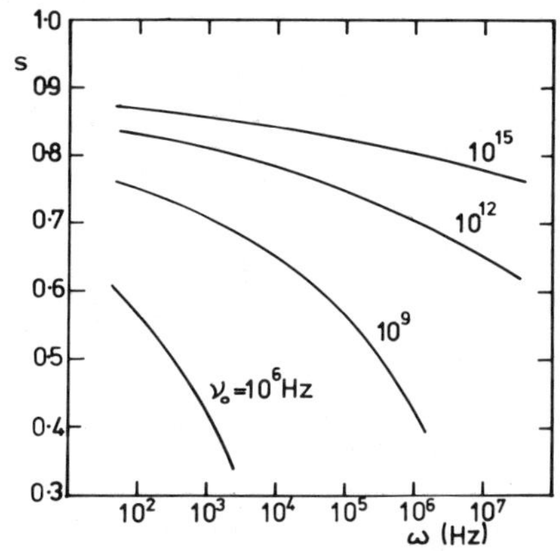

Fig. 2. Plot of s versus ω for various values of ν_o according to the expression $s = 1-4/\ln(\nu_o/\omega)$.

only approximates to 0.8 in the electrical range of frequencies when ν_o is $\sim 10^{12}$ Hz. However, values of s outside the range ~0.4-0.8 cannot occur for physically reasonable values of ν_o.

Experimentally, values of s lying close to unity (or occasionally greater) have been observed. Furthermore s is normally independent of frequency and is often found to be a decreasing function of temperature. These facts cast doubt on the applicability of the quantum-mechanical tunnelling model to the interpretation of a.c. conductivity behaviour in many amorphous semiconductors. Since it is now believed that chalcogenide and oxide glasses do not have a finite density of one-electron states at the Fermi level, this is hardly surprising for these materials. For others, however, one may wonder if the several approximations that are used in the theory are justified. Principally these are the "pair approximation" and neglect of multiphonon and correlation effects. All have been fairly extensively considered; lack of space prevents a discussion of the detailed theoretical work but for completeness a fairly comprehensive set of references is provided.[6–17]

One principal modification to the formulae for quantum-mechanical tunnelling which will be mentioned is that necessitated if small polarons are formed.

Then Eq. (6) becomes

$$r_\omega = (1/2\alpha)\left[\ln(\nu_o/\omega) + \frac{W_H}{kT}\right] \tag{6a}$$

and Eq. (8) becomes

$$s = 1 - 4/\left[\ln(\nu_o/\omega) - \frac{W_H}{kT}\right] \tag{8b}$$

It should perhaps be noted that this last formula predicts that s should *increase* with increasing temperature, a situation which is the opposite to that frequently observed experimentally.

D. AN ATOMIC MODEL

An attempt to link a.c. conductivity with the anomalous behaviour of the thermal conductivity and specific heat observed in glasses at low temperatures was made by Pollak and Pike.[18] They considered the hopping of atoms over the barrier separating states in a two-level system. To obtain coupling with the electric field it was of course necessary to assume that the species carries a charge.

For a barrier height of W (measured from the lower of the two levels) and an energy difference between the levels of Δ, the relaxation time can be approximated, for $W >> kT$, to

$$\tau = \tau_o \exp(W/kT)/[1 + \exp(-\Delta/kT)] .$$

Assuming that Δ and W are random independent variables, the a.c. conductivity can be calculated on the pair approximation in a manner similar to that used for quantum mechanical tunnelling to yield

$$\sigma(\omega) = \left(\pi e^2 r_o^2/6\Delta_o W_o\right) N\omega\, kT \tanh(\Delta_o/2kT) \tag{10}$$

where r_o, Δ_o and W_o refer to maximum values in the distributions and N is the number of hopping species.

This model therefore predicts an accurately linear dependence of $\sigma(\omega)$ on frequency ($s = 1$) and a temperature dependence that is linear for $T << \Delta_o/2k$ and zero for the opposite condition. It is therefore incapable of explaining data for which s is less than unity and temperature dependent. Furthermore the values of N, deduced by fitting to experimental results for which s is close to unity, lie in the range 10^{19}-10^{20} cm^{-3}. This is somewhat higher than estimates deduced from low-temperature thermal experiments.

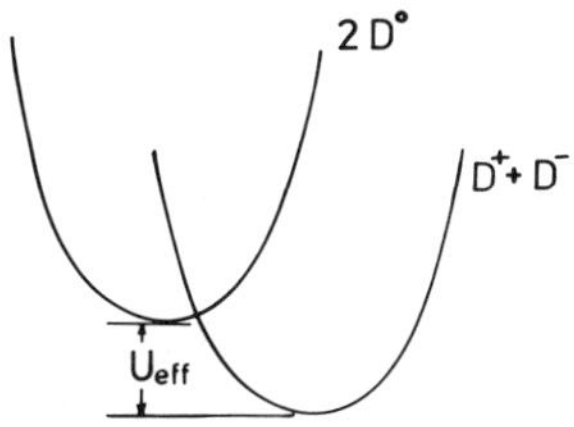

Fig. 3. Configurational-coordinate diagram illustrating the lowering of total energy when two neutral defects, D^0, are converted to charged defects, D^+ and D^-, by spin pairing.

E. CORRELATED BARRIER HOPPING

A model which has had considerable success in explaining data for chalcogenide semiconductors has been proposed by Elliott.[19] The mechanism takes into consideration the nature of the defects known to be present in these materials. From a variety of experiments, particularly luminescence and e.s.r. studies, it has been established that the defects are charged spinless centres which occur in pairs.[20] Two electrons at a dangling bond (undercoordinated) site D^- are compensated by a positively charged defect D^+ which is associated with an overcoordinated chalcogen atom. The repulsive energy required to place two electrons on D^- is more than compensated by the gain in energy at D^+. The two centres are really two charge states of the same defect, and, because of the above property, it is referred to as a negative-U defect, U being the correlation energy. Another way of expressing this feature is to say that the reaction:

$$2D^0 \rightarrow D^+ + D^- \tag{11}$$

is exothermic (Fig. 3). Thus neutral spin-carrying defects D^0 normally occur only by thermal excitation at high temperatures or in a metastable state following optical excitation at low temperatures.

The mechanism of dielectric loss proposed by Elliott is that of two electrons hopping simultaneously over the barrier separating a D^+D^- pair, thereby interconverting the centres. The process has been referred to as bipolaron hopping since the distortion (bonding configuration) around the defects must be transferred along with the charge. Essential to the model is the effective lowering of the Coulomb barrier separating the two sites when they are close together (Fig. 4). The hopping energy is

$$W = W_M - 8e^2/4\pi\epsilon\epsilon_o r \tag{12}$$

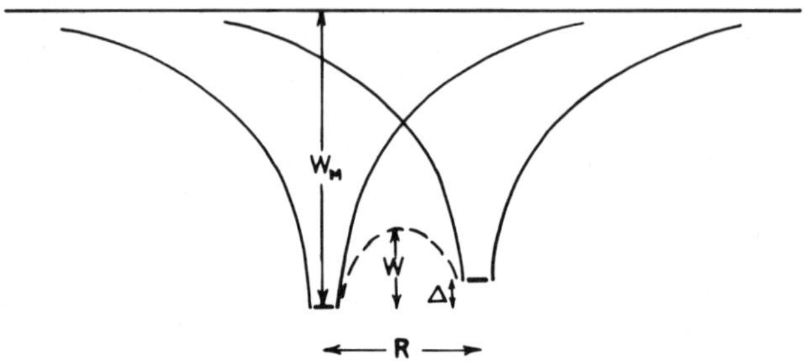

Fig. 4. Lowering of the Coulomb barrier between a pair of charged defects when at a separation R. W_M is the energy to take two electrons from D^- into the conduction band and Δ is the difference in energy between the two sites.

for a separation r and a dielectric constant ϵ. The maximum energy W_M is the energy to transfer two electrons from a D^- to a D^+ in the absence of interaction and, for chalcogenides, is approximately equal to the bandgap of the material.[20] It is also approximately equal to $4W_p$ where W_p is the distortion energy associated with the bipolaron. As mentioned in § II C, the probability for *tunnelling* of the two electrons from D^- to D^+ is proportional to $\exp(-2W_p/kT)$. Thus hopping over the barrier is favoured when $W < 2W_p$, or, using Eq. (12), when

$$4W_p - 8e^2/4\pi\epsilon\epsilon_o r < 2W_p \tag{13}$$

i.e. for a separation r given by

$$r < 4e^2/\pi\epsilon\epsilon_o B$$

where B is the bandgap. For $\epsilon = 10$ and $B = 2$ eV the centres are therefore required to be closer than ~10 Å.

The a.c. conductivity calculated according to this model is, for a random distribution of centres of density N, given by

$$\sigma(\omega) = \frac{N^2}{2}\,\omega \int_{\tau_{min}}^{\tau_{max}} \frac{e^2R^2}{3kT}\,\frac{\omega\,\tau}{(1+\omega^2\tau^2)}\,4\pi R^2 dR \tag{14}$$

where $\tau_{max} = \nu_o^{-1}\exp(W_M/kT)$ and $\tau_{min} = \nu_o^{-1}$

Under certain approximations, this can be evaluated to give[3]

$$\sigma(\omega) = \left[64e^{12}/3\pi^3(\epsilon\epsilon_o)^5\Delta_o\right] B^{-6}N^2\nu_o^\beta\omega^s kT \tanh(\Delta_o/2kT) \quad (15)$$

$$\text{where } s = 1-\beta \text{ and } \beta = 6kT/B \quad (16)$$

The essential new feature of this expression is that it contains a dependence on frequency which departs from linearity and which depends on the temperature and the bandgap.†

Comparison of the above theory with experiment is not easy to make in the absence of independent knowledge of the density of defects N. However, for a fixed value of N, the variation of $\sigma(\omega)$ with bandgap at a fixed temperature can be plotted. This is done in Fig. 5 (the dashed curve marked RANDOM) which also includes representative points (triangles) for three amorphous arsenic chalcogenides. The curve is seen to follow the form of the empirical relation reported by Mott and Davis[1] shown in the inset and illustrating the previously unexpected and unexplained correlation between the d.c. and a.c. conductivity. Different mechanisms are appropriate for each but they are now seen to be linked by their strong dependence on the magnitude of the bandgap.

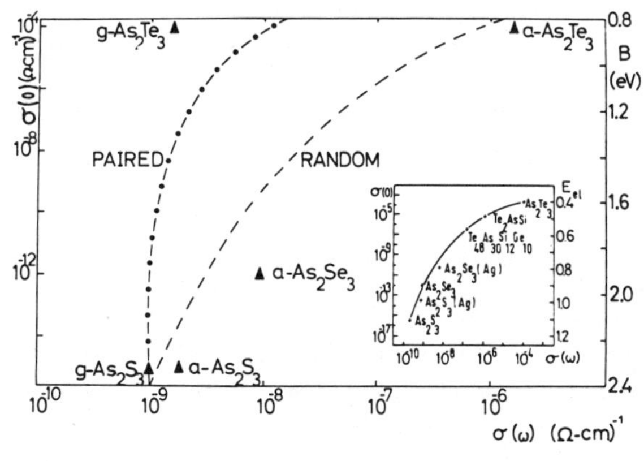

Fig. 5. Predictions of the correlated barrier hopping model for the variation of a.c. conductivity with d.c. conductivity (or bandgap B) for random and paired distributions of defect sites. Results for representative materials are shown. The inset is discussed in the text (from Elliott,[19] (1980)).

† A more accurate expression gives $s = 1 - \beta - \gamma$, where $\gamma = 6(kT/B)^2 \ln(\nu_o/\omega)$).

The curve marked PAIRED in Fig. 5 was obtained by Elliott by relaxing the assumption of a random distribution of charged defect centres. Instead a distribution was used that takes into account the likelihood that, for glasses obtained by quenching from the melt, a high proportion of defects of opposite sign will lie closer together than the average separation, having been subject to Coulomb attraction in the liquid state. The main difference between $\sigma(\omega)$ calculated in this case and that given by Eq. (15) is that the frequency exponent is given by

$$s = 1 - \beta + \delta$$

where $\delta = T/8T_g$ and T_g is the glass transition temperature, below which further spin pairing during cooling cannot occur. Since β is as before, Eq. (16), there is a partial cancellation of the two factors and a value of s close to unity and virtually independent of both temperature and bandgap results. This is in good agreement with data obtained on glassy (melt-quenched) chalcogenides as shown for representative materials in Fig. 5, but more extensively considered by Kočka.[21] For situations such that $\delta > \beta$, a superlinear frequency dependence of $\sigma(\omega)$ is predicted and such a variation has been reported in several instances (see reference 3).

The temperature dependence of $\sigma(\omega)$ according to the correlated barrier hopping model can be obtained from Eq. (15) and obeys

$$\frac{d[\ln \sigma(\omega)]}{dT} = \frac{6k}{W_M} \ln (\nu_o/\omega) \tag{17}$$

$$\text{or} \quad \frac{d[\ln \sigma(\omega)]}{d(1/T)} = - \frac{6kT^2}{W_M} \ln (\nu_o/\omega)$$

Thus, on a semilog/reciprocal temperature plot, the slope increases with increasing temperature or decreasing frequency. A successful comparison with data on chalcogenides has been made by Elliott.[19]

F. VARIANTS ON THE MODELS AND OTHER MATERIALS

One of the conditions employed in the evaluation of the integral in Eq. (14) is that $\omega\tau_{max} >> 1$. If this condition does not hold then $\sigma(\omega)\propto\omega^2$ and a dielectric loss peak occurs at $\omega \sim \tau_{max}^{-1}$ (if $\omega\tau_{max} \simeq 1$). Furthermore the peak should be activated with energy W_M. Such behaviour has been observed by Street and Yoffe[22] in evaporated films of As_2S_3 and As_4S_4. Elliott,[19]

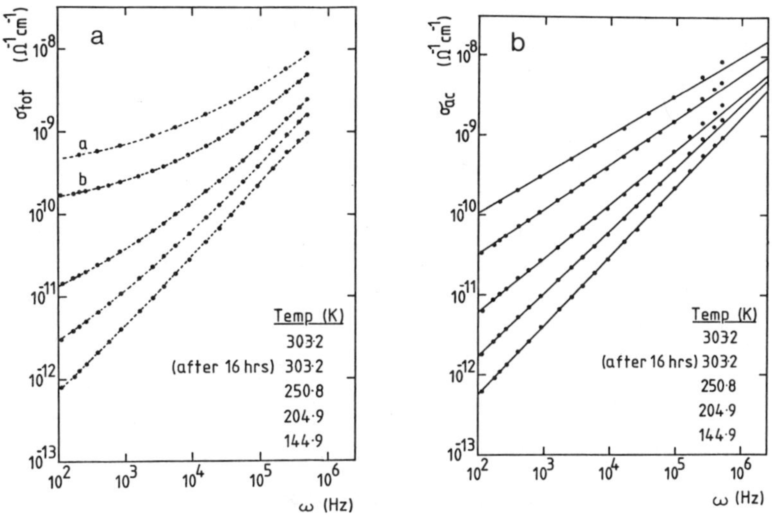

Fig. 6. a) Variation of total a.c. conductivity with frequency for a sputtered film of amorphous phosphorus at various temperatures. b) The same after subtraction of the d.c. component (from Extance, Ph.D. thesis).

recognizing that the values of W_M so obtained were roughly equal to one quarter of the respective bandgaps, has interpreted these data in terms of single-polaron correlated barrier hopping (CBH). He suggested that in evaporated films there exist a high concentration of very close D^+D^- pairs (intimate valence alternative pairs, IVAPs) which might be expected to have some D^0 character and thus behave like single-electron centres.

In amorphous arsenic, D^0 centres are believed to coexist with D^+D^- pairs, there being certain sites in the network where negative-U defects are not favoured energetically.[23] Single-electron CBH has been postulated as contributing to the a.c. conductivity in this material also.[24].

Another Group V material, amorphous phosphorus, appears to exhibit single-polaron CBH in the as-deposited state when prepared by sputtering but bipolaron CBH when annealed or when hydrogen is present in the sputtering gas. Data for as-deposited films are shown in Fig. 6a and b.[25] The slopes of the lines in Fig. 6b, along with those for similar data on annealed or hydrogenated material, are plotted versus temperature in Fig. 7. The solid curves are theoretical fits according to CBH theory and yield values for W_M of 0.7 and 1.6 eV respectively. Although the defect energy levels in a–P have not yet been determined with certainty, the second value is close to twice the activation energy for d.c. conduction and therefore seems likely to be

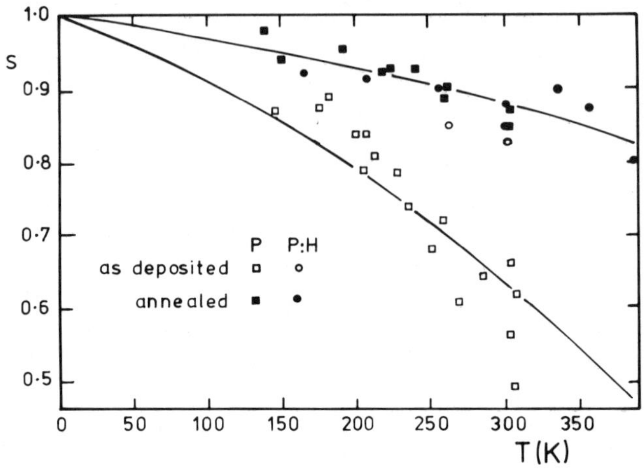

Fig. 7. Lower points: variation of the slope, s, with temperature for the data of Fig. 6b and for other samples of a—P. Upper points: the same for the films after annealing and for a—P:H as indicated. Solid lines: theoretical fits according to Elliott's correlated barrier hopping model [$s = 1 - (6kT/W_M) - 6(kT/W_M)^2 \ln(\nu_o/\omega)$] (from Extance, Ph.D. thesis).

associated with the bipolaron. That the lower value is associated with single polaron hopping is supported by the fact that, in the as-deposited films, the d.c. conductivity obeys the $T^{-1/4}$ variable-range hopping law which suggests neutral, spin-carrying D^0 centres. This conduction process is no longer present after annealing or hydrogenation.

An interesting development of the CBH model has been proposed by Shimakawa[26] to explain the large temperature dependence of the a.c. conductivity observed in certain chalcogenides at highish temperatures. It had been suggested earlier that this was associated with quantum-mechanical tunnelling in band-edge localized states (process B, Fig. 1). Shimakawa considered the thermal creation of D^0 centres by the back reaction of Eq. (11) and postulated that single-electron CBH between D^0 and D^- and between D^0 and D^+ was responsible. An example of fitting to experimental data is shown in Fig. 8 for which the only adjustable parameter was the density of charged defect centres $N(= 4.2 \times 10^{18}\ \text{cm}^{-3})$.

Meaudre and Meaudre[27] have measured $\sigma(\omega)$ as a function of temperature for films of SiO_2 prepared by r.f. sputtering. For $T \lesssim 320\text{K}$, $\sigma(\omega)$ was found to be close to a linear function of frequency and almost independent of temperature. This contribution was interpreted in terms of bipolaron CBH with a paired distribution of defects of density 3×10^{18} cm^{-3} and a value of $W_M = 6.2$ eV. For $T \gtrsim 320K$ Meaudre and Meaudre propose that quantum-mechanical tunnelling of hole-like polarons takes place between D^- and D^0 centres with a hopping energy $W_H = 0.35$ eV. In order to explain the temperature dependence of this contribution to $\sigma(\omega)$ it was necessary to assume that the density of D^0 centres was activated with an effective correlation energy of 0.56 eV. A complete energy level scheme for the defects was proposed. The above authors also compared their data with a theory of a.c. conduction (not described in this paper owing to lack of space) developed by Scher and Lax.[28] This theory, known as continuous time random walk (CTRW), predicts frequency exponents, s, of $\sigma(\omega)$ that can be less than unity, but when $s = 1$ a T^{-1} dependence is expected. This is at variance not only with data for SiO_2 but also for many chalcogenides.

Measurements of $\sigma(\omega)$ on evaporated a–Si have been reported by Rieder[29]

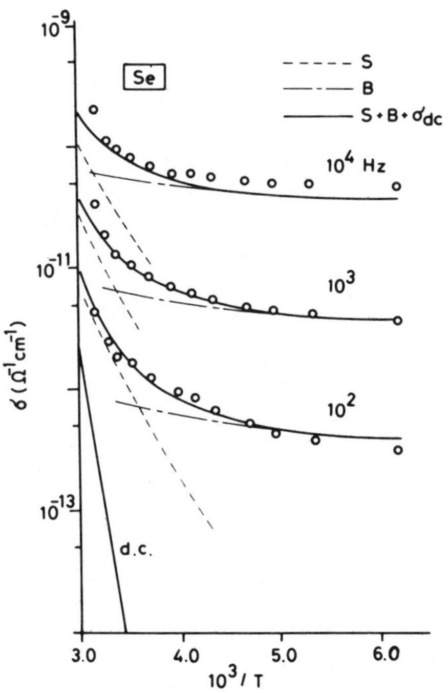

Fig. 8. Temperature dependence of $\sigma(\omega)$ in glassy Se. Experimental points from A. I. Lakatos and M. Abkowitz (Phys. Rev. **B3**, 1791 (1971)). Lines as indicated. S: single-polaron CBH. B: bipolaron CBH. Solid curves: sum S + B + d.c. contributions (after Shimakawa[26]).

and on glow-discharge a–Si:H by Abkowitz et al.[30] and by Nitta et al.[31] The advantage of using the latter material is that the dangling-bond defects are much reduced in density with a concomitant reduction in the $T^{-1/4}$ d.c. contribution which otherwise has to be subtracted. However the mechanisms for $\sigma(\omega)$ in the two types of material are entirely different. Reider observes $\sigma(\omega) \sim T^n\omega^s$ with $n \sim 1.4$ and s decreasing from 0.94 to 0.88 as T increases from 180 to 350K. In the region of $\omega \sim 1$ MHz, he also finds $s \rightarrow 0$ and $n \sim 1$. Reider interprets his data in terms of one-electron CBH with a low-energy cut-off in the distribution of activation energies. However the value of W_M required is 1.5 eV - probably larger than the band gap. Abkowitz et al. found for a–Si:H, $s \sim 1$ and $n \sim 0.8$. In addition, by studying the field-effect on the same samples they were able to show that $\sigma(\omega) \propto N(E_F)^2$. This particular observation is consistent with quantum-mechanical tunnelling (see § 3.2); however, $s \sim 1$ is not. Elliott[32] has applied the bipolaron CBH model to the data, negative-U centres being postulated to exist in special regions of the material - perhaps at void surfaces. He finds $W_M \sim 1.8$ eV (in this case less than the bandgap which in hydrogenated material is ~2 eV); this accounts for the weak temperature dependence of s but predicts a dependence of T somewhat stronger than observed. In their samples, Nitta et al. find n decreases with frequency which is also consistent with CBH (see § III E).

Finally in this section, mention is made of a series of comprehensive measurements of $\sigma(\omega)$ on sputtered films of a–Ge by Long et al.[33,34] These data, which include measurements of $\sigma(\omega)$ down to 1K and also as a function of the a.c. electric field strength, have not yet been fully interpreted, but the latest suggestion is that a single-electron polaron tunnelling model is appropriate. Long suggests that the polarons involved are *large* and that their potential wells overlap. Furthermore, by subtracting the large $T^{-1/4}$ d.c. contribution, he has found evidence for an ω^2 term in $\sigma(\omega)$ at low frequencies which he associates with a loss peak and a low-R cut-off in the distribution of active centres. These concepts overcome some of the earlier problems encountered with fitting quantum-mechanical tunnelling theories to data on a–Ge, such as reconciling parameters deduced from the $T^{-1/4}$ behaviour with the a.c. behaviour, the non-linear variation of s with temperature, and the unphysically large value of ν_0 required.

IV. MICROWAVE ABSORPTION

Measurements in the microwave region of the spectrum are less numerous than the models that have been used to interpret them. As the frequency is increased several possibilities exist. One of these is resonant absorption of photons by two-level systems, electronic or atomic. Such direct transfer leads

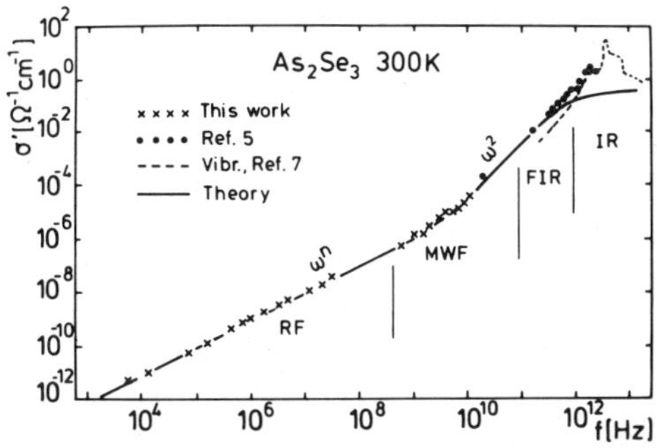

Fig. 9. Variation of $\sigma(\omega)$ with ω for a$-As_2Se_3$ (from Schmidt and Breitschwerdt[35] where further details may be found).

to a variation of $\sigma(\omega) \sim \omega^2$ (see reference 1 for example) and indeed such a dependence is commonly observed on the low-frequency side of the infrared absorption region of the spectrum. Figure 9 illustrates results on As_2Se_3 compiled by Schmidt and Breitschwerdt.[35]

Strom and Taylor,[36] whose results are shown as filled circles in Fig. 9, consider the ω^2 region as arising from disorder-induced charge fluctuations involving coupling to lattice modes with a wide range of wavevectors. On the other hand Schmidt and Breitschwerdt use a generalized model in which the dielectric response is coupled to relaxing structural modes. They obtain the solid line shown in Fig. 9 and hence account for the lower-frequency linear region in terms of the same phenomenological model. It should be stressed that these authors do not claim that their two-step description can distinguish between electronic, ionic, dipolar or structural mechanisms. A similar generalized model has also been developed by Jonscher and co-workers.[37]

Although there are certain problems in extending the CBH model to frequencies comparable to ν_o, the intrinsic relaxation frequency, Shimakawa[38] has recently shown that good agreement can still be reached in the gigahertz region. Figure 10 shows an example of the temperature dependence of $\sigma(\omega)$ in As_2Se_3 at 2.37 GHz. The solid line was calculated using parameters derived from lower-frequency fitting. The only modification made to existing formulae was the inclusion of a Gaussian (rather than a flat) distribution of site energy differences Δ with a width $\sim$15 meV. Similar agreement was found for several other chalcogenide glasses.

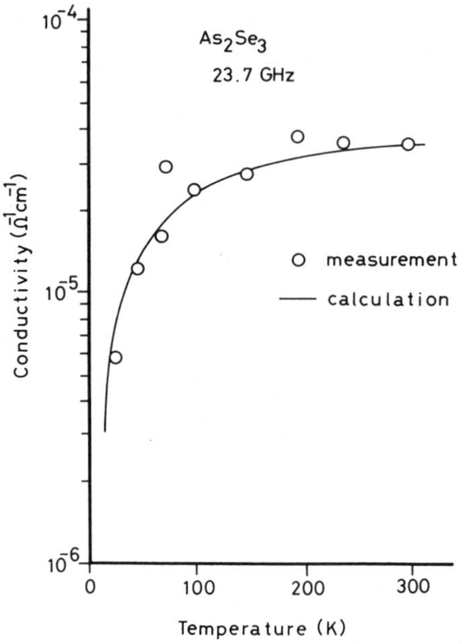

Fig. 10. Temperature variation of microwave conductivity for a$-As_2Se_3$. Experimental points from reference 36; solid curves according to modified CBH theory (from Shimakawa, unpublished).

V. CONCLUSIONS

In spite of the apparent wealth of results on the frequency dependence of conductivity in amorphous semiconductors (by no means all of which has been reviewed here), insufficient information is actually available to decide which, if any, of the various models and theories that have been proposed is preferable. It has been shown that, for the electrical range of frequencies, it is the *departures* from linearity and the details of the temperature dependence that should permit a distinction to be made. Experimentally, problems associated with contacts and sample inhomogeneities often arise, making it difficult to obtain accurate and reliable data for comparison with theory. There is also a danger in trying to make one model fit results on very diverse systems. For the chalcogenides and Group V materials it appears that the model of single- or double-polaron correlated barrier hopping, suitably modified to take account of the different defect distributions, is able to account for most measurements. Further developments of this theory might result in more general applicability to other classes of materials. However it must be freely admitted that in many cases it is not yet possible to decide even between electronic and atomic models, and in fact when polarons are formed such a distinction may not be meaningful.

ACKNOWLEDGEMENTS

The author is grateful to A. R. Long and S. R. Elliott for providing preprints of forthcoming review articles to be published in Advances in Physics. He is also grateful to P. Extance and K. Shimakawa for permission to reproduce (as yet unpublished) Figures 6 and 11 respectively.

REFERENCES

In the references, the following abbreviations are used:

Garmisch (1974): Proceedings of the 5th. International Conference on Amorphous and Liquid Semiconductors, eds. J. Stuke and W. Brenig (Taylor and Francis 1974).

Leningrad (1976): Proceedings of the 6th. International Conference on Amorphous and Liquid Semiconductors, ed. B. T. Kolomiets (Nauka 1976).

Edinburgh (1977): Proceedings of the 7th. International Conference on Amorphous and Liquid Semiconductors, ed. W. E. Spear (University of Edinburgh, 1977).

1. N. F. Mott and E. A. Davis, "Electronic Processes in Non-crystalline Materials" 2nd. edition (Oxford University Press 1979).
2. W. A. Phillips, Phil. Mag. **34**, 983 (1976).
3. S. R. Elliott, to be published.
4. M. Pollak and T. H. Geballe, Phys. Rev. **122**, 1742 (1961).
5. I. G. Austin and N. F. Mott, Adv. in Phys. **18**, 41 (1969).
6. M. Pollak, Phys. Rev. **133**, A564 (1964); Phil. Mag. **23**, 519 (1971); Garmisch, 127 (1974); Leningrad, 79 (1976); Phil Mag. **36**, 1157 (1977).
7. P. N. Butcher and P. L. Morys, J. Phys. C. **6**, 2147 (1973); Garmisch, 153 (1974).
8. H. Böttger and V. V. Bryksin, Phys. Stat. Sol. (b) **78**, 415 (1976).
9. P. N. Butcher and K. J. Hayden, Phil Mag. **36**, 657, (1977); Edinburgh, 234 (1977).
10. P. N. Butcher, K. J. Hayden and J. A. McInnes, Phil. Mag. **36**, 19 (1977).
11. J. A. McInnes, P. N. Butcher and J. D. Clark, Phil. Mag. B. **41**, 1 (1980).
12. B. Movaghar, B. Pohlman and W. Schirmacher, Solid State Comm. **34**, 451 (1980).
13. B. Movaghar, B. Pohlman and G. W. Sauer, Phys, Stat. Sol. (b) **97**, 533 (1980).
14. A. L. Efros, Phil. Mag. B. **43**, 829 (1981).
15. P. N. Butcher and S. Summerfield, J. Phys. C. **14**, L1099 (1981).
16. S. Summerfield and P. N. Butcher, to be published.
17. A. R. Long, N. Balkan, W. R. Hogg and R. P. Ferrier, Phil. Mag. B. XXX (1982).
18. M. Pollak and G. E. Pike, Phys. Rev. Lett. **28**, 1449 (1972); G. E. Pike, Phys. Rev. B **6**, 1572 (1972).
19. S. R. Elliott, Phil Mag. **36**, 1291 (1977); Phil. Mag. B. **37**, 135 (1978); Phil Mag. B. **37**, 553 (1978); B **38**, 325 (1978); Phil. Mag. B. **40**, 507 (1979); J. Non-Cryst. Solids, **35** and **36**, 855 (1980).
20. N. F. Mott, E. A. Davis and R. A. Street, Phil. Mag. **32**, 961 (1975).
21. J. Kočka, A. Tríska and L. Štourač, Leningrad, 249 (1976).
22. R. A. Street and A. D. Yoffe, J. Non-Cryst. Solids **8-10**, 745 (1972).
23. G. N. Greaves, S. R. Elliott and E. A. Davis, Adv. in Phys. **28**, 49 (1979).

24. S. R. Elliott and E. A. Davis, Edinburgh, 637 (1977).
25. P. Extance and E. A. Davis, to be published.
26. K. Shimakawa, J. de Physique **42**, C4-167 (1981); Phil. Mag. B 46, 123 (1982); to be published.
27. M. Meaudre and R. Meaudre, Phil Mag B **40**, 401 (1979); Phil. Mag. B, to be published.
28. H. Scher and M. Lax, J. Non-Cryst. Solids, **8**, 497 (1972); Phys. Rev. B **7**, 4491, 4502, (1973).
29. G. Rieder, Phys. Rev B **20**, 607 (1979).
30. M. Abkowitz, P. G. LeComber and W. E. Spear, Commun. on Phys. **1**, 175 (1976).
31. S. Nitta, K. Shimakawa and K. Sakaguchi, J. Non-Cryst. Solids **24**, 137 (1977); S. Nitta, K. Shimakawa and S. Nonomura, J. Non-Cryst. Solids **35** and **36**, 339 (1980).
32. S. R. Elliott, Phil Mag. B. **38**, 325 (1978).
33. A. R. Long and N. Balkan, J. Non-Cryst. Solids **35**, and **36**, 415 (1980); Phil Mag. B. 41, 287 (1980).
34. A. R. Long, W. R. Hogg, N. Balkan and R. P. Ferrier, J. de Physique, **42**, C4-107 (1981).
35. W. W. Schmidt and K. G. Breitschwerdt, J. de Physique **42**, C4-171 (1981).
36. U. Strom and P. C. Taylor, Phys. Rev. **B16**, 5512 (1977).
37. A. K. Jonscher, Phys. Status Solidi b, **82**, 585 (1977); Phil. Mag. B **38**, 587 (1978).
38. K. Shimakawa, to be published.

DISCUSSION

Transport in Amorphous Semiconductors

In discussion it was emphasized that the qualitative form of $\sigma(\omega,T)$ can be generally explained by distributions of the parameters involved in the relaxation time τ of the charge carriers, working within the framework of exponential relaxation in the Debye theory. A more careful comparison of $\sigma(\omega,T)$ and experiment over a large dynamic range shows deviations which provide searching tests of theoretical self-consistency and require a more careful consideration of the nature of the charge carriers. These deviations include derived values of the pre-exponential factor much larger than is physically plausible, although it was pointed out that this often results from an attempt to explain both a.c. and d.c. conductivity by the same model.

A number of possibilities have been suggested to account for these problems. In addition to the possibility of non-exponential relaxation, a more detailed investigation of the nature of the charge carriers has allowed greater flexibility in fitting theory to experiment. Correlation and polaron effects might be important, and reservations were expressed about attempts to explain the behavior of different classes of amorphous solids using the same model. The relative importance of "intrinsic" and "extrinsic" effects is different in different materials and in only very few cases could the effect of specific charge carriers be identified (e.g. OH in vitreous silica). It was emphasized that there was no clear distinction between electron motion accompanied by distortion of the surrounding medium (polaron motion) and ionic motion. Models involving ionic motion can give the same frequency and temperature dependences for $\sigma(\omega,T)$ as those involving electronic motion and perhaps a more useful distinction is between models assuming variable range hopping and those in which the charge carriers move between two sites of almost uniform separation.

B. Golding, Chairman
W. A. Phillips, Discussion Leader

COMMENTS

Frequency Dependent Loss in Amorphous Semiconductors

These observations are mainly concerned with tetrahedral amorphous semiconductors. The conventional notation in which the AC conductivity σ is expressed in the form $\omega^s T^n$ (ω = frequency, T temperature) will be used.

1. Low (liquid helium) temperatures.

   a. The AC loss in a wide variety of amorphous semiconductors is characterized by similar behavior of the parameters s and n, namely $s \rightarrow 1$ and $n \rightarrow 0$ so that $\sigma \sim \omega$.

   b. The *magnitude* of the loss, characterized by $\sigma/\omega|_{T\rightarrow 0}$ is also remarkably constant for many materials.[1] This is illustrated for tetrahedral semiconductors in Table 1.

TABLE 1

AC Loss of Amorphous Semiconductors

Reference	Material	Preparation	$\frac{\omega}{2\pi}$ (Hz)	T(K)	$2\pi\sigma/\omega(\Omega\text{cm Hz})^{-1}$
2	Ge Ge:H	R.F. sputtering	3.10^3	4.2	7.10^{-14}
3	Ge	Evaporation	2.10^6	4.2	2.10^{-13}
4	Si	Sputtering	1.10^3	2.0	3.10^{-14}
5	Si:H	Glow discharge & sputtering	1.10^4	4.2	2.10^{-14}

Similar examples to those in Table 1 would of course vary in their DC conductivities, and in the AC loss observed at higher temperatures, by many orders of magnitude.

c. These examples are often characterized by strong *AC field effects*[2–5]

d. The s = 1, n = 0 behavior is characteristic of relaxation in a broad distribution of entities whose dipole moments are uncorrelated with their relaxation times. It is however not easy to distinguish between atomic and correlated electronic relaxation on the basis of this evidence alone.[1]

2. Higher temperatures: some results on the a-Ge:H system.
A recent study of AC loss in sputtered hydrogenated a-Ge films[1,6] shows the following points:

a. At higher temperatures than considered above, the losses in a a-Ge:H are likely to be electronic, as they correlate well with the DC conductivity.

b. The loss has similar magnitude, temperature and frequency dependence for all samples. The incorporation of hydrogen merely shifts the temperature scale of the loss to higher temperatures.

A. R. Long
Dept. of Natural Philosophy
University of Glasgow
Glasgow G12, 800, Scotland

REFERENCES

1. A.R. Long, Adv. in Physics *31*, 553 (1982).
2. A. R. Long, W. R. Hogg, N. Balkan and R. P. Ferrier, Journal de Physique *42*, C4-107 (1981).
3. M. H. Gilbert and C. J. Adkins, Phil. Mag. *34*, 143 (1976).
4. B. Golding, J. E. Graebner and W. H. Haemmerle, *Proc. 7th Int. Conf. on Amorphous and Liquid Semiconductors*, edited by W. E. Spear, (Centre for Industrial Consultancy and Liaison, University of Edinburgh, 1977) p. 367.
5. B. Pistoulet, F. Roche and A. Cagna, Journal de Physique *42*, C4-147 (1981).
6. A. R. Long, N. Balkan, W. R. Hogg and R. P. Ferrier, Philos. Mag. *B45*, 497 (1982).

THERMAL AND ACOUSTIC EXPERIMENTS IN LOW TEMPERATURE GLASSES

M. v. Schickfus

Max-Planck-Institut für Festkörperforschung
Heisenbergstr. 1
7000 Stuttgart 80
FRG

The low-temperature thermal, acoustic and dielectric properties of glasses differ largely from those of corresponding crystals. Specific heat is enhanced and varies linearly with temperature, whereas thermal conductivity is quadratic in temperature and lower than in crystals. Ultrasonic and dielectric properties are anomalous as well: Absorption is much higher and the velocity of sound and the dielectric constant exhibit a characteristic temperature dependence in all glasses. These anomalies are explained within the framework of a model which assumes tunneling of groups of atoms of the amorphous network between equivalent sites.

I. INTRODUCTION

One fundamental difference between the structure of a glass and that of the corresponding crystal is the lack of long-range order. In glasses the regular arrangement of neighbouring atoms is lost at distances exceeding a few interatomic spacings. Therefore, for large wavevectors, when the wavelength approaches the interatomic distances, phonons in glasses cease to be plane waves. Therefore they are highly damped like in anharmonic crystals. Consequently, below room temperature, the phonon transport properties of the two modifications (e.g. thermal conductivity) differ strongly. On the other hand, at small wavevectors, a crystal and a glass can be considered as an elastic continuum where phonon properties are only determined by macroscopic quantities. At very low temperatures, where the wavelength of thermal phonons becomes very large, the differences between crystal and glass are therefore expected to disappear. The same should therefore be valid for the low-temperature specific heat: both, crystal and glass should follow a Debye T^3-law with only slight differences in the absolute value caused by the differences in density and elastic constants. Experiments performed in the past ten years have revealed a totally different behaviour. The thermal, acoustic and dielectric low-temperature properties of insulating and metallic glasses show an anomalous behaviour down to the lowest temperatures investigated so far (below 10 mK). These anomalies are attributed to low-energy excitations which are able to interact with phonons, microwave photons and conduction electrons.

In this paper we shall report experiments which have turned out to be important for the understanding of these low-temperature anomalies. For the sake of clarity and conciseness we do not intend to give a complete review of the large number of experiments which have been performed in the past years. Recent reviews on this subject can be found in Refs. 1, 2, and 3.

In the following section the thermal properties of dielectric and metallic glasses will be described first. In Section 3 we shall introduce the widely accepted tunneling model, which gives a phenomenological description of the observed anomalies. Section 4 is devoted to ultrasonic and dielectric experiments which test the predictions of the tunneling model. Finally, in Section 5, experiments are reported which might help in the search for the microscopic nature of the tunneling systems. Such experiments were performed at very high pressure, in polymers, and in crystals with very high defect concentration.

II. THERMAL ANOMALIES AT LOW TEMPERATURES

A. SPECIFIC HEAT

The low-temperature specific heat of vitreous silica was first investigated in 1969[4] and, more systematically, in 1971.[5] These authors found a very pronounced deviation from the Debye behaviour at low temperatures: The specific heat is considerably larger and does not approach the T^3-behaviour of

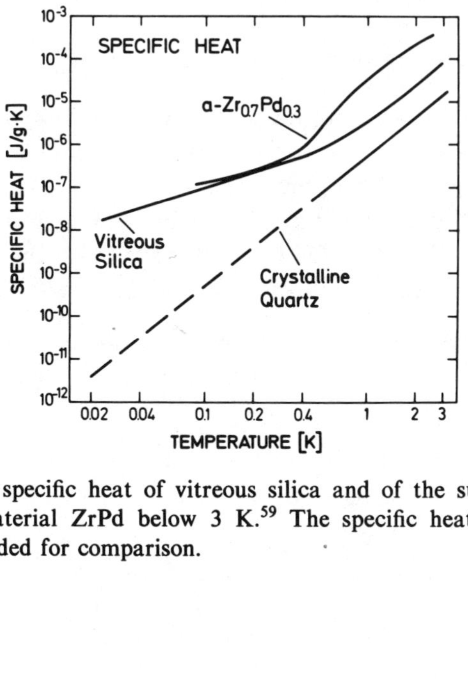

Fig. 1. The specific heat of vitreous silica and of the superconducting amorphous material ZrPd below 3 K.[59] The specific heat of crystalline quartz is included for comparison.

B. THERMAL CONDUCTIVITY

In crystals at low-temperatures the lifetime of thermal phonons and thus the thermal conductivity κ is dominated by phonon scattering at the surfaces of the sample. The temperature dependence of the thermal conductivity is therefore determined by the T^3-dependence of the phonon heat capacity. The same behaviour is expected for amorphous materials because the phonon wavelength is large enough for the elastic continuum approximation. Instead it turns out that the thermal conductivity (Fig. 2) is much lower than in crystals. The thermal conductivity falls steadily and passes a plateau near 10 K, which has been observed in all amorphous materials. Below this temperature κ decreases like aT^α, where α is close to two in all cases. Again this "universal" behaviour is observed in practically all amorphous materials with only slight variations of the prefactor a,[5,12,13] including metallic glasses.

III. THE TUNNELING MODEL

In order to explain the anomaly of the specific heat in glasses, the existence of a wide distribution of low-energy excitations has to be assumed. Because of the quasi-linear temperature dependence of the specific heat, the density of states of these excitations must be almost independent of energy for at least 20 mK $< E/k_B <$ 1 K. To account for the thermal conductivity, a strong interaction of these excitations with phonons is required.

These requirements are met by a model introduced by Anderson, Halperin and Varma,[14] and Phillips.[15] This model assumes that in the amorphous

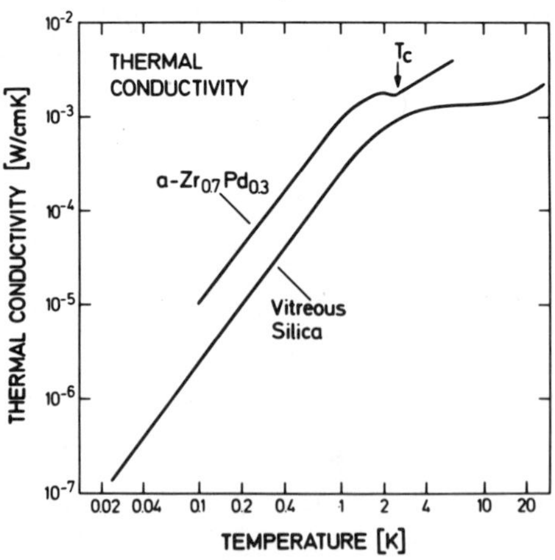

Fig. 2. Thermal conductivity of vitreous silica and of the superconducting metallic glass PdZr below 20 K.[59]

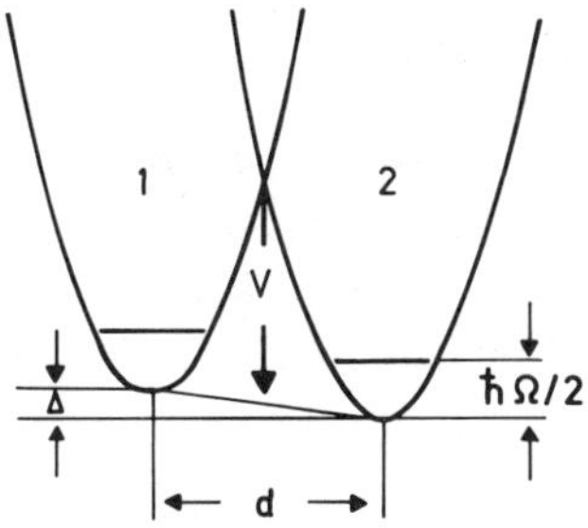

Fig. 3. The double-well potential of the tunneling model. The particle tunnels between sites 1 and 2 separated by a potential barrier V. The wells are separated in configuration space by a distance d. $\hbar\Omega/2$ is the zero-point energy of the potential wells and Δ the asymmetry.

structure there exist atoms or groups of atoms in a double-well potential (Fig. 3). This is plausible in a glass: Due to frozen-in entropy there may be configurations with equivalent free energy minima. If the minima are nearby in configuration space, transitions between them may occur by quantum mechanical tunneling. The tunneling system is described by the following Hamiltonian:

$$H = H_o + H_i = 1/2 \begin{pmatrix} \Delta & \Delta_o \\ \Delta_o & -\Delta \end{pmatrix} + 1/2 \begin{pmatrix} \delta\Delta & 0 \\ 0 & -\delta\Delta \end{pmatrix} \tag{1}$$

where H_o describes the undisturbed tunneling system and H_i the interaction with external fields. An influence of off-diagonal terms in H_i has not been found experimentally (with the possible exception of thermal expansion[16]). Therefore they will not be considered here. Δ is the asymmetry of the potential well minima (Fig. 3), and Δ_o is the tunnel splitting caused by the overlap of the wavefunctions of the two wells. Δ_o is related to the physical parameters of the tunneling defect as follows:

$$\Delta_o = \hbar\Omega \sqrt{\frac{\pi}{\lambda}}\, e^{-\lambda}; \qquad \lambda = \frac{d}{\hbar}\sqrt{2mV} \tag{2}$$

where $\hbar\Omega/2$ is the zero point energy of the particle in one of the potential wells, m the mass of the tunneling particle. The other quantities are defined in Fig. 3. The interaction term $\delta\Delta$ can be written as

$$\delta\Delta/2 = \overleftrightarrow{\gamma}\vec{e} + \vec{p}\vec{F} \tag{3}$$

The two contributions represent the interaction with strain fields $\overleftrightarrow{e}$ through the deformation potential $\overleftrightarrow{\gamma}$, and with electrical fields $\vec{F}$ through the dipole moment $\vec{p}$. Although $\overleftrightarrow{\gamma}$ and $\overleftrightarrow{e}$ are tensor quantities, they are generally treated as scalars for simplicity. After diagonalization of H_o one finds $E = \sqrt{\Delta^2 + \Delta_o^2}$ for the energy of a tunneling system. The interaction term is transformed into a diagonal term $(\Delta/E)\gamma$ and $(\Delta/E)p$ and an off-diagonal term $2(\Delta_o/E)\gamma$ and $2(\Delta_o/E)p$ in the elastic and the electrical case, respectively.

A. SPECIFIC HEAT

In order to explain the linear temperature dependence of the specific heat, a constant density of states of the tunneling system is required. Therefore it has been assumed[14,15] that a uniform distribution exists of the tunneling parameters Δ and λ

$$P(\Delta,\lambda) = \bar{P} = \text{const.} \tag{4}$$

This expression can be transformed into

$$P\,(E,r)\;dEdr = \tfrac{1}{2}\,\bar{P}r^{-1}\,(1 - r)^{-1/2} \tag{5}$$

with $r \equiv (\Delta_o/E)^2$. For a given energy this distribution is shown in Fig. 4. In order to keep the total number of states $N = \int\int P(E,r)drdE$ from diverging, a cutoff $\Delta_{o,min}$ (or λ_{max}) and a maximum energy E_{max} have to be assumed. With the density of states given by (5) the specific heat can be calculated as

$$c_v = \bar{P}k^2T\,\frac{\pi^2}{6}\,(1 + \ln\,(2kT/\Delta_{o,min})) \tag{6}$$

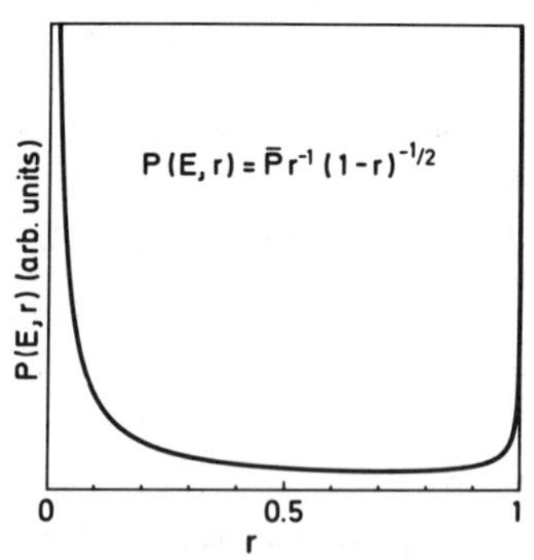

Fig. 4. Tunneling density of states P(E,r) with $r = (\Delta_o/E)^2$. The major contribution comes from tunneling states with small tunnel splitting Δ_o/E.

If the weak temperature dependence of the logarithmic term is neglected, this expression has the required linear temperature dependence. With $\Delta_{o,min} \approx 10^{-5}$ K the value of the logarithmic term is about 11 at T = 1 K. With Eq. (6) one finds $\bar{P} \cong 7 \times 10^{31}$ erg^{-1}cm^{-3} from the specific heat data of vitreous silica. As we will see in Section III C, tunneling systems with a small tunnel splitting Δ_o have long relaxation times. This leads to a dependence of the measured specific heat on observation time, which has indeed been observed and will also be discussed in Section VI. This time dependence, which becomes larger at low temperatures, may also be the origin of the deviation of the specific heat from a linear temperature dependence.

B. THERMAL CONDUCTIVITY

In the tunneling model the anomaly of the thermal conductivity is ascribed to the resonant scattering of phonons from the tunneling defects. Using the golden rule, the mean free path of phonons can be calculated[17] from Eqs. (1), (3) and (4):

$$l_{res}^{-1} = \frac{\pi \bar{P} \gamma^2}{\rho v^3} \omega \tanh \frac{\hbar\omega}{2k_B T} \tag{7}$$

where ρ is the mass density of the material, v the sound velocity and $\hbar\omega$ the energy of the phonon. To calculate the thermal conductivity we use the kinetic formula $\kappa = 1/3 c_{ph} \bar{v} \bar{\ell}$ where c_{ph} is the T^3-dependent specific heat of the phonons and $\bar{v}$ and $\bar{\ell}$ are appropriately weighted averages of all phonon modes. In the dominant phonon approximation the hyperbolic tangent in Eq. (7) is constant, so that the mean free path goes like 1/T. This leads to the observed quadratic temperature dependence of the thermal conductivity. With the deformation potential $\bar{\gamma} \approx 1$ eV,[18] a density of states $\bar{P} = 4 \times 10^{31}$ erg^{-1}cm^{-3} is found for a–SiO_2. This value is very close to the one derived from specific heat.

C. RELAXATION OF THE TUNNELING SYSTEMS

After a perturbation the tunneling systems return into thermal equilibrium within the relaxation time τ. Thermalization is achieved through the interaction with thermal phonons, or electrons in the case of amorphous metals. In the phonon-assisted case this relaxation time is given by[17]

$$\tau^{-1} = \left(\frac{\gamma_l^2}{v_l^5} + 2\frac{\gamma_t^2}{v_t^2} \right) \frac{rE^3}{2\pi\rho\hbar^4} \coth \frac{E}{2k_B T} \tag{8}$$

where the indices l and t refer to the polarization of the phonons. Because of the factor $r = (\Delta_o/E)^2$ in Eq. (8), the distribution of the relaxation rates $P(E,\tau^{-1})$ for a given E is similar to the density of states P(E,r) in Eq. (5). For r = 1, corresponding to a symmetric double-well potential, the relaxation time has its smallest value τ_m. Fig. 4 shows that we have to expect two important regions for the relaxation behaviour of glasses: short times, where $\Delta_o/E \approx 1$, which contribute less than 10% of the specific heat, and long relaxation times with $\Delta_o/E << 1$, which contribute most to the specific heat. This distribution of the relaxation times also explains why the logarithmic factor in the specific heat does not appear in thermal conductivity: Here only strongly coupling tunneling systems ($\Delta_o \approx E$) with short relaxation times contribute. On the contrary, in a specific heat experiment the measuring time is typically some 10 sec, so that slow tunneling systems ($\Delta_o/E \rightarrow 0$) can contribute.

In metallic glasses there exists a very efficient relaxation channel through the coupling between conduction electrons and tunneling systems. It is described in an expression derived in Ref. 19

$$\tau_e^{-1} = \frac{\pi}{4\hbar} (\tilde{\rho}_e V_\perp)^2 \, E \coth \frac{E}{2kT} \tag{9}$$

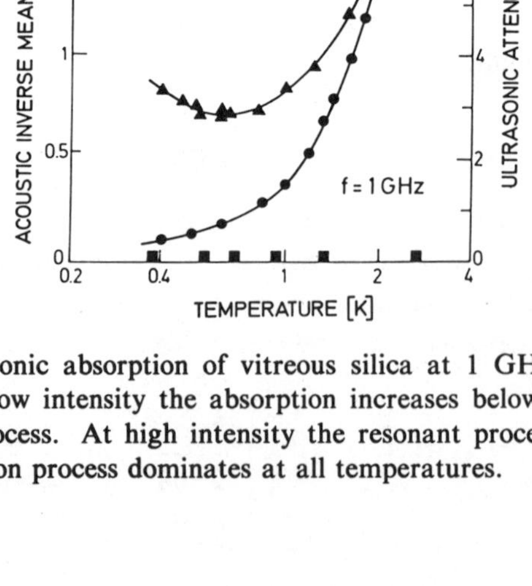

Fig. 5. Ultrasonic absorption of vitreous silica at 1 GHz between 0.4 and 2 K.[59] At low intensity the absorption increases below 0.8 K due to the resonant process. At high intensity the resonant process is saturated and the relaxation process dominates at all temperatures.

where $\tilde{\rho}_e$ is the electronic density of states per atom at the Fermi level and $V_\perp$ an off-diagonal coupling matrix element (corresponding to $\gamma(\Delta_o/E)$). For $(\tilde{\rho}_e V_\perp)$ a value of 0.16 was found for PdSi.[19] At 0.1 K the relaxation rates due to this mechanism are about four orders of magnitude faster than the phonon relaxation rates, whereas above a few K the phonon contribution dominates because of its stronger energy dependence.

IV. ULTRASONIC AND DIELECTRIC EXPERIMENTS

The low-temperature ultrasonic and the dielectric response of amorphous materials is characterized by two processes: One is the resonant interaction with the tunneling systems, the absorption contribution of which has already been given in Eq. (7). The other is caused by the relaxation of the tunneling systems which has been discussed in the preceding section.

A. ABSORPTION

As an illustrative example for the behaviour of dielectric glasses[20,21], polymers[22] and amorphous metals,[19,23] Fig. 5 shows the temperature dependence of the ultrasonic absorption of a–SiO_2 below 4 K and at 1 GHz. At low acoustic intensity the increase of the absorption below 0.5 K is caused by the resonant process in accordance with Eq. (7). In this expression, however, the influence of the magnitude of the external field onto the population difference of the tunneling systems has not been taken into account: If the intensity approaches a critical value J_c, the tunneling systems are driven out of thermal equilibrium towards a smaller population difference and the absorption is thus reduced. This process is described by a modification of Eq. (7)

$$l_{res}^{-1} = l_{res,0}^{-1}\,(1 + J/J_c)^{-1/2} \tag{10}$$

with $l_{res,0}^{-1}$ given by Eq. (7) and J the intensity. The critical intensity J_c depends on the linewidth of the tunneling excitations, inversely on their relaxation time τ and their coupling $(\Delta_0/E)\gamma$ to the ultrasonic strain field. In normal conducting metallic glasses τ is very short, so that the saturation threshold is much higher than in dielectric glasses.[19,23]

An analogous behaviour has been observed in the microwave dielectric absorption of glasses.[24] Here the deformation potential γ has to be replaced with the dipole moment p and we find $4\pi^2\bar{P}p^2/\left(3c_0\sqrt{\epsilon}\right)$ for the prefactor in Eq. (7). At 10 GHz and 0.4 K the dielectric absorption of vitreous silica can be lowered by a factor of 100 by increasing the electrical field (Fig. 6).

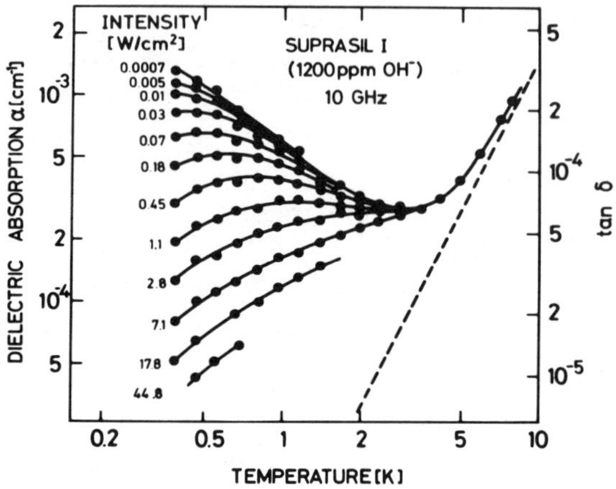

Fig. 6. Dielectric absorption of vitreous silica at 10 GHz between 0.4 and 8 K.[24] Below 3 K the absorption increases due to the resonant process. At high intensity the resonant absorption is saturated. The dashed line indicates the contribution of the relaxation process.

Above the (weakly frequency dependent) temperature of the absorption minimum, the absorption increases with temperature (Figs. 5 and 6). This contribution is caused by the relaxation of the tunneling systems whose energy is slightly changed by the applied field through the diagonal part of the deformation potential $(\Delta/E)\gamma$ and $(\Delta/E)p$, respectively. The system returns into a new thermal equilibrium within the relaxation time τ, resulting in an absorption contribution of an individual tunneling system:[17]

$$l_{rel}^{-1} = (1 - r)\,\frac{\gamma^2}{\rho v^3}\,\frac{\partial}{\partial E}[f(E,T)]\,\frac{\omega^2\tau}{1 + \omega^2\tau^2}\,, \tag{11}$$

in the dielectric case the quantity $\gamma^2/\rho v^3$ has to be replaced by $4\pi p^2/(3c_0\sqrt{\epsilon})$. τ is given by Eqs. (8) and (9) and $f(E,T) = -1/2\tanh(E/2kT)$. To arrive at the macroscopically observed absorption, one has to integrate Eq. (11) over all relaxation times and over all energies - i.e. over all tunneling states. In general this integration has to be done numerically, but limiting cases can be found analytically: In the acoustic case the absorption is proportional to $\bar{P}\gamma^4\omega^0T^3$ for $\omega\tau_m >> 1$ and to $\bar{P}\gamma^2\omega^1T^0$ for $\omega\tau_m << 1$, if a linear relation between γ_t and γ_l is assumed. τ_m is the shortest relaxation time at a given energy $E(r = 1)$. In the dielectric case these expressions are changed to $\bar{P}\gamma^2p^2\omega^0T^3$ for $\omega\tau_m >> 1$ and to $\bar{P}p^2\omega^1T^0$ for $\omega\tau_m << 1$. This difference is due to the fact that the external field is elastic in the former case and electric in

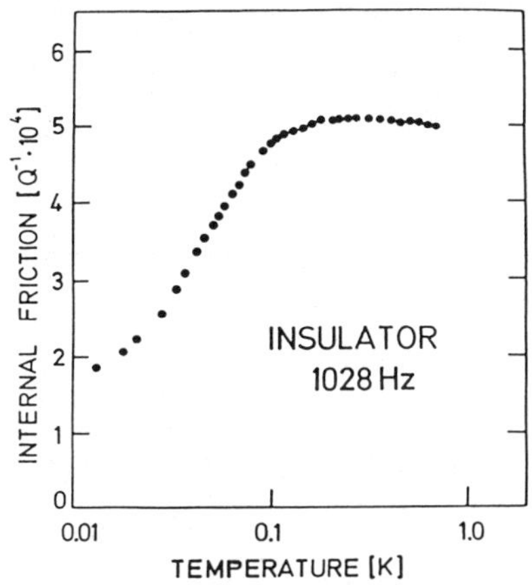

Fig. 7. Internal friction of a silica-based glass at 1 KHz in a vibrating reed experiment. Above 0.2 K the absorption in nearly temperature independent.[25]

the latter, whereas for dielectric glasses the relaxation time is determined by the coupling to phonons in both cases. The temperature independence of the relaxation absorption for $\omega\tau_m \ll 1$ is a consequence of the linear increase of the density of states P(E,r) predicted by the tunneling model for decreasing relaxation rates and $r \ll 1$ (Fig. 4 and Eq. (5)).

This behaviour is most clearly illustrated in an elastic experiment on a silica-based glass carried out at the very low frequency of 1 kHz by a vibrating reed technique[25] (Fig. 7). The absorption first rises (because of an additional background absorption it is difficult to verify the T^3-dependence)

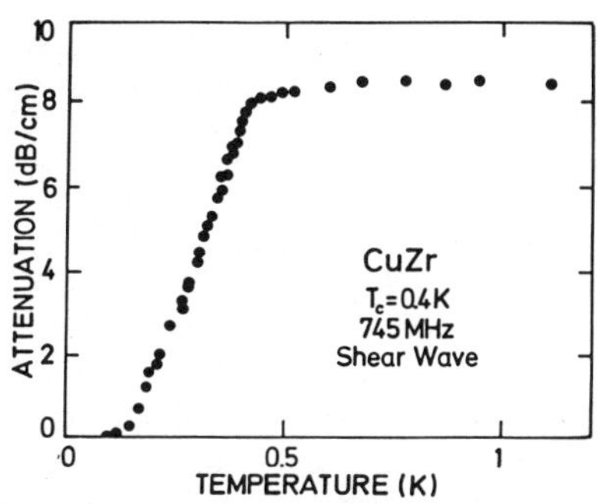

Fig. 8. Ultrasonic absorption of the superconducting metallic glass CuZr at 745 MHz.[27] The absorption becomes temperature independent above $T_c = 0.4$ K.

and changes into a temperature independent behaviour above 0.2 K. A similar behaviour has been observed in polymers at frequencies around 20 MHz.[26] Another example is the acoustic absorption of the amorphous metal CuZr at 745 MHz (Fig. 8).[27] Here we find again the increase of the absorption with temperature ($\omega\tau_m >> 1$) until 0.4 K. Then the sample becomes normal conducting and the relaxation rate increases rapidly so that $\omega\tau_m << 1$ because of the interaction with conduction electrons. Therefore above $T_c = 0.4$ K the absorption is again temperature independent.

B. SOUND VELOCITY AND DIELECTRIC CONSTANT

The influence of the tunneling systems on the velocity of sound and the dielectric constant can be obtained from the absorption through the Kramers-Kronig relation. For the resonant contribution a simple logarithmic law is found for $\hbar\omega << kT$:[28,29]

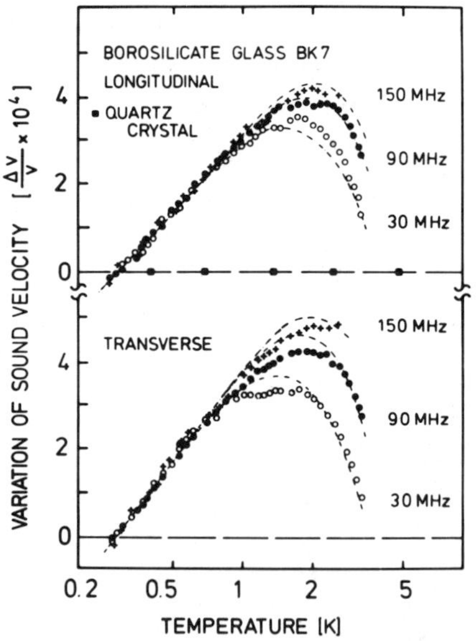

Fig. 9. Low-temperature sound velocity of a borosilicate glass between 30 and 150 MHz for longitudinal and transverse waves (S. Hunklinger, M. v. Schickfus in Ref. 1). Below 1 K the resonant interaction with the tunneling systems produces a logarithmic temperature dependence of the sound velocity. In crystalline quartz the sound velocity is constant in this temperature range.

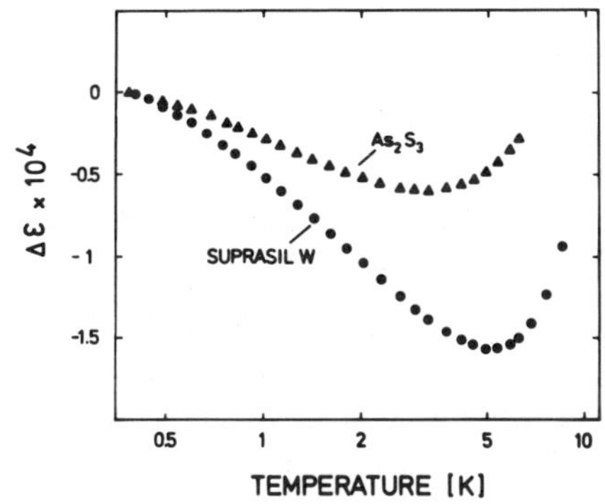

Fig. 10. Microwave dielectric constant of vitreous silica and of amorphous As_2S_3 below 10 K.[29] The logarithmic temperature dependence of ϵ due to the resonant process is modified below 0.5 K because $\hbar\omega \cong k_BT$.

$$\left(\frac{\Delta v}{v}\right)_{res} = A_{ph} \ln (T/T_o); \quad \left(\frac{\Delta \epsilon}{\epsilon}\right)_{res} = A_{el} \ln (T/T_o) \qquad (12)$$

where T_o is an arbitrary reference temperature and $A_{ph} = \bar{P}\gamma^2/(\rho v^2)$ in the acoustic and $A_{el} = -8\pi\bar{P}p^2/3$ in the dielectric case. In ultrasonic or microwave dielectric experiments this behaviour has been observed below temperatures of a few K in a large variety of amorphous dielectrics and metals[28,29,26,30] (Figs. 9 and 10). In the case of normal conducting amorphous metals, however, it has turned out later that this effect is not caused by the resonant contribution alone (see below).

In general, the contribution of the relaxation process has to be calculated numerically. However, in most cases the conditions $\omega\tau_m >> 1$ and $\omega\tau_m << 1$ are valid, where analytical solutions can be found. A $\omega^{-2}T^6$-dependence of the velocity of sound and dielectric constant (with a negative sign like in Eq. (12)) is expected for $\omega\tau_m >> 1$. Because of the strong frequency dependence this contribution is very small in ultrasonic and microwave dielectric experiments. It is therefore masked by the resonant process. But also at low frequencies the T^6-dependence is not observable, because other processes seem to dominate the high-temperature behaviour.

In the case $\omega\tau_m << 1$ a temperature dependence is predicted which depends on the nature of the relaxation process: In the case of phonons with an ω^2-density of states one finds[31]

$$\left(\frac{\Delta v}{v}\right)_{rel} = -\frac{3}{2} A_{ph} \ln (T/T_o); \left(\frac{\Delta \epsilon}{\epsilon}\right)_{rel} = -\frac{3}{2} A_{el} \ln (T/T_o) \qquad (13)$$

where A has already been defined for the resonant process. For the relaxation by conduction electrons with a constant density of states the prefactor is changed to $-1/2$.[31] This behaviour is again most clearly observable at low frequencies, where the condition $\omega\tau_m << 1$ is reached at low enough temperatures, so that the relaxation process is not modified by higher-order processes. Fig. 11 shows the temperature dependence of the velocity of sound of a silica-based glass at 1 kHz.[25] Below 0.05 K the condition $\omega\tau_m >> 1$ is fulfilled (implying a τ_m of $\approx$150 μsec) and the logarithmic contribution of the resonant process dominates (the T^6-dependent contribution of the relaxation process is negligible here). Above 0.1 K the relaxation process contributes and leads to a logarithmic decrease of the sound velocity with $-½$ of the slope of the resonant process (Eqs. (12) and (13)). The corresponding behaviour of the dielectric constant has been observed,[32,33] but not interpreted in terms of this simple formalism. In normal conducting metallic glasses the condition $\omega\tau_m >> 1$ has not been reached experimentally. The observed logarithmic temperature dependence of the velocity of sound[30,34,19] is therefore the sum of both, the resonant and the relaxation contribution (i.e. ½ of the slope of the resonant contribution

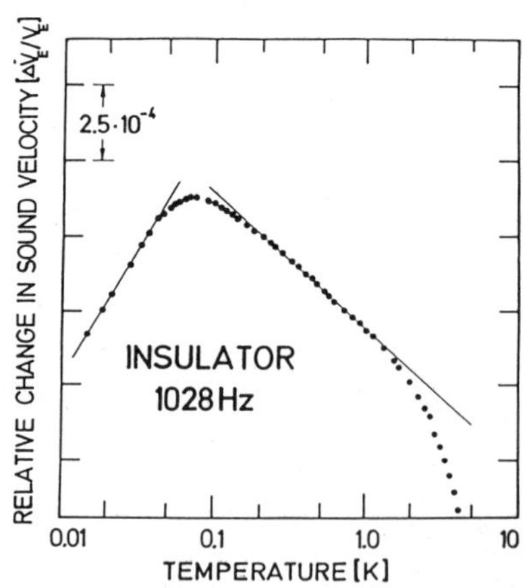

Fig. 11. Low-frequency velocity of sound in a vibrating reed experiment on a silica-based glass.[25] Below 0.08 K the resonant process causes a logarithmic variation of the velocity of sound. Above 0.1 K the relaxation contribution sets in and the sound velocity decreases with half the slope of the resonant part.

alone). Only in the case of a metallic glass undergoing the superconducting transition, a change of the slope from A to A/2 has been observed:[27] Above T_c the electronic contribution reduces τ so that $\omega\tau_m << 1$. The same observation can be made if superconductivity is suppressed with a magnetic field.[35]

From the sound velocity measurements which can be performed with high accuracy, values for the coupling constant $\bar{P}\gamma^2$ can be derived with much higher accuracy than from absorption measurements. These can be compared with the values obtained from measurements of the thermal conductivity and agree very well in general.

An important difference between the behaviour of the velocity of sound and of the dielectric constant should be pointed out: Whereas the acoustic properties depend only very weakly on the chemical composition of the sample, the dielectric behaviour is very sensitive to impurities. In the case of a$-SiO_2$ the coupling constant $\bar{P}p^2$ depends linearly on the content of OH^--groups and only a small residual value can be ascribed to the "intrinsic" tunneling systems mainly probed in ultrasonic experiments.[29] This difference between the ultrasonic and the dielectric experiments in vitreous silica is mainly caused by the different dipole moments of the "intrinsic" (0.6 Debye) and the "extrinsic" (3.7 Debye) tunneling systems.[18] Another important difference is that the "extrinsic" tunneling systems are more weakly coupled to the lattice.[18]

V. NATURE OF THE TUNNELING SYSTEMS

In the preceding sections the tunneling model has been introduced as a phenomenological model for the description of the low-temperature anomalies of glasses. Two questions remain open in this context: are two-level tunneling systems really the origin of the low-temperature anomalies, and with which entities in the amorphous network can they be identified? In this section experiments will be reported which were performed to answer some of these questions.

A. LONG-TIME RELAXATION

One of the predictions of the tunneling model is the distribution of relaxation times (Eq. (5)). According to this distribution the largest contribution to the specific heat comes from tunneling systems with relaxation times $\tau >> \tau_m$. Therefore a logarithmic time dependence of the specific heat is expected, which has indeed been observed on a millisecond time scale in recent experiments.[36,37] Another experiment which confirms the existence of the very long relaxation times predicted by the tunneling model has been performed by rapidly cooling a sample of vitreous silica from a few

Kelvin to about 0.2 K.[38] It was observed that the sample emitted heat even at times exceeding one hour. The authors found their results to agree with the predictions of the tunneling model. A third experiment which gives evidence for the long-lived tunneling states has been performed by applying stress to a sample of vitreous silica or of the amorphous metal PdSiCu. In both cases heat evolved after a change of stress at times up to several minutes at temperatures around 2 K.[39] This experiment can be explained in terms of the relaxation absorption in a milli-Hz "acoustic" experiment.

Although these experiments agree with the predictions of the tunneling model, they give no quantitative information about the distribution of the relaxation times and thus of the tunneling parameter λ (Eq. 8). This, however, is the case for the relaxation contribution to the absorption and the velocity of sound in the limit $\omega\tau_m \ll 1$. The temperature independence of the absorption and the logarithmic temperature dependence of the sound velocity are consequences of the distribution $p(E,r) \propto r^{-1}$ for $r \ll 1$. The fact that this behaviour has indeed been observed (Section IV) strongly supports the tunneling model.

B. CRYSTALS WITH HIGH DEFECT CONCENTRATION

In the search for the microscopic origin of the tunneling systems it may be very useful to compare with crystalline systems with defects which are known to be tunneling systems. A well-known class of such systems are the alkali-halides with impurities substituting a halide atom. A theoretical concept predicting a glass-like behaviour of these substances because of an elastic mutual interaction of these defects has been developed.[40] Experimental investigations have only been performed[41] in NaBr:F. A Schottky-like peak was found in the specific heat around 1 K followed by a linear decrease at lower temperatures. The magnitude of the linear term was found to depend on the F^- concentration. The temperature dependence of the dielectric constant[41] might be interpretable in terms of the "glassy" tunneling model; unfortunately the data do not extend to low enough temperatures.

In the ionic conductor Li_3N a low-temperature behaviour has been found which is perfectly glass-like. The specific heat is linear in temperature,[42,43] and there is a logarithmic temperature dependence of the velocity of sound and of the dielectric constant.[43,44] In this system the tunneling particle has been identified with an H^+-impurity which cannot be totally eliminated for technical reasons.[44] Interestingly, the magnitude of the acoustic and the thermal anomaly did not depend on H^+-concentration. This behaviour agrees with the interaction picture.[40]

Many other ionic conductors have been investigated, one of the most studied is Na–β alumina, where it is assumed that the Na^+-ion can tunnel between two nearly equivalent sites. Because of the non-stoichiometry of this material, the configurational disorder in the conducting planes leads even to a "two-dimensional" disorder. Again, the typical "glassy" low-temperature behaviour is found in the thermal, acoustic and dielectric properties.[45,46,47]

Finally we should mention experiments on irradiated quartz. These have been performed with electrons,[48,49] neutrons,[50,51,52] and γ-radiation.[53]. The "glassy" anomalies become observable in the case of neutron irradiation, but with a dose-dependent magnitude. It is important to note that the magnitude of the deformation potential γ is the same as in the glass. This result is explained with the fact that in neutron irradiated quartz small amorphous regions are created by the absorbed neutrons. In the case of γ- or electron irradiation, localized defects are created. Again, "glass-like" low-temperature anomalies are observed in the thermal properties. It has turned out, however, that these defects couple weakly to the lattice and that for energies exceeding 1 K they do not exhibit the constant density of states which is typical for glasses.

In all of these examples the question arises, whether the defects investigated have the typical distribution $P(\Delta,\lambda)$ of the tunneling parameters as in glasses. Is there a constant distribution of λ also for well defined defects, if only the mutual interaction between them is strong enough? If so, is this static interaction also important in the case of glasses? Generally it would be more plausible that the interaction only produces a distribution of the asymmetry Δ. An answer to this question should be found in a low-frequency acoustic or dielectric experiment similar to those described in Section IV, which would test the distribution of relaxation times for $\omega\tau_m << 1$.

C. EXPERIMENTS UNDER HIGH PRESSURE

With the thermal, acoustic and dielectric experiments cited above it is practically impossible to probe the density of states of the tunneling systems beyond energies of a few Kelvin. Above this temperature the rapidly increasing phonon density of states dominates the thermal properties and higher-order processes and other relaxation mechanisms make a unique analysis of the ultrasonic and dielectric data impossible. The question whether the density of states of the tunneling systems stays constant up to higher energies has been investigated in an acoustic experiment performed at pressures up to 8 kbar.[54] Although, with a deformation potential $\gamma \approx 1$ eV of the tunneling systems, this pressure should cause an energy shift of more than 100 K, only a weak influence on the low-temperature acoustic anomalies could be observed. Even if a strain dependence of γ is assumed, this means that the constant density of states should extend to energies beyond some 10 K, because the tunneling systems "shifted up" from the low-energy region have to be replaced by an equal number "shifted down" from high energies.

D. INFLUENCE OF THE COORDINATION NUMBER

A detailed analysis reveals that there are some exceptions from the typical low-temperature behaviour of glasses. The specific heat of amorphous arsenic and of amorphous germanium layers exhibits at most a much reduced linear term in the specific heat.[55,56] Furthermore, no tunneling contribution to the acoustic properties comparable with that in vitreous silica has been found in amorphous silicon[57] and germanium.[58] These observations have been connected with the coordination number of these materials: a-Arsenic is threefold and a—Ge and a—Si are fourfold coordinated. This means that the existence of tunneling systems is based on a low coordination number and the corresponding softer structure. In a—SiO_2, for instance, the two-fold coordinated bridging oxygen atoms are easily subject to a bending motion with two potential minima. A plausible candidate for a tunneling system would then be the rotation of a SiO_4 tetrahedron. Of course, this idea is limited to glasses with a structure similar to that of silica. It would be a tremendous task to identify the microscopic nature of the tunneling systems in all other amorphous systems.

E. EXPERIMENTS ON POLYMERS

Recently a number of acoustic experiments have been performed on amorphous polymers. As already mentioned, a "glassy" behaviour is observed. The magnitude of the anomaly hardly depends on chemical details. If the methyl side group in PMMA ("Plexiglas") is replaced with a heavier ethyl side group, almost no change of the coupling factor $\overline{P}\gamma^2$ is observed.[26] It is therefore concluded that the origin of the tunneling systems is not associated with a motion of a single side group or a rotation of the backbone.

VI. CONCLUSION

At low temperatures the thermal, acoustic and dielectric properties of amorphous materials differ largely from those of their amorphous counterparts. Below, say 5 K these anomalies are almost independent of the chemical composition. Only in amorphous metals the additional contribution of the conduction electrons needs to be considered. The low-energy excitations causing these anomalies are quite successfully explained in a model which assumes the tunneling of atoms or groups of atoms between nearly equivalent sites in a double-well potential. This model assumes in a plausible way that there are excitations with an energy-independent density of states and very low energy splittings.

Acoustic and dielectric experiments which are sensitive to the relaxational behaviour of the tunneling systems have confirmed predictions on the distribution of these relaxation times and thus a very important and basic detail of the tunneling model. Nevertheless, until now it has not been possible to correctly predict the existence and the properties of the tunneling systems from known properties of the amorphous structure. Attempts have been made to attack this problem by gradually introducing disorder in crystalline materials, or to deliberately modify details of the structure of amorphous polymers.

Finally it should be mentioned that there remain open questions in the tunneling picture itself: For instance the facts that the thermal conductivity does not exactly go like T^2, that there is an enhanced T^3 - term in the specific heat, the plateau in the thermal conductivity around 10 K, and deviations of the ultrasonic and dielectric response from the theoretical expectation need to be reconciled.

A. K. Raychaudhuri's and S. Hunklinger's critical and helpful remarks during the preparation of this manuscript are gratefully acknowledged.

REFERENCES

1. Amorphous Solids (Topics in Current Physics, Vol. 24); W. A. Phillips, ed. (Springer, Heidelberg 1981).
2. Glassy Metals I (Topics in Applied Physics, Vol. 46); H. -J. Güntherodt, H. Beck eds. (Springer, Heidelberg 1981).
3. H. v. Löhneysen, Phys. Reports **79**, 161 (1981).
4. E. W. Hornung, R. A. Fisher, G. E. Brodale, and W. F. Giauque, J. Chem. Phys. **50**, 4878 (1969).
5. R. C. Zeller and R. O. Pohl, Phys. R. v. B **4**, 2029 (1971).
6. J. C. Lasjaunias, A. Ravex, M. Vandorpe, and S. Hunklinger, Solid State Comm. **17**, 1045 (1975).
7. R. B. Stephens, Phys. Rev. B **13**, 852 (1976).
8. J. E. Graebner, B. Golding, R. J. Schutz, F.S.L. Hsu, and H. S. Chen, Phys. Rev. Lett **39**, 1480 (1977).
9. J. C. Lasjaunias, A. Ravex, and D. Thoulouze, J. Phys. F **9**, 803 (1979).
10. H. J. Schink, H. v. Löhneysen, W. Sander, and K. Samwer, Physica B+C **108**, 389 (1981).
11. A. Ravex, J. C. Lasjaunias and O. Béthoux, Solid State Comm. **40**, 853 (1981).

12. J. R. Matey and A. C. Anderson, J. Non-Cryst. Solids **23**, 129 (1977).
13. H. v. Löhneysen, D. M. Herlach, E. F. Wassermann, and K. Samwer, Solid State Comm. **35**, 591 (1981).
14. P. W. Anderson, B. I. Halperin, and C. M. Varma, Phil. Mag. **25**, 1 (1972).
15. W. A. Phillips, J. Low-Temp. Phys. **7**, 351 (1972).
16. K. G. Lyon, G. L. Salinger, and C. A. Swenson, Phys. Rev. B **19**, 4231 (1979).
17. J. Jäckle, Z. Physik **257**, 212 (1972).
J. Jäckle, L. Piché, W. Arnold, and S. Hunklinger, J. Non-Cryst. Solids **20**, 365 (1976).
18. B. Golding, M. v. Schickfus, S. Hunklinger, and K. Dransfeld, Phys. Rev. Lett. **43**, 1817 (1979).
19. B. Golding, J. E. Graebner, A. B. Kane, and J. L. Black, Phys. Rev. Lett. **41**, 1487 (1978).
20. S. Hunklinger, W. Arnold, S. Stein, R. Nava, and K. Dransfeld, Phys. Lett. A **42**, 253 (1972).
21. B. Golding, J. E. Graebner, B. I. Halperin, and R. J. Schutz, Phys. Rev. Lett. **30**, 223 (1973).
22. M. Schmidt, R. Vacher, J. Pelous, and S. Hunklinger, Proc. 5th Int. Conf. - The Physics of Non-Crystalline Solids (1982).
23. P. Doussineau, J. de Physique Lett. **42**, L-83 (1981).
24. M. v. Schickfus and S. Hunklinger, Phys. Lett. A **64**, 144 (1977).
25. A. K. Raychaudhuri and S. Hunklinger, Proc. 5th Int. Conf. - The Physics of Non-Crystalline Solids (1982).
26. G. Federle and S. Hunklinger, Proc. 5th Int. Conf. - The Physics of Non-Crystalline Solids (1982).
27. W. Arnold, A. Billmann, P. Doussineau, and A. Levelut, f. de Physique C6 **42**, 37 (1981).
W. Arnold, A. Billmann, P. Doussineau, and A. Levelut, Proc. 5th Int. Conf. - The Physics of Non-Crystalline Solids (1982).
28. L. Piché, R. Maynard, S. Hunklinger, and J. Jäckle, Phys. Rev. Lett. 32, 1426 (1974).
29. M. v. Schickfus and S. Hunklinger, J. Phys. C**9**, L439 (1976).
30. G. Bellessa, J. Phys. C **10**, L 285 (1977).
31. F. L. Black and P. Fulde, Phys. Rev. Lett. 43, 453 (1979).
G. Weiss and S. Hunklinger, private communciation.
32. P. J. Anthony and A. C. Anderson, Phys. Rev. B **20**, 763 (1979).
33. A. K. Raychaudhuri, Phonon Scattering in Condensed Matter, H. J. Maris ed., Plenum N.Y. 1980.
34. G. Bellessa, P. Doussineau, and A. Levelut, J. de Physique Lett. **38**, L-65 (1977).
35. G. Weiss, S. Hunklinger, and H. v. Löhneysen, Phys. Lett. **85**, 84 (1981).
36. M. T. Loponen, R. C. Dynes, V. Narayanamurti, and J. P. Garno, Phys. Rev. B. **25**, 1167 (1982).
37. M. Meissner and K. Spitzmann, Phys. Rev. Lett. **46**, 265 (1981).
38. J. Zimmermann and G. Weber, Phys. Rev. Lett. **46**, 661 (1981).
39. H. Tietje, M. v. Schickfus, and E. Gmelin, Proc. 5th Int. Conf. - The Physics of Non-Crystalline Solids (1982).
40. M. W. Klein, B. Fischer, A. C. Anderson, and P. J. Anthony, Phys. Rev. B **18**, 5887 (1978).
B. Fischer and M. W. Klein, Phys. Rev. Lett. **43**, 289 (1979).
41. R. J. Rollefson, Phys. Rev. B **5**, 3235 (1972).
42. E. Gmelin and K. Guckelsberger, J. Phys. C**14**, L21 (1981).
43. D. A. Ackermann, D. Moy, R. C. Potter, A. C. Anderson, and W. N. Lawless, Phys. Rev. B **23**, 3886 (1981).
44. T. Baumann, M. v. Schickfus, S. Hunklinger, and J. Jäckle, Solid State Comm. **35**, 587 (1980).
45. P. J. Anthony and A. C. Anderson, Phys. Rev. B **16**, 3827 (1977).
46. P. Doussineau, C. Frénois, R. G. Leisure, A. Levelut, and J. Y. Prieur, J. de Physique **41**, 1193 (1980).
47. U. Strom, M. v. Schickfus, and S. Hunklinger, Phys. Rev. B **25**, 2405 (1982).
48. C. Laermans, A. M. de Goer, and M. Locatelli, Phys. Lett. A **80**, 331 (1980).
49. M. Hofacker and H. v. Löhneysen, Z. Physik B **42**, 291 (1981).

50. J. W. Gardner and A. C. Anderson, Phys. Rev. B **23**, 474 (1981).
51. C. Laermans, Phys. Rev. Lett. **42**, 250 (1979).
52. B. Golding and J. E. Graebner, in "Phonon Scattering in Condensed Matter", H. J. Maris ed., Plenum N.Y. 1980.
53. J. Chaussy, J. Le G. Gilchrist, J. C. Lasjaunias, and M. Saint-Paul, J. Phys. Chem. Sol. **40**, 1073 (1979).
54. U. Bartell and S. Hunklinger, Proc. 5th Int. Conf. - The Physics of Non-Crystalline Solids (1982).
55. D. P. Jones, N. Thomas, and W. A. Phillips, Phil. Mag. B **38**, 271 (1978).
56. H. v. Löhneysen and H. J. Schink, Phys. Rev. Lett. **48**, 1121 (1982).
57. M. v. Haumeder, U. Strom, and S. Hunklinger, Phys. Rev. Lett. **44**, 84 (1980).
58. M. v. Haumeder, unpublished.
59. S. Hunklinger, J. de Physique C6-39, **1444** (1978).

DISCUSSION

Low Energy Excitations

After initial comments on the difficulty of identifying small concentrations of impurities in glasses, a comparison was made between electric dipoles in real glasses and those in alkali halides. In $KTaO_3$:Li the dipoles may give rise to a spin glass, but this does not appear to happen in real glasses. It was suggested that the difference was a result of the distribution of local parameters present in the glass, contrasting with the relatively well-defined local sites in $KTaO_3$:Li.

The discussion then concentrated on two topics, the nature of the tunneling states and a broader discussion of coherent phenomena.

W. A. Phillips, Chairman
B. Golding, Discussion Leader

COMMENTS

Low-Temperature Specific Heat of Amorphous Films

Disorder-induced excitations which can be described as two-level systems (TLS) exist in both insulating and metallic glasses.[1] In tetrahedrally bonded amorphous semiconductors their density might be smaller because of the rather "closed" network structure inhibiting atomic tunneling.[2] In order to check this conjecture we have measured the specific heat of electron-beam evaporated a-Ge films below 1K.[3] Our results also provide some clues concerning the spatial distribution of, and interactions between, specific defects, i.e. dangling bonds, which are known to be present in a large number ($10^{19} - 10^{20}cm^{-3}$) in vapor-deposited a-Ge films.

The specific heat C of three different a-Ge films is shown vs. temperature T in Figure 1. Above 1K, C exhibits a T^3 dependence in good agreement with previous work.[4] Below 1K, an anomaly centered at 0.4K is observed which

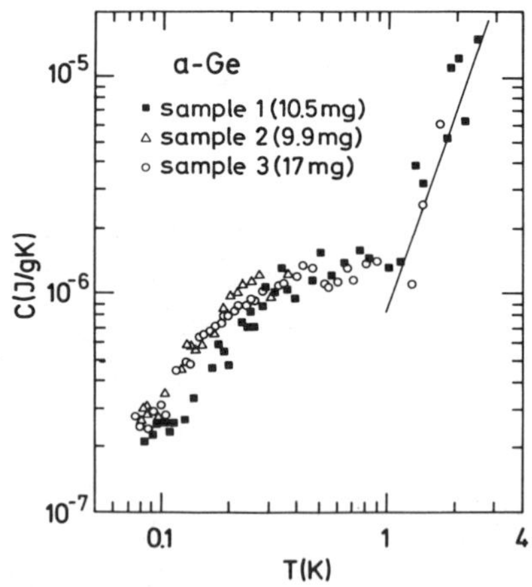

Fig. 1. Specific heat C of amorphous Ge films as a function of temperature T. The solid line is T^3 dependence as measured[(4)] above 1K.

falls off roughly as C = aT towards lower temperatures, with amplitude a between 2.9 and 3.9×10^{-6} J/g–K^2 for the different samples, i.e. three times larger than the TLS-induced linear term in vitreous silica. The maximum near 0.4K corresponds to a very small high-energy cut-off of the distribution of the underlying excitations, which is not observed for TLS in insulating glasses. From the results of an annealed film, for which C is strongly reduced and the maximum is shifted to ~0.1K, one can conclude that the density of TLS in our films is six times smaller than in vitreous silica.[3] Our estimate of a TLS$\lesssim 2.10^{-7}$ J/g–K^2 for the linear specific-heat coefficient due to TLS is compatible with some phonon scattering by TLS observed earlier in a-Ge.

The large specific heat reported in Figure 1 probably arises from dangling bonds whose density in our films agrees with the above-mentioned values as checked by ESR. However, the observed anomaly cannot be due to isolated dangling bonds, because the dipolar interaction between these and the hyperfine interaction are both two to three orders of magnitude weaker than the energy range corresponding to 0.1 and 0.4K. Therefore, it must be attributed to exchange-coupled clusters of dangling bonds. Because of the disordered structure, a broad distribution of exchange interactions J is very plausible, and in fact is supported by the linear T behavior of C below 0.3K. The temperature of the maximum corresponds to a cut-off at $J/k_B \sim 0.5$K. From entropy considerations,[3] the number of dangling bonds in clusters is ~10% of their total number. This ratio is larger than that expected from a random spatial distribution of dangling bonds, indicating that clustering is preferred. Finally, the reduction of C in the annealed film arises from the annealing of dangling bonds in clusters.

This work has been performed in collaboration with H. J. Schink.

H. v. Lohneysen
Physikalisches Institut
der Rhein.-Wesf. Hochschule
D-1500 Aachen, FRG

REFERENCES

1. M. von Schickfus, Chapter II.
2. W. A. Phillips, J. Low Temp. Phys. *7*, 351 (1972).
3. H. v. Lohneysen and H. J. Schink, Phys. Rev. Lett. *48*, 1121 (1982).
4. C. N. King, W. A. Phillips, and J. P. de Neufville, Phys. Rev. Lett. *32*, 538 (1974).

Soap Film Networks - A Model System for Amorphous Solids?

Tunneling modes must be a very general feature of disordered structures. *Macroscopic* disordered systems should show analogous features. An interesting example is the case of *2d soap film networks,* originally studied by Cyril Stanley Smith as a model for grain growth. For purposes of demonstration, these can be formed in a few seconds by squeezing some detergent foam between two transparent sheets on an overhead projector. If the random structure is disturbed, its relaxation towards equilibrium can be seen to include sudden rearrangements, as shown in Fig. 1. It is easy to see that the system is making a local transition, between the two levels of a two-level system. It is, of course, not a matter of quantum tunneling or thermal excitation, but rather the instability of one of the two levels. Apart from this, the system may be used to illustrate many aspects of the TLS, such as the interaction of individual TLSs.

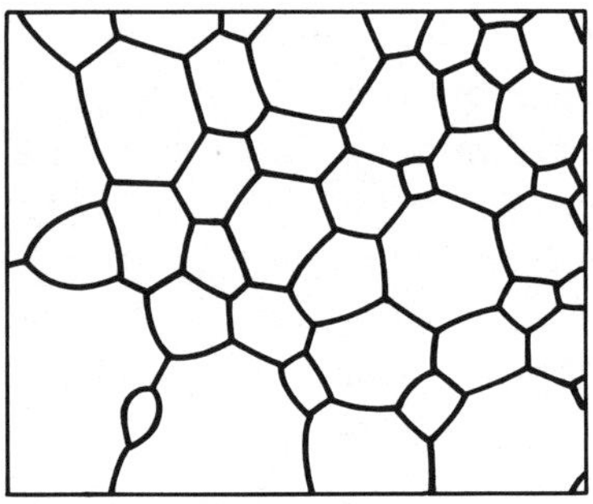

Fig. 1. Example of a soap film network in a computer simulation with periodic boundary conditions.

Such structures have been computer simulated by Weaire and Kermode (to be published), with a view to studying this and other essential features of random structures, less accessible in 3d systems. It is suggested that they might play the same role for amorphous solids as did Bragg's soap bubble rafts (*not* the same thing) in demonstrating the properties of crystals.

D. Weaire
University College, Dublin
Dublin, Ireland

COHERENT PHENOMENA IN GLASSES

J. Joffrin

Laboratoire de Physique des Solides
Université de Paris-Sud
Centre d'Orsay, 91405 Orsay, France

Concepts underlying and experiments revealing coherent phenomena in glasses and other disordered solids are reviewed. The description of low energy excitations in glasses in terms of "two level systems" (TLS) permits analogy with the rich variety of coherent experiments familiar in electron spin- and nuclear magnetic resonance. Both elastic and electric field pulses have been employed to reveal saturation of absorption, free induction decay, self induced transparency and both spontaneous and stimulated echoes in glasses at low temperatures. Information provided by these experiments on the nature of TLS in glasses is reviewed.

I. INTRODUCTION

"Coherence" in physics has a very general significance which applies to many different problems. In every case, "coherence" means appearance of a macroscopic quantity made up by addition of elementary or atomic quantities of the same nature. The counterpart of "coherence" is "incoherence" or randomness, brownian motion or statistical independence.

When a sound wave propagates in a liquid, one can define a local quantity like the pressure which oscillates or propagates coherently. The local velocity is also a non-zero statistical average over the different velocities of the molecules of the liquid. The phase of the propagating wave is perfectly defined and may be confirmed by interference experiments or phase-matching conditions.

More subtle is the case of superfluid helium: the Bose-Einstein condensation results in a non-zero mean value for a non-diagonal (boson) operator, the amplitude of condensation. It manifests itself in the Josephson-equations, where the phase appears as the conjugate of the amplitude of condensation in Hamilton-type equations.[1]. However, in that case, the phase can be understood from a more general and powerful point of view.[2] There is a relation between the phase ϕ and the circulation Γ of the vector potential (the velocity $\overline{v}$)

$$\Gamma = \int_{(c)} \overline{v} \cdot \overline{d\ell} \tag{1}$$

which is

$$\phi = \frac{m}{\hbar} \Gamma \,. \tag{2}$$

The circulation Γ is quantified as a consequence of the quantization of the phase in the wave function which includes the factor $e^{i\phi}$. Vortex lines result from this. The quantum of circulation is

$$\Gamma = \frac{h}{m} \,. \tag{3}$$

The same argument applies to superconductivity.

Very useful for what we need to discuss later is the case of an assembly of almost identical spins placed in an external applied magnetic field. A longitudinal magnetization appears. This is a diagonal macroscopic and static quantity when equilibrium is reached. However, it is easy to induce a non-diagonal or transverse macroscopic magnetization by a pulsed oscillating magnetic field. The magnetization oscillates afterwards according to the Larmor equations and one can observe "echoes" when later similar pulses are applied. In that case the coherence injected in the movement of the whole assembly of spins appears for a limited period of time but it can be restored at will as long as irreversible phenomena have not destroyed it.[3]

This last remark illustrates the usefulness of the echo-phenomenon and more generally of coherence. By observing the decay of a macroscopic signal, one can access the variety of the elementary processes which govern the irreversible relaxation of a population of spins or of an assembly of oscillators.

In recent years, many experiments of this type have been performed on glasses or on disordered systems. This paper intends to make a presentation of the subject.

II. AN ECHO EXPERIMENT

An echo experiment explores the relaxation properties of a population of "oscillators", whatever they are. The simplest example is a nuclear or an electronic spin population. In an external field there is a discrete number of levels, and as long as they are well separated one can consider only two levels whose energy separation is adjusted to be almost equal to the frequency of an exciting field (or vice versa). This alternating field can be equivalently an electromagnetic, acoustic or crystalline field if some indirect coupling exists between spin variables and strain. The assembly of spins forms a population distribution because, in general, they have not exactly the same energy. Local strains or inhomogeneities of the environment build-up a more or less narrow line. In a crystalline material, the line is considered to be narrow if the width is smaller than the energy separation of the two levels in question. In a disordered material, this condition may not be satisfied even for nuclear spins. For electronic spins or for impurity levels, one should speak of bands or densities of states instead of lines.

It can be shown in general that if one starts from an equilibrium situation where the static magnetization M_z is oriented parallel to the z axis and if one applies successively two "pulses" one can always get a signal at a time later i.e. an echo resulting from the coherence of the population of oscillators introduced by the first pulse and revealed by the second one.[4] If one wants to extract valuable information from the measurements of the echo the applied pulse-sequence must fulfill particular conditions. Let us call ω_o the resonance frequency of the spins, ω the frequency of the exciting field whose intensity is $h = \dfrac{\Omega}{\gamma}$ and is oriented perpendicular to z, $\Delta t_1 = \dfrac{1}{\delta\omega_1}$ the width in time of the first pulse, $\Delta t_2 = \dfrac{1}{\delta\omega_2}$ the width of the second and τ their separation in time. It can be shown that the oscillating transverse magnetization appearing at a time τ after the second pulse is given by

$$M_y(\omega_o) = -n_{lx}^3 \sin\phi_1 \sin^2\frac{\phi_2}{2} - 2n_{lx}^3 n_{1z} \sin^2\frac{\phi_1}{2} \sin^2\frac{\phi_2}{2}$$

where

$$\begin{cases} \omega - \omega_o = \Delta\omega \\ \phi_1 = (\Omega^2 + \Delta\omega^2)^{1/2}\, \Delta t_1 \\ \phi_2 = (\Omega^2 + \Delta\omega^2)^{1/2}\, \Delta t_2 \\ n_{1x} = \sin\theta \\ n_{1z} = \cos\theta \\ \mathrm{tg}\,\theta = \dfrac{\gamma h}{\Delta\omega} \end{cases} \tag{4}$$

For a narrow resonance line, (that is for a population of spins where $\Delta\omega$ remains much smaller than Ω) one gets the familiar result

$$M_y(\omega_o) \simeq \sin\phi_1 \sin^2\frac{\phi_2}{2} \tag{5}$$

which can be maximized for

$$\phi_1 = \frac{\pi}{2} = \frac{\phi_2}{2}\,. \tag{6}$$

This is by definition a $\left(\frac{\pi}{2}, \pi\right)$ pulse sequence (see Fig. 1). It means that at low intensity the amplitude of the signal observed should be proportional to h^3. If h is known in absolute value, the position of the maximum of intensity of the signal provides a value of the constant γ which is the coupling constant between the field and the spin e.g. Bohr magneton for spins, electric dipole moment for optical transitions. The maximum of amplitude of the signal provides a quantity which gives the number of spins in resonance.

On the other hand, when the population of spins has a large width, the answer is not as simple. It would require a sum over the quantities $M_y(\omega_o)$ with a weighting factor $p(\omega_o)$ and an additional phase factor in order to calculate the time evolution of the echo. Usually for disordered materials or glasses we are in the extreme case of a very broad band of oscillators with a smooth distribution $p(\omega_o)$ at a scale given by $\delta\omega_1$ or Ω. The cubic law in h^3 is preserved at low exciting power, and a signal maximum always occurs when h is progressively increased allowing estimates of the magnitude of γ and for the density of states.

What has been said of spins can be carried over without modifications to

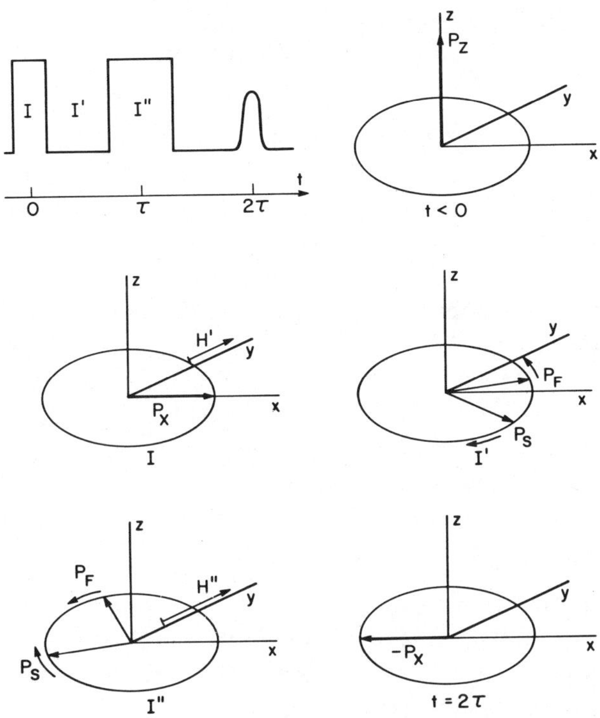

Fig. 1. The echo generation can be understood by considering the rotation of the static polarization P_z in rotating frame. A ($\pi/2$; π) pulse sequence works as indicated below: initially P_z lies along the z axis; during the phase I it is rotated 90° onto the x-axis; phase I' shows the effect of inhomogeneous broadening; some packets precess faster (p_F) than other (p_S). A second pulse, phase I" causes a 180° rotation of the x components of the P vectors; fast dipoles are then lagging slow dipoles; all meet at time 2τ to form an echo resulting from a macroscopic magnetization aligned along x.

any quantum system with a discrete number of levels including non-harmonic oscillators. Echoes have been observed from molecular vibrational modes of molecules in the gas state,[6] on plasma modes in a tokamak-machine[7] and also on various paraelectric impurities[8] in crystals.

Of major importance for us is the case of the so-called T. L. S. (two-level systems) in glasses. Other authors in this volume have explored these T. L. S. in relation to the intrinsic nature of glasses or disordered materials, and how they govern the low temperature properties of these materials: heat capacity, thermal conductivity, dielectric response, and propagation of sound. These T. L. S. have both an electric and an elastic dipole moment; thus echoes may be triggered by an oscillating electric or acoustic field. By extension we shall speak here also of spin-echoes, although more precisely one could call them "T. L. S. - echoes".

For glasses the difficulty of interpretation of the results is one step greater than for spin echoes because to define the physical properties of a T. L. S, one needs more than one parameter. That is, their resonant energy does not characterize completely their kinematics as it does for spins. In the previous chapter the T. L. S. were specified not only by E, their energy of excitation, but also by an additional parameter r which is a measure of the coupling of the T. L. S. to the lattice. For a given E, r is distributed. If a good model is available for the T. L. S. it would inevitably lead to a distribution of associated quantities. For instance, the electric induced dipole moment is distributed,[9] i.e. the coupling term γ in equation (4). Thus, the calculation of $M(y_o)$ requires[5] at the onset an integration with a weighting factor p(r). All T. L. S. do not satisfy the condition (6) simultaneously.

Not only is the induced moment distributed but so is the permanent electric dipole moment. It would be desirable to confirm certain microscopic models by systematically exploring the correlations between these different dipoles. Thus far only one attempt[10] has been made in that direction, by breaking the time-reversal properties of evolution of the echo during the interval of time $0 - \tau$ or $\tau - 2\tau$. Unfortunately, the answer was not definitive.

Thus, in considering T. L. S. in disordered materials, the simplicity of the narrow nuclear spin resonance lines is lost. Conversely, if the maximum of the echo signal is broad (as a function of the applied power) or even possesses a multi-peaked structure, it may provide quantitative information about the glass. For example, in silica glass two echo maxima have been observed: one of them due to OH^- molecules, the other considered to be intrinsic[35] (see Fig. 2). In γ-irradiated quartz two plateau have been clearly distinguished, corresponding to two kinds of electronic defects[8] (see Fig. 3).

In contrast to the usual spin echo situation an additional selection rule occurs when the propagation of the exciting pulses has to be considered. For instance, when the width in time of a laser beam pulse is a few tens of picoseconds and excites a population of molecules at resonance in a cell, the echo propagates and is strongly peaked in the forward direction. If $\vec{q}_e$ is the wave vector of the echo and $\vec{q}_1$ and $\vec{q}_2$ the wave vectors of the two applied pulses the selection rule is

$$\vec{q}_e = 2\vec{q}_2 - \vec{q}_1 \tag{7}$$

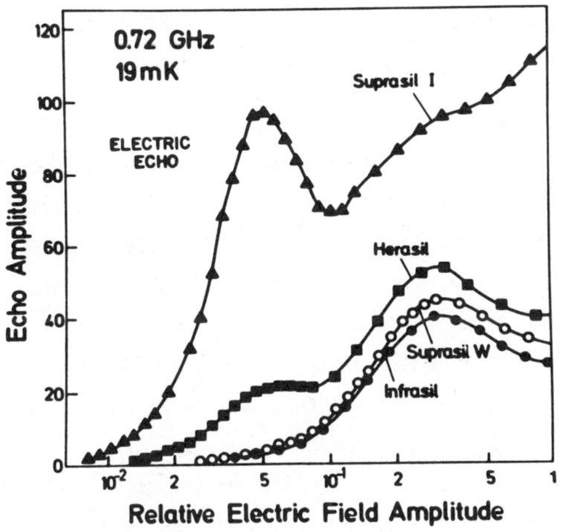

Fig. 2. The figure presents the echo amplitude versus the pulsed electric field[35] amplitude for four different glasses: Suprasil W and infrasil, free of OH^- show only one maximum; Herasil and Suprasil I show a second echo maximum proportional to the OH^- constant, at lower field (i.e. for higher induced electric dipole moment; compare with equations (4) and (6)). The authors normally infer that the first peak has an intrinsic character or at least that it is not due to OH^- impurities.

Clearly, it would be very difficult to obtain such echoes in an anisotropic medium because the laws of dispersion of the two input pulses and of the echo have to be satisfied simultaneously; in glasses this is not a very stringent condition.

Propagation effects also occur when the exciting fields are acoustic fields. The acoustic transit time in a piece of material can easily be made larger than the pulse length. Such an experiment[11] is more delicate to perform than the excitation of T. L. S. in glasses by electric fields. The sample has a finite size and the acoustic waves reflect back and forth, leading rapidly to many interfering signals. As in acoustics, a practical useful rule is to work only with the first observed echo to avoid spurious effects resulting from interferences or from non-exact plane waves (see Fig. 4). In some cases there is no other field than the strain field which couples with T. L. S. In metallic glasses, for example, because the electric field does not penetrate into the sample. In fact, even so the experiment in metallic glasses is difficult because the coherence relaxation time is very short.

A priori, echoes obtained with electric fields or strain fields should lead to complementary information on the T. L. S. For instance, the induced electric and elastic dipoles may have different distributions, but they are correlated, and this knowledge would provide a stringent test for any model.

On the other hand, these two ways of producting echoes should provide the same information concerning the relaxation behaviour of the T. L. S. (see below) furnishing a test of consistency of the two types of experiments. This last remark is valid because we consider T. L. S. in glasses or more generally quantum systems with only two levels so that the four operators 1, S_x, S_y, S_z describe completely the evolution of the two-by-two matrix characterizing the spin - 1/2. If, on the contrary, one faces spin - 1 or three-level systems, the transition induced by a magnetic field or an electric field must still be expressed using 1, S_x, S_y, S_z but for those induced by an acoustic field one must introduce the quadratic operators S_x^2, S_y^2 ··· $(S_xS_y + S_yS_x)$, ···

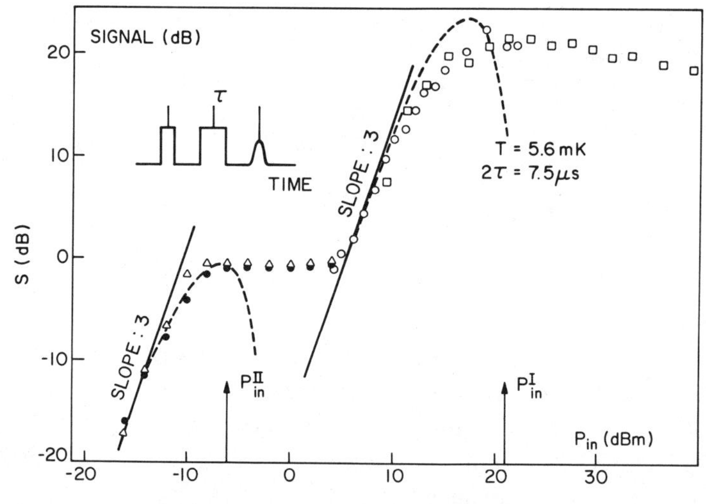

Fig. 3. This Figure gives the echo signal as a function of the power of the applied pulses in e-irradiated quartz. Different features must be noticed; at low power the slope of the curve goes to 3; two independent defects are observed leading to two "plateaux" reacting their value at two very different incident powers; the dotted curves result from a theoretical calculation of the signal corresponding to a $(\pi/2, \pi)$ pulse sequence applied to a narrow line of resonating electronic dipoles. The two corresponding electric defects have density of states differing by a factor 220 and a dipole moment by 1/22 (see Ref. 8): the first one has been attributed to holes trapped at aluminum impurities, and the second one to electrons trapped by crystal imperfections.

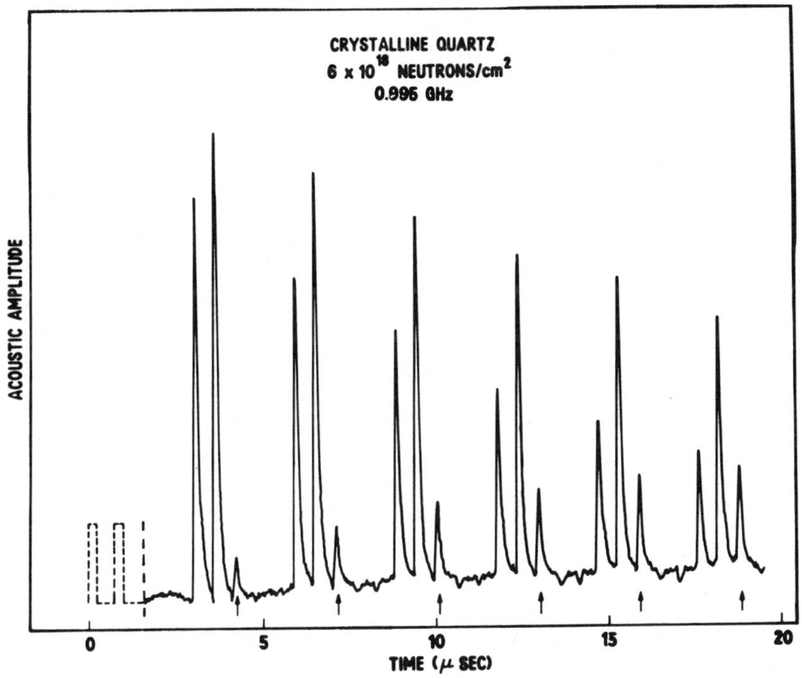

Fig. 4. Acoustic reflections and phonon echoes generated by a two-pulse sequence in neutron irradiated quartz: echoes are indicated by arrows. Each sequence of reflections and echoes is separated by 3 μsec, the round-trip transit time of the longitudinal waves in quartz.[37]

Consequently the relaxation times observed by triggering with an acoustic field or an electromagnetic field may not be equal.[36]

At this point it is probably worthwhile to mention that "echoes" have been obtained not only on T. L. S. or on electronic defects but also on a population of phonons excited by pulses of uniform electric field.[12] "Population" in that case means phonon-modes of almost the same frequency but with very different wave vectors. The procedure is the following: The first electric pulse at time $\tau = 0$ excites at the surface of the crystal (even if it is very rough) all the acoustic modes within the bandwidth of the pulse. At high frequency, a few GHz, the efficiency of such a process is not bad even if it is non-resonant. During the interval of time $0-\tau$ the different modes propagate in the crystal almost independently and at any time it would be impossible to detect any vibration at a given point in the crystal because all the modes combine destructively. Their wave vectors and elastic polarizations are different. The second electric pulse, at time τ, reverses the wave-vector of the incoming phonon-modes, each mode by each mode, using a term in the internal energy of the form

$$V = E^n s^m \tag{8}$$

where s is the strain and E represents the electric field; n and m are integers. The simplest term to be used as the origin of this parametric process is n = 2, m = 2. This is a four-wave interaction, the electric field of the second pulse being taken twice, together with the forward and backward (or reversed) modes. The phonon-photon selection rules are still given by (7)

$$\omega_e = 2\omega_2 - \omega_1 \tag{9}$$

$$\vec{q}_e = 2\vec{q}_2 - \vec{q}_1$$

But here $(\vec{q}_e, \omega_e)$ and $(\vec{q}_1, \omega_1)$ refer to phonon-modes, and $(\vec{q}_2,\omega_2)$ to photon modes (or electromagnetic modes).

To obtain an efficient parametric process we must have

$$\omega_e = \omega_1 = \omega_2 \tag{10}$$

$$\text{and} \qquad \vec{q}_e = -\vec{q}_1 \qquad \text{because } \vec{q}_2 = 0$$

During the interval of time $(\tau,2\tau)$ each mode propagates backward and at a time 2τ they all refocus in phase at the surface of the crystal where they had been generated. The signal is detected by the inverse process responsible for the excitation at time t = 0. This is the "phonon-echo". For this type of echo, the decay of the amplitude is determined by the acoustic attenuation in the sample i.e. by dissipative processes which cannot be wave-vector reversed.

Phonon-echoes of this type have been observed in many crystalline materials but not as yet in glasses. It is important to remember that here, the coherence was introduced into the phonon population.

For completeness we note another type of echo-experiment has proved valuable in glasses: they are called "backward-wave-phonon echoes":[13] They are to be distinguished from the "phonon-echoes" mentioned in the previous paragraph. Here, the first pulse is an ultrasonic wave and the second, a standing-wave electric field. The effect is thus very similar to "phonon-echoes" because the echo is detected as an acoustic wave, and it is as well due to a parametric process. However in glasses the important non-linearity is associated with the presence of the T. L. S. The coherence of the T. L. S.

population plays no role. More precisely, if the T. L. S. have a response to both the applied acoustic and electric fields, it is of no consequence to know their behavior after the second pulse has been applied. The T. L. S. follow a driving force and their relaxation properties do not play a role. The T. L. S. instead play the role of an indirect coupling between the phonon field and the electric field because they have both an electric and an elastic dipole moment. The coherence is attached to the exciting phonon field and not to the T. L. S. population. This is in contrast to T. L. S.-echoes where their free-precession during the interval of times $0-\tau$ and $\tau-2\tau$ is the source of incoherence (or irreversible relaxation). To summarize, the "backward-wave-phonon-echoes" appear as a particular case of "phonon-echoes". Wave-vector reversal is produced as well by a parametric process but only a single phonon mode is concerned.

III. RELAXATION OF THE ECHO SIGNAL

The main advantage of the echo-method is to pick-up only the influence of the non-reversible decay processes on the amplitude of the signal.

The irreversible processes are all those which correspond to a thermal fluctuation of the resonance parameters of each oscillator. In a gas of molecules the collisions have such an effect, including transitions between different states, exciting the vibration up to anharmonic vibrations.

In a disordered solid one must distinguish between two types of relaxation processes for the T. L. S. echoes: a jump in energy with a modification of the quantum numbers corresponding to an exchange of energy with the thermal bath due to inelastic collisions which occur with phonons or free electrons. The population of the excited levels is changed or, in an other language, the longitudinal "magnetization" of the T. L. S. is modified. The associated relaxation time is called T_1. It varies with T, the temperature of the bath and can be explored by an echo-method using three unequally spaced pulses.[5–11] This has been done carefully in glasses leading in general to the conclusion that the phonons, at low temperature, control T_1. T_1 is rather well proportional to T^{-1}, a signature of a one phonon direct process. From the numerical value of T_1 one can extract a coupling constant between the T. L. S. and the strain of the local environment. This constant is a deformation potential and appears to be very large in comparison with true spins, a few electron volts.[11] Incidently, this answer is more valuable than the coupling strength deduced from standard acoustic measurements because in the later, the deformation potential is always multiplied by the density of states of the spins, a quantity rather difficult to measure independently. Experimentally T_1 is obtained when the echo signal obtained after three applied pulses is measured as a function of the interval of time t between the second and third pulse. The decay pattern is not as perfectly exponential as one would like to have it, in particular at long times (see Fig. 5), possibly an effect of the distribution of T_1 is, but it is good enough to confirm one-phonon relaxation processes.

The second type of relaxation is characterized by T_2. A most important problem is to evaluate the relaxation time T_2, the coherence time of the excited population of T. L. S. T_2 is obtained by measuring the echo amplitudes as a function of τ following a two-pulse sequence. There is no doubt that a diffusion process is responsible for its behavior. The number of non-resonant T. L. S. (unlike-spins) in a glass is very large. The T. L. S. are all coupled together by pseudo-Ising exchange forces through the virtual phonon bath.[14] Thus any thermal fluctuation of the unlike-T. L. S. will be perceived by the resonant T. L. S. as a thermal fluctuation of the local field. Some experiments seem to prove directly that diffusion occurs.

The idea of diffusion was introduced several years ago[15] and it was

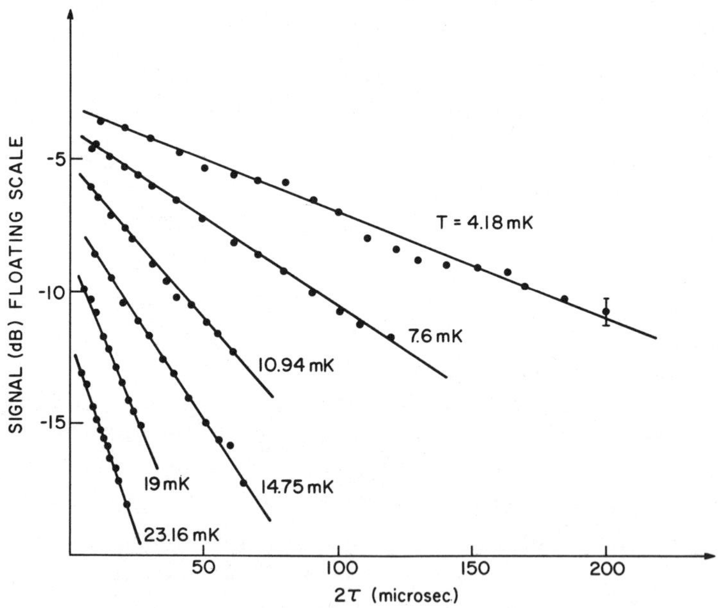

Fig. 5. Experimental results: intensity of the echo signal obtained in silica-glass versus 2τ; these data are taken[8] at low power in a region where the law in h^3 is fulfilled (see (5)); for that very reason the dynamic of the signal is not very large. It is a matter to taste to decide if the decay is exponential or not.

suggested that T_2 would vary according to T^{-2} where T is the temperature and that the decrease of the echo as a function of τ should be given by:

$$\text{signal}\ (T,\tau) \cong \exp(-aT^4\,\tau^2) \tag{11}$$

Experimentally some results[11] are in favor of the T^{-2} law, some are not. The $\exp(-a\tau^2)$ decay has never been observed unambiguously, probably because in disordered materials one always faces a distribution of relaxation times. The calculation leading to (11) supposed in fact that the fluctuation time of the unlike-T. L. S. was larger, for any of them, than the interval of time τ. In other words:[16]

$$\tau \ll \{T_1\} \tag{11}$$

However, since in a glass the properties of the T. L. S. are widely distributed, some of the T. L. S. are strongly coupled with strain fluctuations and others only weakly. As a result there is a distribution in T_1 for the unlike-T. L. S. and the diffusion theory[17] leads to the law:

$$\text{signal}\ (T,\tau) \cong \exp(-bT\tau) \tag{12}$$

if τ falls within the distribution of T_1 of the whole T. L. S. assembly $\{T_1,\tau\}$. Some authors claim to have observed that law (12) instead of (11) and in addition to have determined the value of the constant b in reasonable agreement with the calculated one, possibly fortuitously. Let us mention here that the law given by (12) has been observed in spin-glasses[25] which in common with glasses seems to have a hyperbolic distribution of relaxation times for the whole assembly of T. L. S. and a constant density of states (see Figs. 5 and 6).

The same refined theory[17] explains also why in the other extreme case, τ larger than any of the T_1, the signal should be:

$$\text{signal}\ (T,\tau) \cong \exp(-cT^{1/2}\tau^{1/2}) \tag{13}$$

with

$$\tau \gg \{T_1\}$$

Experimentally it is rather confusing [18] that in a partially disordered linear polymer, a substance where the coupling of the T. L. S. with the phonons is

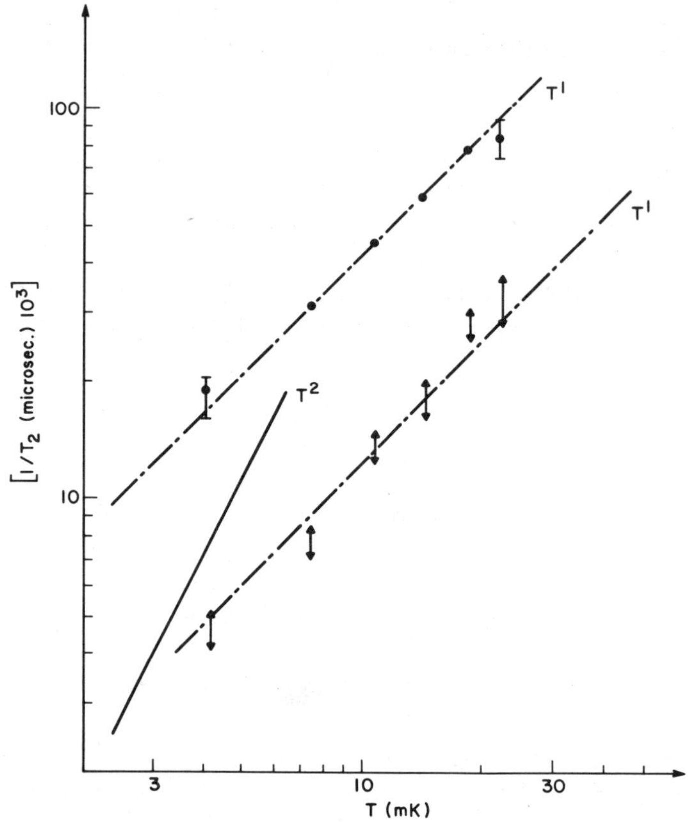

Fig. 6. Two experimental results giving the line width $\frac{1}{T_2}$ versus the temperature[8]; the upper curve is obtained for higher power than the lower curve; the last one corresponds to the experimental points of Figure 6. A straight line which corresponds to a T^2 variation is drawn for comparison.

weak,[19] thus implying $\{T_1\}$ large the law (13) has been observed (see Fig. 7). One would expect instead to observe (11).

At this stage, it is probably useful to mention a similar mechanism for T^{-2} dependence of the coherence relaxation time. It was proposed to explain a completely different experiment. If, in a disordered system, one introduces optical centers (like Eu^{3+} in silicate glass[20] or Pr^{3+} in BeF_2 or GeO_2 glass[21]) the homogeneous line-width of the optical centers is proportional to T^2. The mechanism responsible for this behaviour is the following: the T. L. S. and the optical centers are coupled and the T. L. S. have their energy modulated by the phonons of the thermal bath. Consequently the optical centers feel indirectly the thermal fluctuations, if the temperature is low in comparison with the energy of the unlike-T. L. S., the homogeneous line-width of the optical centers is proportional to T^2. This has been experimentally observed for almost two decades in T.

For our echo problem the same mechanism may be applied replacing the optical centers by the resonant-T. L. S. and we recover the situation described previously. Several papers in this volume (see D. Huber, W. Yen and R. Brewer, R. Macfarlane) consider these effects for optical transitions observed by F. L. N. or F. I. D. techniques. However, it is worthwhile mentioning that these authors describe a case where T_2 varies as T^{-1} (Pr^{3+} in silica glass) and that they also have observed non-exponential decay. One conclusion is probably that it is too naive to hope to observe only one behavior for $T_2(T)$, either at low temperature for the T. L. S. or for optical centers in another range of temperature.

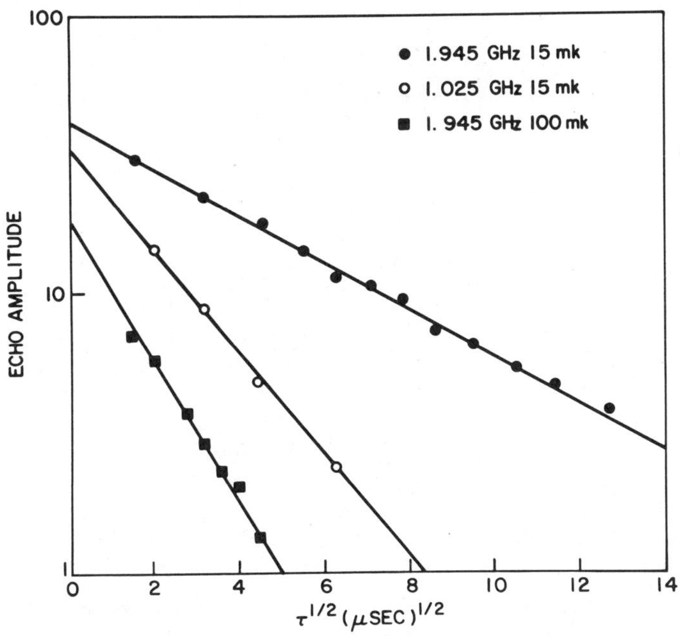

Fig. 7. Decay of the echo signal in polyethylene versus the parameter $\tau^{1/2}$ (Ref. 18). Under various conditions of frequency and temperature (13) is verified.

IV. OTHER COHERENT EFFECTS: SELF-INDUCED TRANSPARENCY

We return to propagation effects by considering the conditions for high transparency to acoustic waves propagating in glass. At any given frequency there are T. L. S. on speaking-terms. Thus, at low incident power, the medium is opaque, providing a large absorption. But above a critical threshold for a given pulse-width[22] the wave propagates with anomalously low energy loss. This is the phenomenon of self-induced transparency which occurs when the "area" of the pulse is "2π" (see condition (6)). Here one needs:

$$\phi_1 = 2\pi n \qquad (n = 1,2,3 \ldots) \tag{14}$$

The main effect of self-induced transparency for plane waves is an decrease of the velocity in comparison with low power conditions (see Fig. 8). At the same time, the width of the pulse should narrow because the continual absorption from the pulse leading edge and emission of energy into the pulse training edge stops. A detailed measurement of the increase of the intensity

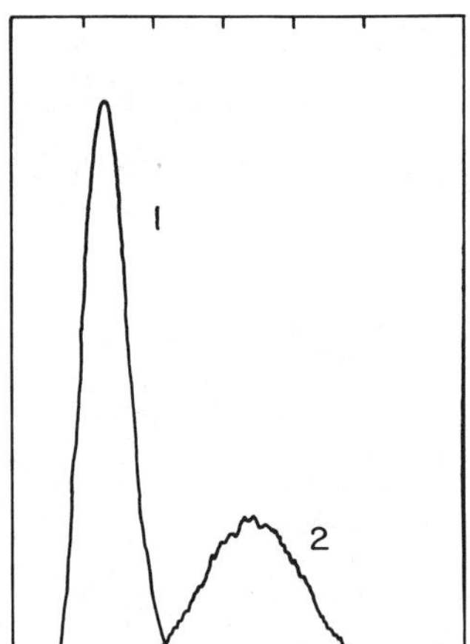

Fig. 8. Propagation of an acoustic pulse in a glass in the coherent regime i.e. when the time for a round trip 2τ is smaller than T_2. At low energy (ϕ_1 and ϕ_2 in (5) smaller than $\frac{\pi}{2}$) the velocity is rather small and the pulse is broadened; when the energy is increased (pulse labelled 1) the transit time 2.2 μs. becomes shorter. In this experiment,[23] the change in velocity propagation is as high as 10%.

of the acoustic pulse in such conditions should provide information on the nonresonant relaxation mechanism in glasses.

The phenomenon has been beautifully demonstrated[23] in glasses, surprisingly more clearly than in acoustic paramagnetic resonance in a crystalline material[24] Similar effects could be observed in optics using picosecond laser pulses, particularly on the T. L. S. if they can be directly excited.

V. CONCLUSION

A wide variety of materials can become glasses. If glass is really a new phase, as the crystalline phase is, the question remains open. Without any doubt the majority of glass compounds exhibit a moderately sharp transition from the liquid state. We may ask: what are the dynamic characteristics of that transition?

It is clear that most of the experiments reported previously have no connection with the glass-transition, and probably not with the glass-state. However, the T. L. S. are the typical excitations which govern the thermodynamic and transport properties at low temperature and the disorder is essential, leading to a constant density of states, distributed properties, etc. But the phenomena reported here could also have been descriptive of highly distorted crystalline materials. So, in conclusion, I would prefer to enlarge the class of materials envisaged to the wider category of disordered systems even though no experiments using coherence has been published so far.

The two important groups of disordered materials to be considered are, on one the hand, crystalline materials with a large number of impurities or defects, and on the other hand, those which have to a true glass state: spin-glass systems (a world in itself) and spin-glass like systems (dipolar electric glasses, molecular glasses, ...).

First, it seems worthwhile noting that in crystalline semiconductor materials, electronic non-ionized impurities may be studied by most of the techniques used in glasses at low temperature. In fact, here also they govern the transport properties as a result of their strong interactions with phonons, the concepts used to understand their properties are the same as for glasses.[26] In the same category are crystalline quartz-crystals irradiated with neutrons[27] or γ-rays.[8] Bands of defects appear which possess large electric and elastic dipole moments. A beautiful example of what can be obtained by echo experiments combined with other techniques has been provided in this volume. In the amorphous semiconductor a—As_2S_3, it has been shown[34] that the T. L. S. can be excited by light pumping, thus decreasing the echo signal.

On the other hand, possibly more fruitful for the future and eventually for the understanding of the glass transition are materials where molecular defects with dipole or quadrupole electric moments lead to a glassy state. The two most studied are $KTaO_3$:Li[28] and the quadrupolar-glass phase in solid hydrogen.[29]

Although not yet so well advanced, two other examples should be mentioned: the mixed compounds KBr–KCN[30] and KCl–OH. KCl–OH is well known[32] at low T and at low frequency to have a dielectric constant which exhibits a maximum depending on the OH^- content. As in the case of spin glasses, there exists a thermal remanent polarization. Recent careful measurements at very low temperature ($0.01K < T < 0.5K$) of both the real and imaginary parts of the dielectric constant at different frequencies show a behaviour very similar to $KTaO_3$:Li around 50K. In addition the relaxation time T_1 deduced from echo-measurements at 300 MHz are in very good agreement[31] with those deduced from dielectric relaxation. On the other hand, dielectric constant measurements at 9 GHz where the authors observe[33] a log T behaviour for $\Delta\epsilon/\epsilon_0$ between 1K and 10K is only a signature of a broad band density of T. L. S. These results are not considered as proof of the existence of a glassy state. At $T > 1K$ the situation is very similar to OH^- in silica-glass: there is no collective freezing of the polarized impurities. The advantage of KCL–OH^- over $KTaO_3$:Li or even KBR–CN^- is the low concentration of impurities (0.1%) resulting in simplification of the models.

These last examples indicate that the subject of coherence in disordered materials will continuously move to new aspects. Concerning the glass transition, the dynamical properties are of clear concern. Is there any critical slowing-down or only a progressive freezing? Echo-techniques may be particularly well adapted to help to resolve such questions.

REFERENCES

1. P. W. Anderson, Rev. of Mod. Physics **38**, 298 (1966).
2. A. Blandin (private communication).
3. E. L. Hahn, Phys. Rev. **80**, 580 (1950).
4. A. Bloom, Phys. Rev. **98**, 1105 (1955).
   W. B. Mims, Electron paramagnetic resonance, editor: S. Geschwind (Plenum Press 1972).
5. L. Bernard, L. Piché, G. Schumacher, J. Joffrin, J. Graebner, J. de Physique, Lettres **39**, L 126 (1978).
   L. Bernard, L. Piché, G. Schumacher, J. Joffrin, J. of Low Temp. Phys. **35**, 411 (1979).
6. I. D. Abella, N. A. Kurnit, S. R. Hartmann, Phys. Rev. **141**, 391 (1966).
7. G. Hermann, R. Hill, D. Kaplan, Phys. Rev. **156**, 118 (1967).
8. L. Bernard, M. Saint-Paul, J. Joffrin, J. de Physique Lettres **40**, L 593 (1979).
   M. Saint-Paul, J. Joffrin, J. of Low Temp. Phys. **49**, 195 (1982).
9. W. A. Phillips, Phil. Mag. B **43**, 747 (1981).

10. L. Bernard, L. Piché, G. Schumacher, J. Joffrin, J. de Physique, **39** C6, 957 (1978).
11. B. Golding, J. E. Graebner, Phys. Rev. Letters **37**, 852 (1976).
J. Graebner, B. Golding, Phys. Rev. **B 19**, 964 (1979).
12. S. Popov, N. Krainik, Sov. Phys. Solid State **12**, 2440, (1971).
J. Joffrin, A. Levelut, Phys. Rev. Letters **29**, 1325 (1972).
A. Billmann, C. Fresnois, J. Joffrin, A. Levelut, S. Ziolkiewicz, J. de Physique, **34**, 747 (1973).
R. Melcher, N. Shiren, Phys. Rev. Letters **34**, 731 (1975).
13. N. S. Shiren, W. Arnold, T. Kazyaka, Phys. Rev. Letters **39**, 239 (1977).
14. J. Joffrin, A. Levelut, J. de Physique **36**, 811 (1975).
15. J. Black, B. Halperin, Phys. Rev. B **16**, 2879 (1977).
16. W. Mims, Phys. Rev. **168**, 370 (1968).
S. McCall, E. Hahn, Phys. Rev. **183**, 457 (1969).
R. Rammal, R. Maynard, J. de Physique, Lettres **39**, L 195 (1978).
17. R. Maynard, R. Rammal, R. Suchail, J. de Physique Lettres **41**, L 291 (1980).
18. B. Golding, J. Graebner, W. Haemmerle, Phys. Rev. Letters **44**, 899 (1980).
19. G. Frossati, J. Gilchrist, J. of Phys. C **10**, L 150 (1977).
W. A. Phillips, Proc. Roy. Soc. **A 319**, 565 (1970).
20. P. Selzer, D. Huber, D. Hamilton, W. Yen, M. Weber, Phys. Rev. Letters **36**, 813 (1976).
21. J. Hegarty, W. Yen, Phys. Rev. Letters **43**, 1127 (1979).
22. S. McCall, E. Hahn, Phys. Rev. **183**, 457 (1969).
23. J. E. Graebner, B. Golding, in "Lattice Dynamics" ed. M. Balkanski (Flammarion, Paris, 1977) p. 464.
24. N. Shiren, Phys. Rev. B **2**, 2471 (1970).
25. D. Lewitt, R. Waldstedt, Phys. Rev. Letters **3B**, 178 (1977).
26. H. Zeile, O. Mathani, K. Lassmann, J. de Physique Lettres **40**, L 53 (1979).
27. C. Laermens, Phys. Rev. Letters **42**, 250 (1979).
28. U. Höchli, Phys. Rev. Letters **48**, 1494 (1982).
U. Höchli, H. Weibel, L. Boatner, J. Phys. **C 12**, L 563 (1979).
29. M. Devoret: J. Phys. C **13**, 2257 (1980).
N. Sullivan, M. Devoret, J. de Physique Lettres **40**, L 559 (1979).
30. A. Loidl, R. Feile, K. Korr, Phys. Rev. Letters **48**, 1263 (1982).
S. Bhattacharya, S. Nagel, L. Fleishmann, S. Susman, Phys. Rev. Letters **48**, 1267 (1982).
31. G. Bonfait, H. Godfrin, J. Joffrin, D. Thoulouze, to be published.
32. A. T. Fiory, Phys. Rev. **B 4**, 614 (1971).
33. M. Saint-Paul, M. Mesa, R. Nava (submitted to Phys. Letters).
34. B. Golding: this Volume.
35. B. Golding, M. Schickfus, S. Hunklinger, K. Dransfeld, Phys. Rev. Letters **43**, 1817 (1979).
36. N. Shiren, T. Kazyaka, Phys. Rev. Letters **28**, 1304 (1972).
37. B. Golding, J. Graebner, *Phonon Scattering in Solids*, edited by H. J. Maris (Plenum, New York, 1980) p. 11.

COMMENTS

Coherent Resonance Studies of Tunneling Processes in Glasses

The past decade has seen significant advances in understanding the low temperature behavior of glasses in terms of two-level atomic tunneling processes.[1] Perhaps the most satisfying aspect of this work is the way in which it has emerged as a distinct sub-branch of resonance physics, in much the way that optical resonance grew from the foundations of nmr and esr. A pivotal question remains to be satisfactorily answered, however: what is the microscopic structure of a tunneling center? Indeed, is one description sufficient to explain the apparently universal appearance of two-level atomic tunneling in the presence of structural disorder? To set the stage for discussion, Table I exhibits some of the diverse materials in which atomic tunneling has been observed.

Theory has been little help in solving this puzzle. No "first principles" calculation which derives the existence of tunneling systems in glass exists. Computer modeling of disordered structures has attempted to illustrate the plausibility of the low-energy modes, but no calculation has done more than to hint at the presence of a low energy tail to a broad distribution of localized states.

Experimentalists, on the other hand, have been actively attacking the problem on a variety of fronts. Their approaches may be grouped into the following categories.

(1) Classification. Low temperature properties of standard and novel types of disordered media are measured and tabulated with the hope of discovering correlations or unique phenomena.

(2) Selective disorder. An increasing number of defects are introduced into a nearly perfect crystal, until "glass-like" properties emerge. A variant is the use of particle bombardment to create disordered regions in crystals or clusters of point defects.

TABLE I

Classes of Solids Exhibiting Tunneling Centers

Inorganic insulators:	SiO_2, GeO_2, B_2O_3 ...
Organic polymers:	Polystyrene, polyethylene, PMMA
Amorphous semiconductors:	As_2S_3, Se, Si:H
Amorphous metals:	PdSi, NiP, superconducting ZrPd
Superionic conductors:	Na β-alumina, ZrO_2
Defective crystals:	n-irradiated quartz ω-ZrNb Polycrystalline Cu

(3) Correlation with other classes of defects. Tunneling centers are linked to electronic centers which can be studied by transport, esr, or by their optical response.

(4) New probes. Scattering methods eg. neutrons, x-rays, or light scattering.

In spite of our lack of a detailed picture of the tunneling centers, the phenomenology of resonance as applied to glasses has demonstrated the subject's beauty and elegance. In silica glass, for example, because of the strongly temperature dependent relaxation times (T_1 and T_2), it is possible to span the range from incoherent to coherent regimes by varying the temperature by less than 1 degree![2] This remark is literally correct, but it must be admitted that the region referred to lies between 0.001 - 1K. The reason for the sensitivity of tunneling system properties to temperature results from the coincidence of tunneling energies to k_BT in the spectral region below 50 GHz. The necessity for low temperatures requires that most experiments be performed in the environment of a dilution refrigerator.

Although we have referred to the use of resonance methods in the study of tunneling processes, the most common types of resonance, nuclear magnetic and electronic spin resonance have not been particularly helpful. Rather, novel and unorthodox resonance methods using rf and microwave acoustic and electric field resonant processes have provided the most powerful probes of two-level systems. This arises (1) from the large values of

TABLE II

Resonance Frequency and Wavelength Scales

	f	λ
nmr	10-100 MHz	m
esr	10-30 GHz	cm
optical	10^5 GHz	μm
acoustic	0.1-10 GHz	μm
electric	0.1-10 GHz	cm

the deformation potential linking the tunneling systems to long wavelength phonons, and (2) from the presence of electric dipole moments on tunneling systems.

Table II compares frequency and wavelength regimes for spin, optical, and glass resonance experiments. Although the spectral region and excitation field will obviously determine the particular experimental method, it is the wavelength of the resonant radiation field relative to the sample dimensions d which determines the character of the system's response. In this respect, it can be seen that optical and microwave acoustic resonance share a common wavelength region, one in which λ is much less than d. Such conditions imply that resonant emission, to be observable, must satisfy non-trivial wavevector conservation conditions in addition to energy conservation. This in turn results in directional emission of spontaneous or stimulated resonant radiation in a coherent regime. The following discussion will illustrate these points by describing coherent resonance experiments on glasses using microwave phonons and photons at low temperatures.

Coherent Acoustic Resonance

Microwave phonon experiments are often carried out in the geometry shown in Fig. 1. A coherent propagating sound wave is generated by a thin piezoelectric film transducer e.g. ZnO sputtered onto a flat polished surface of the sample. A short electric field pulse across the ZnO launches a sound

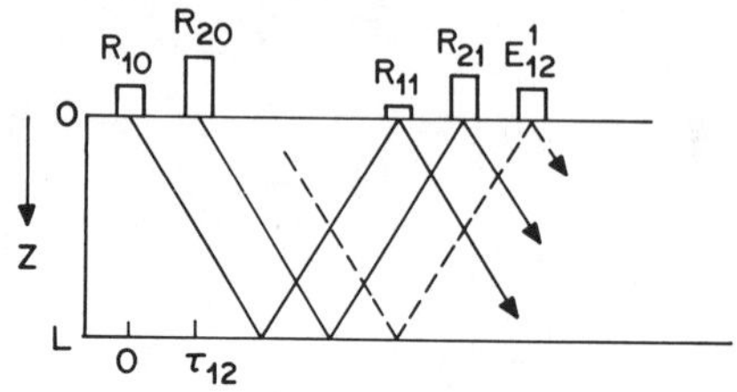

Fig. 1. Illustration of space and time relationships for pulsed acoustic echo experiment.

wave which propagates perpendicular to the surface, is totally reflected from the opposite parallel surface, and returns to the ZnO receiver. For collinear propagation the phase matching condition $\vec{k}_{echo} = 2\vec{k}_2 - \vec{k}_1$ yields $\vec{k}_{echo} = \vec{k}_1 = \vec{k}_2$. The echo therefore is a pulse traveling at time τ_{12} after the second generating pulse.

Since the acoustic pulses are trapped in the medium, they undergo multiple reflections with a decay time characteristic of the effective attenuation of the system. A particularly clear example of phonon echo generation in a medium with a relatively small number of absorption lengths $\alpha_o L$ is neutron-irradiated quartz (NIRQ)[3] as shown in Fig. 2. Light bombardment (integrated flux $\lesssim 10^{19}$ neutrons/cm^2) creates small (<20Å) disordered regions in a crystalline matrix.[4] The clusters themselves are a relatively dense form of disordered SiO_2 which contain tunneling centers. The effective density of tunneling states (per unit volume) is nevertheless only about 5% that of silica glass which results in a relatively small $\alpha_o L$. In Fig. 2 the two generating pulses decay slowly with time but we observe that the phonon echo grows with propagation distance and time. At 20 mK all tunneling lifetimes are long compared to τ_{12} and the growth of the echo arises from the coherent transfer of energy from the generating pulses into the tunneling systems with subsequent re-radiation into the echo. The growth of the echo with propagation distance is a consequence of the conservation of total pulse area $A = (\gamma/\hbar) \int e(t)dt$ [γ, deformation potential coupling; e, strain field envelope] and is a manifestation of phonon self-induced transparency.

Under conditions in which $\alpha_o L$ is large ($\approx$10) and the generating pulses are appreciably absorbed, echo generation can be very effective with the echoes themselves generating more echoes. Fig. 3 illustrates this situation in fused silica at 5 mK for 1.0 GHz phonons. Although multiple phonon echoes have been reported previously,[5] in this situation the first echo is larger than either of the generating pulses after one pass through the glass. At least five phonon echoes are generated from two pulses.

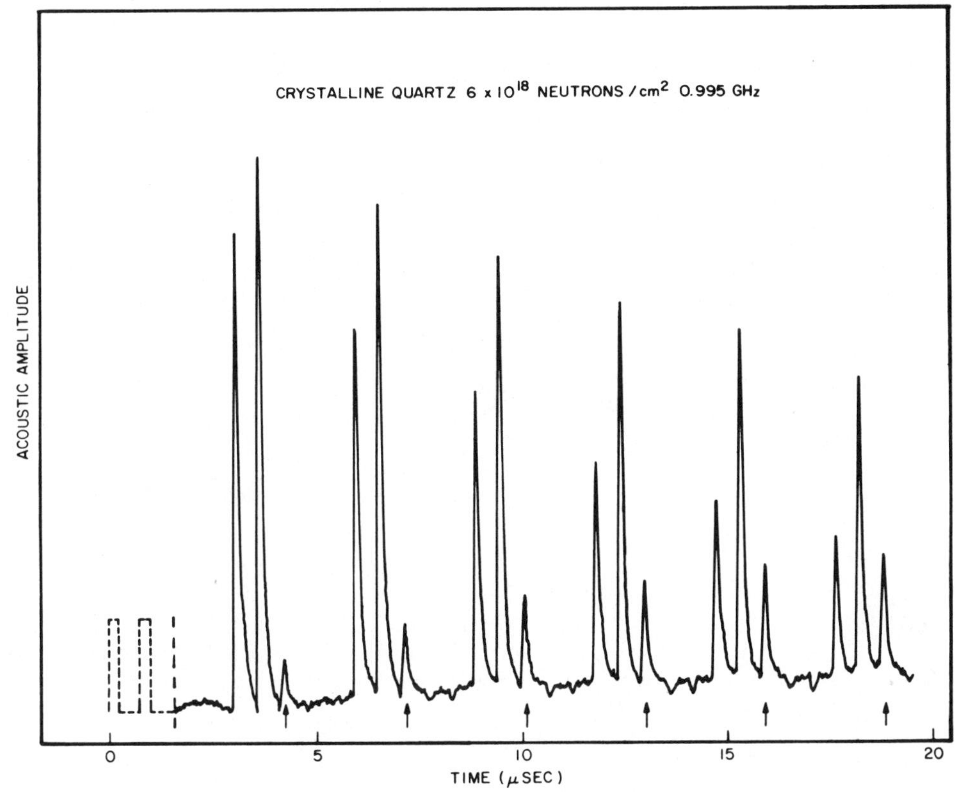

Fig. 2. Phonon echoes in neutron-irradiated quartz. The echoes are indicated by small arrows.

The above examples of phonon echoes have been exact analogs of spin echoes[6] or, more precisely, photon echoes.[7] Such echoes are a manifestation of free induction decay. Solomon first showed, in nmr, the existence of an echo arising from spin nutation, usually referred to as a rotary echo. In the vector representation shown in Fig. 4, the polarization is driven by a multiple π pulse to nutate along the x-axis. After the system is "wound-up", it is forced to "unwind" by reversing the phase of the driving field. In cases in which there is a distribution of matrix elements, as the xy plane becomes uniformly populated after a long driving pulse, the phase reversal of the spins restores a state of maximum polarization when the spins align along $\pm y$. The rotary echo appears while the system is being driven by the resonant field.

Glasses appear to be nearly ideal systems in which to study rotary echoes since there is a broad distribution of matrix elements. This distribution leads

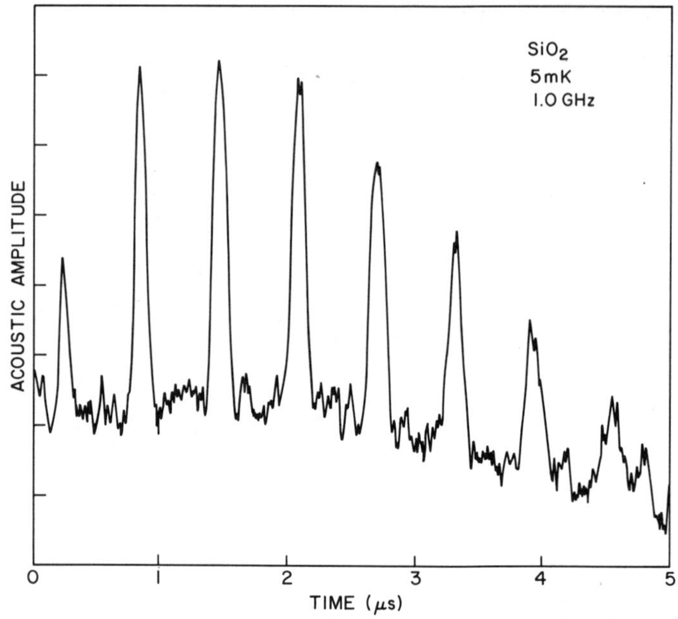

Fig. 3. Multiple phonon echoes in Suprasil W glass. The first two pulses are the excitation pair, attenuated by one round-trip pass through the 0.6 cm sample.

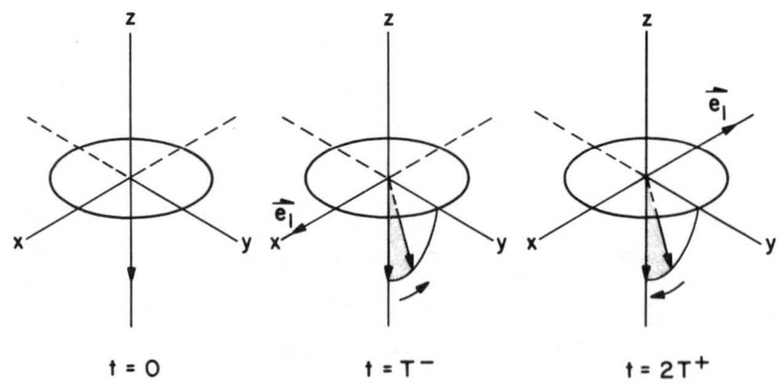

Fig. 4. Bloch-vector description of rotary echo formation. The "pseudo-polarization" is initially oriented along −z. Upon application of a transverse e field, the vectors nutate at different rates in the initial direction and dephase. After the e field is reversed (π phase shift), the Bloch vectors nutate at the same rates but in the opposite direction. Rephasing along −z creates the echo.

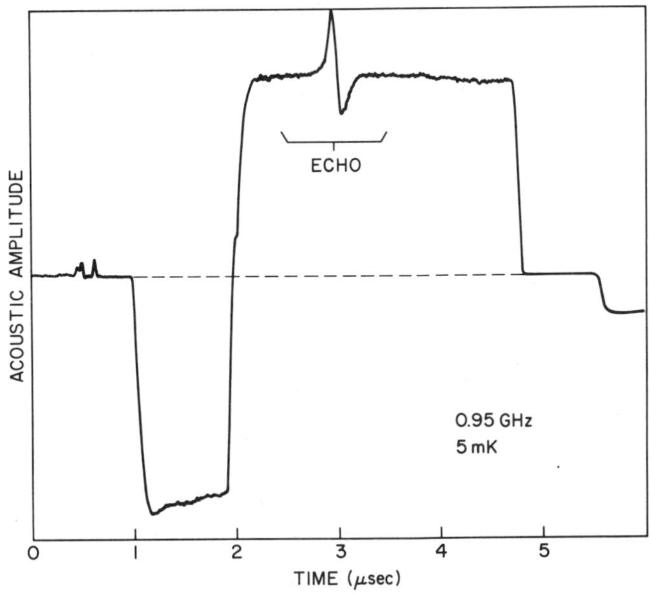

Fig. 5. Rotary phonon echo in Suprasil W silica glass. The echo is the modulation of the second, positive-going part of the excitation pulse.

to a distribution in T_1 which is responsible for the time-dependent tunneling specific heat at low temperature. Here the distribution of phonon couplings is largely cancelled out in rotary echoes.

The experimental method is straightforward. A multiple-π pulse is generated for a time τ at which point a phase shift of π radians is applied to the driving field, and the system is rephased along z at 2τ. The echo appears at $2\tau \pm (\pi/2\omega_1)$ where $\omega_1 = (\gamma/\hbar)e$, and therefore gives a direct measurement of the Rabi frequency, ω_1. Fig. 5 shows the first observation of a rotary phonon echo in SiO_2 glass at 6 mK and 0.95 GHz. Phase sensitive detection is utilized so that a π phase shift is reflected in the sign change of the excitation field. The rotary echo appears as a modulation atop the $\phi = \pi$ pulse. Only one cycle of the echo is seen as a result of inhomogeneous broadening which rapidly damps the Rabi oscillations.[9]

Coherent Electric Resonance: Photon Echo

Electric resonance methods in the microwave and rf region have been particularly useful in examining the role that dipolar impurities play in tunneling processes in glasses. In these experiments the glass forms a capacitive element in a cavity[10] or in a lumped element resonant circuit.[11]

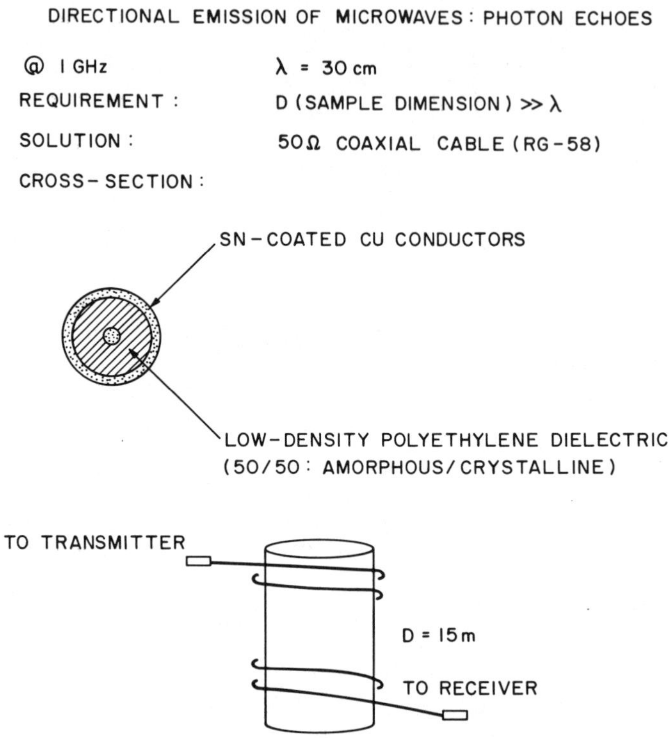

Fig. 6. Conditions for the experimental observation of a microwave photon echo. The guide is RG-58 coaxial cable wound on a copper drum and attached to the mixing chamber of a dilution refrigerator.

The experiment consists of exciting a standing-wave mode of the structure so that $k \cong 0$ i.e. the field does not propagate.

In one instance,[12] it has proved possible to excite a propagating electromagnetic mode at microwave frequencies in a disordered system. Fig. 6 illustrates how a coaxial cable filled with partially disordered polyethylene satisfies these requirements. The TEM mode of a coaxial waveguide produces a radial electric field which excites tunneling systems contained in the polyethylene dielectric insulation. The mode propagates along the guide with speed $c_o/\sqrt{\epsilon}$, so that a 15m cable provides an easily resolvable propagation delay ~100 nsec.

Fig. 7 illustrates the directional photon echoes which traverse the coaxial cable after excitation by three generating pulses. E_1 is a stimulated echo, whereas E_2, E_3, and E_4 are spontaneous echoes generated by combinations of P_1 and P_2 with P_3. Pulse powers in this geometry are very low; approximately 100 μW is required to generate π pulses.

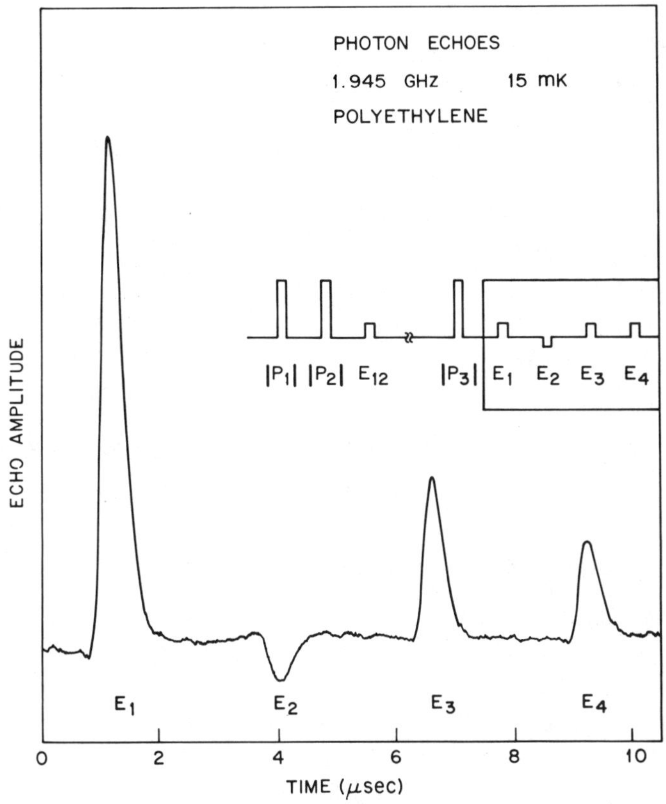

Fig. 7. Microwave photon echoes emitted by tunneling systems contained in the polyethylene dielectric of a coaxial cable. The inset shows the pulsing sequence.

Summary

The above discussion has illustrated resonance phenomena in glasses utilizing coherent acoustic and electric fields which result in directional emission of echo radiation. Analogs to coherent optical resonance in the visible and infrared are essentially exact. At very low temperatures, $k_BT \ll \hbar\omega$, energy decay occurs by the spontaneous emission of phonons, corresponding to photon spontaneous emission which dominates decay in many atomic and molecular systems. We have not touched upon the processes by which tunneling systems decay and dephase.[2] Such phenomena give a much more complete description of the way in which tunneling systems interact with one another, with phonons, and with extended and localized electronic states.

B. Golding
Bell Laboratories
Murray Hill, New Jersey 07974 USA

REFERENCES

1. See, for example, *Amorphous Solids: Low Temperature Properties,* edited by W. A. Phillips (Springer-Verlag, Berlin, 1981).
2. B. Golding and J. E. Graebner, ibid., p. 107.
3. B. Golding and J. E. Graebner, *Phonon Scattering in Solids,* edited by H. J. Maris (Plenum, New York, 1980) p. 11.
4. H. Grasse, O. Kocar, H. Peisl, S. C. Moss and B. Golding, Phys. Rev. Letters. *46,* 261 (1981).
5. B. Golding and J. E. Graebner, Phys. Rev. Lett. *37,* 852 (1976); J. E. Graebner and B. Golding, Phys. Rev. B*19,* 964 (1979).
6. E. L. Hahn, Phys. Rev. *80,* 580 (1950).
7. I. Abella, N. A. Kurmit, S. R. Hartmann, Phys. Rev. *141,* 391 (1966).
8. I. Solomon, Phys. Rev. Lett. *2,* 301 (1959).
9. N. C. Wong, S. S. Kano and R. G. Brewer, Phys. Rev. A*21,* 260 (1980).
10. B. Golding, M. v. Schickfus, S. Hunklinger and K. Dransfeld, Phys. Rev. Lett. *43,* 1817 (1979).
11. L. Bernard, L. Piché, G. Schumacher and J. Joffrin, J. Low Temp. Phys. *35,* 411 (1979).
12. B. Golding, J. E. Graebner and W. H. Haemmerle, Phys. Rev. Lett. *44,* 899 (1980).

THEORETICAL STUDIES OF OPTICAL ENERGY TRANSFER IN GLASSES AND OTHER DISORDERED SYSTEMS*

D. L. Huber

Department of Physics
University of Wisconsin
Madison, Wisconsin 53706, USA

We discuss various theoretical problems arising in the analysis of the transfer of optical excitation in glasses and other disordered systems. Attention is focused on four areas: the temperature dependence of the homogeneous linewidth, the influence of structural and thermal disorder on the transfer process, calculations of the microscopic transfer rates, and the interpretation of time-dependent fluorescent spectra. In the discussion of the linewidth problem we assess the relative merits of the mechanisms which have been proposed to account for the anomalous temperature dependence observed in glasses. The competing effects of inhomogeneous broadening and ion-ion interactions in determining the eigenstates of multi-ion arrays are pointed out. We emphasize the role of thermal disorder in generating effective single-ion eigenstates by destroying the coherence between

* Research supported by the National Science Foundation.

neighboring ions. The various types of phonon-assisted incoherent energy transfer processes are outlined. It is shown how the rate equation formalism provides a connection between the microscopic transfer processes and the decay of the fluorescence following pulsed excitation.

I. INTRODUCTION

The transfer of excitation between optically active ions is an important phenomenon affecting the performance of lasers and other optical devices. In this chapter we will discuss in detail a number of problems arising in the analysis and interpretation of optical energy transfer in glasses and other disordered systems. We will emphasize the similarities and differences between energy transfer in glasses, where the material is structurally disordered, and energy transfer in crystalline arrays with a small fraction of the sites occupied by optically active ions. The host systems we consider are inorganic insulators which can be regarded as being transparent in the frequency region of interest. The active ions generally belong to the rare earth, actinide, or transition metal series.

With respect to their optical properties there are two major differences between glasses and dilute crystalline arrays. First, the topological disorder inherent in the glass structure generates a broad distribution of crystal field splittings for the active ions and hence a comparatively large inhomogeneous linewidth (linewidths in excess of 10 cm^{-1} are not uncommon[1]). Second, in addition to coupling to phonons the ions also interact with the two-level systems[2,3] (TLS) which are a unique feature of amorphous hosts. This interaction appears to make a significant contribution to the homogeneous linewidth at low temperatures. In this chapter we will discuss the impact of these differences on energy transfer.

We will begin with an analysis of the single-ion homogeneous linewidth which focuses on the current controversy surrounding the interpretation of the anomalous temperature dependence. This section is followed by brief comments on how the structural and thermal disorder determines the effective eigenstates participating in the energy transfer process. The analysis of the energy transfer is separated into two parts. First, we consider various microscopic mechanisms for inter-ion transfer. This section is followed by a discussion of the rate equations characterizing the dynamics of the transfer in which we illustrate how the equations provide a bridge between the microscopic transfer processes and experimental studies of fluorescent decay. This chapter is intended to be a general discussion of energy transfer in glasses. Experimental data for specific systems are reviewed in the following chapter.

II. HOMOGENEOUS LINEWIDTH

The origin of the homogeneous linewidths of the optical transitions in glasses is a difficult problem which as yet has no satisfactory solution. In a general sense the linewidths reflect a rapid modulation of the crystalline field caused by the interaction between the ion and the various thermal excitations in the system. Since the number of thermal excitations taking part in the dynamics depends on the temperature the linewidth is temperature-dependent. In the majority of experimental studies reported to date[4–8] it has been found that the linewidths varied as T^a with $1.7 \lesssim a \lesssim 2.1$ for $T>10K$. In crystalline hosts T^2 behavior is associated with the high temperature ($T \gtrsim 0.5\,T_D$, where T_D denotes the Debye temperature) limit of the two-phonon or raman process where the modulation of the crystalline field arises from the elastic or inelastic scattering of phonons by the optically active ion. At lower temperatures ($T \lesssim 0.5\,T_D$) the linewidth begins to vary more rapidly, ultimately crossing over to a T^7 behavior in the low temperature limit (assuming a negligible contribution from radiative and non-radiative transitions to lower levels).

In glasses the observation of an approximate T^2 behavior when $T \ll T_D$ has been interpreted as evidence for an additional linewidth mechanism involving the TLS.[4,9–12] The most comprehensive analysis of the TLS mechanism is given by Lyo in Ref. 12. In Lyo's model the Hamiltonian takes the form

$$H = \sum_{\nu=o}^{1} E(\nu)\psi_\nu{}^+\psi_\nu + \tfrac{1}{2}e\sigma_z + \sum_{\nu=o}^{1} \textstyle\sum_a V_\nu^\alpha \psi_\nu{}^+\psi_\nu\sigma_\alpha$$

$$+ \sum_q \hbar\omega_q(n_q + \tfrac{1}{2}) + \tfrac{1}{2}\textstyle\sum_\alpha \epsilon\, f^\alpha\sigma_\alpha\,. \tag{1}$$

Here ψ_ν^+ and ψ_ν are the creation and annihilation operators associated with the optical transition between levels o and ν and the σ_α (α=z,+,−) are Pauli matrices associated with the TLS. The third term in (1) is the interaction between the ion and the TLS, while the fourth and fifth terms are the phonon Hamiltonian and the interaction of the TLS with the dynamic strain field ϵ, respectively. In his calculation Lyo diagonalizes the first three terms of Eq. (1); the modulation then arises from the interaction with the phonons induced by the fifth term.

Lyo finds that the dominant effect is associated with the diagonal coupling, V_ν^z, between the ion and the TLS. Assuming a density of states for the TLS varying as E^{μ} and an r^{-s} dependence for V_{ν}^{z}, r being the separation between the ion and the TLS, Lyo obtains a linewidth $\Delta\omega$ behaving as

$$\Delta\omega \propto T^{\mu+4-9/s} \tag{2}$$

With $\mu=0$ (constant density of states) and $s=4$ (dipole-quadrupole interaction) one has $\Delta\omega \propto T^{1.75}$, in reasonable agreement with experiment. However as discussed in Ref. 12 the difficulty with the theory is that while it provides a plausible explanation for the linewidth at low temperatures, it predicts a weaker temperature dependence for $T>50K$ than is observed experimentally. Although the point is made that the raman mechanism is likely to dominate in this regime the experimental data show no evidence of a cross-over from one mechanism to another (see Fig. 7 of Yen).

Because of the apparent absence of a cross-over we have undertaken a reassessment of the raman mechanism with emphasis on the applicability of the standard approximations in glassy hosts.[13] The (elastic) raman linewidth takes the form

$$\Delta\omega_R = C \int_0^{\infty} |\langle \mathbf{H}^{OL} \rangle|_{\omega}^{4}\, \rho(\omega)^2 \frac{e^{\hbar\omega/k_BT}}{\left(e^{\hbar\omega/k_BT}-1\right)^2}\, d\omega\ , \tag{3}$$

where C is a constant, $\rho(\omega)$ is the phonon density of states, and $|\langle \mathbf{H}^{OL} \rangle|_{\omega}$ is the average of the modulus of the matrix element of the orbit-lattice interaction taken over the modes of frequency ω. In crystalline systems it is reasonable to use the Debye approximation, $\rho(\omega) \propto \omega^2$, for the density of states and the long wavelength approximation, $|\langle \mathbf{H}^{OL} \rangle|_{\omega}^{2} \propto \omega$, for the matrix element *over the entire acoustic phonon spectrum.* In this case one finds

$$\Delta\omega_R \propto T^7 \int_0^{T_D/T} \frac{x^6 e^x}{(e^x-1)^2}\, dx\ . \tag{4}$$

Thus according to (4) $\Delta\omega_R$ varies as T^2 only for $T \gtrsim 0.5\ T_D$. However in Ref. 13 it was pointed out that light scattering data taken in glasses[14] show that the acoustic peak in the phonon density of states is found at energies much below k_BT_D. [For the BeF_2 glass studied in Ref. 5 the peak occurs at 50K whereas $T_D = 380K$.] Moreover the approximation $|\langle \mathbf{H}^{OL} \rangle|_{\omega}^{2} \propto \omega$ is valid only for phonon modes of extremely low frequency where the wavevector is a "good" quantum number. At higher frequencies the approximation $|\langle \mathbf{H}^{OL} \rangle|_{\omega}^{2} \propto \omega^{-1}$ is more appropriate. Combining these results we obtain a linewidth of the form

$$\Delta\omega_R = C_1 \int_0^{\omega_c} \frac{\omega^6 e^{\hbar\omega/k_BT}}{(e^{\hbar\omega/k_BT}-1)^2} d\omega + C_2 \int_{\omega_c}^{\Delta_p} \frac{\omega^2 e^{\hbar\omega/k_BT}}{(e^{\hbar\omega/k_BT}-1)^2} d\omega , \qquad (5)$$

neglecting the contribution of the modes with $\omega>\Delta_p$ and assuming a quadratic density of states for $0\leqslant\omega\leqslant\Delta_p$. In (5) Δ_p is the frequency of the acoustic peak and ω_c is the frequency above which the long wavelength approximation for $|<\mathbf{H}^{OL}>|^2_\omega$ is no longer valid. From this equation we obtain a linewidth which varies as T^2 for $T\gtrsim0.5\hbar\Delta_p/k_B$, as T^3 for $\hbar\omega_c/k_B \lesssim T \lesssim 0.5\hbar\Delta_p/k_B$, and T^7 for $T<<\hbar\omega_c/k_B$.

The point of this analysis is that the raman contribution to the linewidth is expected to vary more rapidly than T^2 for temperatures $<0.5\hbar\Delta_p/k_B$, not $0.5\ T_D$, as one would infer from a naive application of the theory. Such a result is consistent with the data reported to date, most of which fall in the range $T\gtrsim0.3\hbar\Delta_p/k_B$. From this it is apparent that additional measurements are needed to provide a complete picture of the linewidth over the full range from liquid helium to room temperature (see note added in proof).

III. EFFECTIVE EIGENSTATES

Before calculating microscopic energy transfer rates one must determine the states involved in the transfer process. In translationally invariant systems the appropriate eigenstates of an array of interacting ions are Frenkel excitons.[15] These are propagating modes labelled by a wavevector $\vec{K}$ with equal amplitude on all sites. In such systems the energy transfer process can be interpreted as phonon-assisted inelastic exciton scattering. An exciton in state $\vec{K}$ with energy $E_{\vec{K}}$ is scattered into a state $\vec{K}'$ with energy $E_{\vec{K}'}$. If the translational symmetry is destroyed either by the addition of impurities or through the mechanism of topological disorder the wavevector is no longer a good quantum number for labelling the eigenstates. When the disorder is "weak" (e.g. a low concentration of impurities) the spectrum is divided into regions of localized and extended states (see Chap. XIII).[16] As the strength of the disorder increases the regions of localized states grow at the expense of the extended states until at some point all of the states are localized.[17] It should be stressed that localized here refers to states having a significant amplitude on a finite number of ions; they are not necessarily single-ion eigenstates.

In arrays of optically active ions there are two principal types of disorder: diagonal disorder arising from the randomness in the distribution of surrounding ligands which produces a distribution in crystal field splittings, and off-diagonal disorder associated with the various configurations of neighboring optically active ions. Both types of disorder can lead to the disappearance of extended states. In the case of diagonal disorder when the ions are arranged on lattice with z nearest neighbors the criterion for the disappearance of extended states takes the form[18]

$$\frac{\Delta\Omega}{zV} \approx 2\text{–}3 , \tag{6}$$

where V is the nearest-neighbor interaction. The symbol $\Delta\Omega$ denotes the width in the distribution of crystal field splittings or equivalently the inhomogeneous linewidth [it is assumed that there is no correlation in crystal field splittings from site to site]. In a dilute array with exponential interactions ($V_{nn'} \sim \exp[-\alpha|\vec{r}_n - \vec{r}_{n'}|]$) and a negligible inhomogeneous linewidth localization is achieved solely through the influence of off-diagonal disorder. The corresponding criterion for the disappearance of extended states is[19]

$$\alpha n^{-1/3} \approx 3 , \tag{7}$$

where n is the number of optically active ions per unit volume and α^{-1} is the range of the interaction.

In crystalline systems with small inhomogeneous linewidths it may be possible to shrink the region of extended states to zero by dilution. In contrast, in the case of glasses the inhomogeneous linewidth is so large that Eq. (7), with zV replaced by the mean interaction, $<\sum_j V_{ij}>$, is generally satisfied. As a consequence the electronic states are localized.

The criteria for localization discussed above are obtained with the assumption that the interactions between ions fall off more rapidly than r^{-3}. While this is true for exchange and quadrupolar coupling it leaves open the possibility that the dipole-dipole interaction may establish extended states.[20] We argue that this is not the case,[21] the point being that at finite temperatures the interaction with the thermal excitations which causes the homogeneous linewidth also destroys the coherence between neighboring ions which is a characteristic feature of multi-ion eigenstates. If the off-diagonal matrix element of the dipolar interaction is denoted by d and the homogeneous linewidth by $\Delta\omega$ then as long as $nd^2/\hbar\Delta\omega << 1$ the thermal disorder is sufficiently strong to prevent the establishment of coherence between dipolar-coupled ions.

The picture that emerges from this analysis is that aside from dipolar effects the electronic states in glasses are localized largely because of the magnitude of the inhomogeneous linewidth. Moreover in the case of energy transfer the single-ion picture is appropriate for those ions which satisfy the criterion

$$\frac{V_M}{\hbar\Delta\omega} << 1 , \tag{8}$$

where V_M is the largest of the interactions with neighboring optically active ions.

IV. INTER-ION ENERGY TRANSFER

In this section we discuss the transfer of excitation between optically active ions. We will consider two types of processes: transfer within the inhomogeneous line, which we refer to as donor-donor transfer, and one-way transfer to traps (donor-acceptor transfer). The trapping centers can either be impurity ions or the donor ions themselves can act as traps through cross-relaxation. In view of the comments on the effects of thermal disorder made previously we will treat the transfer as an incoherent, single-ion process. In the case of donor-donor transfer an ion in an excited state ν undergoes a transition to the ground state while a neighboring ion in the ground state is promoted to the same level. Any energy imbalance is made up by the phonons or other thermal excitations in the system.

The Hamiltonian appropriate to phonon-assisted transfer processes takes the form[22–24]

$$\mathbf{H} = V_1^C + V_2^C + \mathbf{H}_1^{OL} + \mathbf{H}_2^{OL} + \mathbf{H}^{PH} + V_{12} \,. \tag{9}$$

Here V_i^C denotes the crystal field Hamiltonian of the ith ion, $\mathbf{H}_i^{OL}$ is the orbit-lattice interaction coupling the ith ion to the phonons, $\mathbf{H}^{PH}$ is the phonon Hamiltonian and V_{12} is the coupling between the two ions, which arises from exchange and multipolar interactions. The transfer rates can be calculated by the standard methods of time-dependent perturbation theory with initial and final states of the form

$$|\text{initial}> \;=\; |1;\nu>|2;o>|\{n_q\}> , \tag{10}$$

$$|\text{final}> \;=\; |1;o>|2;\nu>|\{n_q\}'> . \tag{11}$$

in which $|i;o>$ and $|i;\nu>$ denote the ground and excited states of the ith ion, respectively, and $\{n_q\}$ and $\{n_q\}'$ are the sets of occupation numbers characterizing the initial and final phonon distributions. In one-phonon processes $\{n_q\}$ and $\{n_q\}'$ differ in the occupation number of one mode,

$n_q' = n_q \pm 1$, whereas in the raman processes two modes are involved: $n_q' = n_q + 1$, $n_p' = n_p - 1$ corresponding to the creation of a phonon in mode q and the destruction of a phonon in mode p. As mentioned, the phonons make up the energy difference between $E_1(\nu)$ and $E_2(\nu)$, where $E_i(\nu)$ denotes the energy of the νth level of the ith ion relative to its ground state.

In the case of one-phonon processes the transfer rate from 1 to 2 takes the form[24]

$$W_{12} = A(r_{12},|\Delta E|)\left(\frac{1}{e^{|\Delta E|/k_BT}-1} + 1\right), \tag{12}$$

if $\Delta E = E_1(\nu) - E_2(\nu)$ is positive and

$$W_{12} = A(r_{12}, |\Delta E|)\left(\frac{1}{e^{|\Delta E|/k_BT}-1}\right), \tag{13}$$

if ΔE is negative. Equation (12) describes a creation process, $n_q' = n_q + 1$ whereas (13) corresponds to phonon destruction, $n_q' = n_q - 1$. In both instances $(\exp[|\Delta E|/k_BT] -1)^{-1}$ is the mean number of phonons present in the mode involved in the transition.

The multiplicative factor $A(r_{12},|\Delta E|)$ depends both on the energy mismatch and the relative separation between the two ions. It has the form

$$A(r_{12},|\Delta E|) \sim V_{12}^2|\Delta E|^{-2}|<\mathbf{H}^{OL}>|^2_{|\Delta E|/\hbar}\,\rho(|\Delta E|/\hbar)\,, \tag{14}$$

where $V_{12} \sim \exp[-\alpha r_{12}]$ (isotropic exchange) or $V_{12} \sim r_{12}^{-s}$ (multipolar). As in Sec. II $|<\mathbf{H}^{OL}>|^2_\omega$ denotes the square of the matrix element of the interaction with modes of frequency ω and $\rho(\omega)$ is the phonon density of states. Assuming $|\Delta E|/\hbar$ is in the range where the long wavelength approximation for the density of states and $|<\mathbf{H}^{OL}>|^2_\omega$ is appropriate we have $A(r_{12},|\Delta E|) \propto |\Delta E|$. Finally, the temperature dependence displayed in Eqs. (12) and (13) reflects a detailed balance symmetry which holds for multi-phonon processes as well

$$W_{12}\, e^{-E_1(\nu)/k_BT} = W_{21}\, e^{-E_2(\nu)/k_BT}\,. \tag{15}$$

The calculation of the two-phonon transfer rates is conceptually straightforward but algebraically complicated.[24] One has to distinguish between processes where the phonon creation and destruction occur on the same site and those where the two processes occur on different sites. The general expression takes the form

$$W_{12} = V_{12}^2\, g(|\Delta E|;\, k_BT)\,, \tag{16}$$

where $V_{12} \sim \exp[-\alpha r_{12}]$ or r_{12}^{-s}, as in the one-phonon transition rates. The function $g(|\Delta E|;k_BT)$ also depends implicitly on the phonon density of states and $|<\mathbf{H}^{OL}>|_\omega^2$. As long as $k_BT >> |\Delta E|$ one has $g \propto T^x$ with $2 \lesssim x \lesssim 7$, in contrast to the one-phonon rate which varies linearly with temperature in the same limit.

The relative importance of the one and two-phonon transfer rates depends on temperature. Typically, when $k_BT >> |\Delta E|$ the two-phonon processes dominate, whereas when $k_BT << |\Delta E|$ and ΔE is positive transfer accompanied by the spontaneous emission of a single phonon is the most important process. It must be emphasized that the transfer processes we have been discussing are present in both glasses and crystalline systems, the only difference being in the frequency dependence of $|<\mathbf{H}^{OL}>|_\omega^2$ and the density of phonon modes. Analogous to the homogeneous linewidth we may also expect transfer processes in glasses in which the TLS play a role. In particular it is plausible that the TLS make a significant contribution to the transfer at low temperatures.

The calculation of the donor-trap transfer rate in an amorphous system is similar to the corresponding analysis in a crystalline host.[25] The trapping involves a transition between initial and final states of the form

$$|\text{initial}> \;=\; |1,\nu>|T,o>|\{n_q\}>\,, \tag{17}$$

$$|\text{final}> = |1,\mu>|T,\rho>|\{n_q\}'>\,. \tag{18}$$

where $|T,o>$ and $|T,\rho>$ denote the ground and excited states of the trap, respectively. Depending on the magnitude of the energy difference, $E_1(\nu) - E_1(\mu) - E_T(\rho)$, one or more phonons will be emitted in the trapping process. Since the detailed balance relation, Eq. (15), applies equally well to trapping backtransfer from the traps will be relatively unimportant as long as $E_1(\nu) - E_1(\mu) - E_T(\rho) >> k_BT$. Because the mismatch in electronic energy is usually large it is unlikely that the TLS play a significant role.

V. FLUORESCENCE

In this section we will discuss the decay of the optical fluorescence following pulsed excitation. We are interested in two classes of experiments:

fluorescence line narrowing[26] and measurements of the decay of the intensity integrated over the inhomogeneous line.[27] In fluorescence line narrowing experiments ions occupying a small segment of the inhomogeneous line are excited by a narrow band source such as a laser. Following the excitation the spectrum consists of two components: a narrow line with a width equal to the bandwidth of the exciting light, and a broad, slowly rising background. The narrow component is the fluorescence from the ions which were initially excited while the background comes from ions which became excited through the incoherent transfer processes discussed in Sec. IV. By monitoring the time development of the fluorescence one can extract information about the microscopic transfer rates. In the broadband experiments one uses a source which covers the entire inhomogeneous line. The decay of the integrated intensity provides information about the rate at which excitation is being transferred to traps.

Both classes of experiments can be analyzed utilizing a formalism based on rate equations for the set of functions $\{P_n(t)\}$ where $P_n(t)$ is the probability that ion n is excited at time t all other ions being in the ground state.[28] These probabilities satisfy the equations

$$\frac{dP_n(t)}{dt} = -(\gamma_R + X_n + \textstyle\sum_{n'} W_{nn'})P_n(t) + \textstyle\sum_{n'} W_{n'n} P_{n'}(t)\,. \tag{19}$$

Here γ_R is the radiative decay rate, X_n is the transfer rate from ion n to all of the traps in the system, and $W_{nn'}$ is the rate of transfer from ion n to ion n′. The transfer rate depends on the temperature and energy mismatch and is related to the rate for the inverse process, $W_{n'n}$, through the detailed balance condition, Eq. (15).

In our analysis of the line narrowing experiments we will make the assumption that the trap concentration is sufficiently low that there is negligible trapping on the time scale of interest. Thus X_n can be taken to be zero. We are interested in the time-dependent ratio of the intensity in the narrow component (with the interpolated background subtracted) to the integrated intensity,

$$R(t) = \frac{\text{intensity in the narrow component}}{\text{integrated intensity}}\,. \tag{20}$$

Various approximations for R(t) (which can be identified with the conditional probability that an ion excited at $t=0$ is also excited at time t) have been developed and tested.[29,30] In crystalline hosts with small concentrations of optically active ions ($\lesssim$10 at.%) a reasonable approximation for the regime $1 \geqslant R(t) \gtrsim 0.01$ takes the form

$$R(t) = \prod_{\ell} (1-c + ce^{-W_{o\ell}t}\cosh(W_{o\ell}t)) \; , \tag{21}$$

where the product is over all lattice sites $\ell \neq o$, and c is the probability that a lattice site is occupied by an optically active ion. In (21) it is assumed that the transfer rates are symmetric, $W_{o\ell} = W_{\ell o}$, and independent of the energy mismatch $E_o - E_\ell$.

As a first step in modifying (21) to bring it into a form appropriate for glasses we write the product as a sum which we evaluate to first order in c

$$R(t) = \exp[\textstyle\sum_\ell \ell n(1-c + ce^{-W_{o\ell}t}\cosh(W_{o\ell}t))] \; ,$$

$$\approx \exp[-c\textstyle\sum_\ell (1-e^{-W_{o\ell}t}\cosh(W_{o\ell}t)] \; ,$$

$$\approx \exp[-n \int d\vec{r}(1-e^{-W(r)t}\cosh(W(r)t))] \; , \tag{22}$$

where n is the concentration of active ions and W(r) is the donor-donor transfer rate at separation r. As long as $k_BT \lesssim$ inhomogeneous linewidth Eq. (22) has to be generalized to allow for the asymmetry in the transfer rates as well as the dependence on energy mismatch. In place of (22) one obtains[31]

$$R(t,E) \exp\left[-n \int d\vec{r} \int dE'P(E') \left(1 - \frac{1}{1+e^{-(E'-E)/k_BT}} - \frac{e^{-W_Tt}}{1+e^{(E'-E)/k_BT}}\right)\right] \; , \tag{23}$$

for the decay following excitation at frequency $E/\hbar$. In (23) one has

$$W_T = W(\vec{r}, E \rightarrow E') \left(1+e^{(E'-E)/k_BT}\right) \; , \tag{24}$$

where $W(\vec{r}, E \rightarrow E')$ is the transfer rate from an ion with energy E to one with energy E′ separated by a distance r while the symbol P(E) denotes the normalized inhomogeneous lineshape function.

The early-time development of the background fluorescence can also be described by an approximation similar to (23).[31] The normalized intensity for light emitted at frequency $E'/\hbar$ following excitation at frequency $E/\hbar$ is given by

$$f(t,E';E) = \frac{(1-R(t,E))P(E')}{(-n^{-1}\ell nR(t,E))} \int d\vec{r} \left[1 - \frac{1}{1+e^{-(E'-E)/k_BT}} - \frac{e^{-W_Tt}}{1+e^{(E'-E)k_BT}} \right] . \quad (25)$$

Like (23) Eq. (25) is valid only for times such that $1 \geqslant R(t,E) \gtrsim 0.01$.

In the analysis of the decay of the integrated intensity one is in effect studying the relative fraction of ions excited at time t.[32] We write this fraction as

$$\text{rel. fraction.} = e^{-\gamma_R t} f(t). \quad (27)$$

The function f(t) describes the loss due to trapping. In treating this function we make the simplifying approximation that the donor-trap transfer rate depends only on their relative separation and is independent of the position of the donor in the inhomogeneous line. Even with this assumption we can obtain exact results only in two limiting cases.[32] The first of these corresponds to there being negligible donor-donor transfer on the time scale of interest so that excitation is transferred directly to the traps. When this is the case the equation of Inokuti and Hirayama[33] is applicable

$$f(t) = \exp\left[-n_T \int d\vec{r}\,(1-e^{-X(r)t})\right] . \quad (28)$$

Here n_T is the trap concentration and X(r) is the donor-trap transfer rate at separation r. In the case of a dipole-dipole process, $X(r) = \beta r^{-6}$ Eq. (28) reduces to

$$f(t) = \exp[-2.4\, n_T(\beta t)^{1/2}] . \quad (29)$$

The second limit pertains to a situation where the donor-donor transfer rate is (infinitely) rapid in comparison with the rate for donor-trap transfer. When this is the case the probability of a donor site being occupied at $t>0$ is governed by the Boltzmann distribution. Under these conditions we have[32,34]

$$f(t) = \exp[-n_T t \int d\vec{r}\, X(r)] , \quad (30)$$

so that the decay is exponential. With dipole-dipole transfer f(t) reduces to

$$f(t) = \exp[-4.2 n_T \beta t r_{min}^{-3}] , \quad (31)$$

where r_{min} is a measure of the minimum separation between donor and trap.

The behavior of f(t) in between the limits of zero and rapid donor-donor transfer is a complicated problem even in crystalline systems with symmetric transfer rates. The analogous problem in glasses is further complicated by the large inhomogeneous linewidth. As a result one can not in general make the simplifying assumption of mismatch-independent transfer rates. Because of the added degree of complexity it is difficult to make direct contact with the underlying donor dynamics. In such a situation one must resort to phenomenological models. When the donor-donor transfer is slow in comparison with the rate of transfer to traps it is appropriate to use the diffusion model of Yokota and Tanimoto.[35] In the case of dipole-dipole transfer one obtains

$$f(t) = \exp[-8.5 n_T t \beta^{1/4} D^{3/4}] , \tag{32}$$

where D is the diffusion constant for the donor array.

In the opposite limit where the donor-donor transfer is rapid in comparison with the rate of transfer to traps one can make use of the hopping model of Burshtein[36] in which f(t) has the asymptotic behavior

$$f(t) = \exp[-n_T t \int \frac{d\vec{r}\, X(r)}{1+X(r)\tau}] . \tag{33}$$

Here τ, a measure of the time the excitation resides on a particular, donor is given by[32]

$$\tau = \int dE P(E) \int_0^\infty R(t,E) dt , \tag{34}$$

where R(t,E) is approximated by Eq. (23). With the dipole-dipole mechanism one obtains

$$f(t) = \exp\left[-6.6 n_T (\beta t^2/\tau)^{1/2} \left[1-(2/\pi)\arctan\left[r_{min}^6 (\beta\tau)^{-1}\right]^{1/2}\right]\right] , \tag{35}$$

which reduces to (31) in the limit $\beta\tau r_{min}^{-6} \rightarrow 0$.

VI. DISCUSSION

In this chapter we have examined a number of problems connected with the transfer of optical excitation in glasses doped with optically active ions. In considering transfer it is important to keep in mind the similarities and differences between energy transfer in amorphous materials and energy transfer in crystalline systems. The glasses are distinguished by the presence of TLS which probably contribute to the homogeneous linewidth at low temperatures and which may play a role in inter-ion transfer. In addition there is a broad distribution of crystal field splittings which is reflected in the large inhomogeneous linewidths. Because of the breadth of the lines the standard approximation $k_BT \gg$ inhomogeneous linewidth can only be made at high temperatures. At lower temperatures the transfer rates are asymmetric (cf. Eq. (15)) and depend both on k_BT and the energy mismatch. Because of the complications associated with asymmetric transfer rates experimental studies of the dynamics of the fluorescence should first focus on the high temperature regime where theoretical techniques developed for the symmetric transfer problem can be employed in the analysis of the data.

Aside from the consideration of specific systems, we can identify two major areas where additional work is needed. The first of these is the problem of the homogeneous linewidth. More measurements are required, particularly at low temperatures ($T < 10K$), to confirm the existence of mechanisms other than the conventional one and two-phonon processes. A second area is the theory of energy transfer in systems with asymmetric, energy-dependent transfer rates. The problem in glasses is doubly complicated in that it involves both the off-diagonal disorder due to the random placement of the optically active ions and the diagonal disorder associated with the crystal field splittings. In this regard the recent work by Movaghar and Schirmacher[37] is especially noteworthy. They have developed a formalism which allows one to determine the diffusion constant for arbitrary line shapes and transfer rates. Using their formalism we have calculated the diffusion constant associated with a one-phonon transfer process obtaining the result[38]

$$D(T) = D_o(T)\exp\left[-a\Delta^2/(k_BT)^2\right], \quad (36)$$

with $a \approx 0.4$. The first factor in (36) is the limiting form at high temperatures, $k_BT \gg \Delta$, where Δ is the inhomogeneous linewidth. The second factor, which suppresses the diffusion at low temperatures, reflects a decrease in the effective number of ions participating in the transfer process.

REFERENCES

1. M. J. Weber, *Laser Spectroscopy of Solids*, Vol. **49** *Topics in Applied Physics*, W. M. Yen and P. M. Selzer, eds. (Springer-Verlag, Berlin, 1981) Ch. 6.
2. W. A. Phillips. J. Low Temp. Phys. **7**, 351 (1972).
3. P. W. Anderson, B. I. Halperin, and C. M. Varma, Phil. Mag. **25**, 1 (1972).
4. P. M. Selzer, D. L. Huber, D. S. Hamilton, W. M. Yen, and M. J. Weber, Phys. Rev. Lett. **36**, 813 (1976).
5. J. Hegarty and W. M. Yen, Phys. Rev. Lett. **43**, 1126 (1979).
6. P. Avouris, A. Campion, and M. El-Sayed, J. Chem. Phys. **67**, 3397 (1977).
7. J. M. Pellegrino, W. M. Yen and M. J. Weber, J. Appl. Phys. **51**, 6332 (1980).
8. P. M. Selzer, D. L. Huber, D. S. Hamilton, W. M. Yen, and M. J. Weber, *Structure and Excitation of Amorphous Solids*, AIP Conf. Proc. **31**, (American Inst. of Physics, New York, 1976) p. 328.
9. T. L. Reinecke, Solid State Comm. **32**, 1103 (1979).
10. S. K. Lyo and R. Orbach, Phys. Rev. B **22**, 4223 (1979).
11. I. S. Osadko, Pi̒sma Zh. Eksp. Teor. Fiz. **33**, 640 (1981) (JETP Lett. **33**, 626 (1981)); I. S. Osadko and S. A. Zhdanov, Opt. Comm. **42**, 185 (1982).
12. S. K. Lyo, Phys. Rev. Lett. **48**, 688 (1982).
13. D. L. Huber, J. Non-Cryst. Solids **51**, 241 (1982).
14. D. Heiman, R. W. Hellwarth, and D. S. Hamilton, J. Non-Cryst. Solids **34**, 63 (1979).
15. J. Frenkel, Phys. Rev. **37**, 17 (1931).
16. N. F. Mott and E. A. Davis, *Electronic Processes in Non-Crystalline Materials*, Second Ed. (Clarendon Press, Oxford, 1979).
17. P. W. Anderson, Phys. Rev. **109**, 1492 (1958).
18. B. Kramer, A. Mackinnon and D. Weaire, Phys. Rev. B **23**, 6357 (1981).
19. W. Y. Ching and D. L. Huber, Phys. Rev. B **25**, 1096 (1982).
20. P. V. Elyutin, Fiz. Tverd. Tela **22**, 3533 (1980) (Sov. Phys. Solid State **22**, 2070 (1980)).
21. D. L. Huber and W. Y. Ching, Phys. Rev. B **25**, 6472 (1982).
22. T. Holstein, S. K. Lyo, and R. Orbach, Phys. Rev. Lett. **36**, 891 (1976).
23. T. Holstein, S. K. Lyo, and R. Orbach, Phys. Rev. B **15**, 1693 (1977).
24. T. Holstein, S. K. Lyo, and R. Orbach, Ref. 1, Ch. 2.
25. D. L. Huber, J. Lumin. **27**, 333 (1982).
26. W. M. Yen and P. M. Selzer, Ref. 1, Ch. 5.
27. J. Hegarty, D. L. Huber, and W. M. Yen, Phys. Rev. B **23**, 6271 (1981); ibid. B **25**, 5638 (1982).
28. D. L. Huber, Ref. 1, Ch. 3.
29. D. L. Huber, D. S. Hamilton, and B. Barnett, Phys. Rev. B **16**, 4642 (1977).
30. W. Y. Ching, D. L. Huber, and B. Barnett, Phys. Rev. B **17**, 5025 (1978).
31. D. L. Huber and W. Y. Ching, Phys. Rev. B **18**, 5320 (1978).
32. D. L. Huber, Phys. Rev. B **20**, 2307 (1979).
33. M. Inokuti and H. Hirayama, J. Chem. Phys. **43**, 1978 (1965).
34. D. Fay, Phys. Rev. B **25**, 4245 (1982).
35. M. Yokota and I. Tanimoto, J. Phys. Soc.(Japan) **22**, 1779 (1967).
36. A. I. Burshtein, Zh. Eksp. Teor. Fiz. **62**, 1695 (1972) (Sov. Phys. JETP **35**, 882 (1972)).
37. B. Movaghar and W. Schirmacher, J. Phys. C **14**, 859 (1981).
38. D. L. Huber, J. Chem. Phys. **78**, (1983).

Recent work by R. M. Macfarlane and R. M. Shelby has established that the $T^{1.8}$ temperature dependence observed for the homogeneous linewidth of Eu^{3+} in silicate glass holds down to at least 1.6K whereas a linear temperature dependence is observed from 1.6K to 20K for the ${}^1D_2 \rightarrow {}^3H_4$ transition of a Pr^{3+}-doped silicate glass (see R. M. Macfarlane, these proceedings.)

DISCUSSION

Theory of Optical Energy Transfer

The suggestion that two level systems (TLS) are responsible for the T^2 behavior of optical linewidths over a wide temperature range was called into question. While TLS can in principle participate, it was pointed out that at high temperatures the linewidth in the glass was suspiciously close to that in the crystals where there are no TLS. The Raman process discussed by Huber can explain the results in both crystal and glass at high temperatures but not the T^2 dependence at low T in the glass. If TLS give this low T behavior it is puzzling that the magnitude should be the same as that given by the Raman process at high T. In addition, TLS will not give a T^2 dependence of the linewidth at high T. The possibility of distinguishing between TLS and Raman processes by saturating the former was discounted because the requirement of saturating such states over a large bandwidth would necessarily cause a large increase in sample temperature. Further, the diagonal part of the atom-TLS interaction is not saturable. The validity of the assumption of Debye-like phonons was questioned. The need for a more realistic phonon model was expressed, and a study of frequency dependence of the scattering matrix elements was now felt to be important.

J. Ryan, Chairman
J. Hegarty, Discussion Leader

COMMENTS

Raman Scattering in Disordered Solids

Invocation of Raman spectra to (1) support or distinguish among various proposed amorphous structures or (2) to provide phonon densities of states more reliably than simple Debye temperature approaches has been frequently made in the literature. It is worth emphasizing that there are severe limitations and rather drastic approximations inherent in either case. In systems with long range order (i.e. crystals) momentum conservation ($\bar{k}_i - \bar{k}_s = \bar{q}$) determines that first order Raman scattering (one phonon) measures the frequencies ω_q^l, where l designates a Raman-active phonon on the lth branch. For visible light $|k_i| \approx |k_s| \approx 10^5 cm^{-1}$, so only the long wavelength or Brillouin zone center frequencies are accessed. Phonons throughout the zone ($q \leqslant \pi/a \simeq 10^8 cm^{-1}$) can be accessed in combinations, however, through second order Raman scattering. In this case the momentum difference, $\bar{k}_i - \bar{k}_s$, is equated to $\bar{q}_1 + \bar{q}_2$ the vector sum of two phonons' momenta. For $\bar{q}_1 \approx -\bar{q}_2$ the pair can couple to light even though the individual momenta q_1 or q_2 may be very large. In the harmonic approximation the scattered spectrum may be expressed as:

$$I(q,\omega) \cong \text{Const.} \sum_{l,m} \int dq' G_q{}'^{lm}\, S_l(q-q', \omega_{q-q'}) S_m(q', \omega_q{}') \tag{1}$$

where $S_m(q',\omega')$ is the Fourier transform of the one phonon eigenvector correlation function: $\langle \delta U_m(r,t) \delta U_m^*(0,0) \rangle$, for the mth branch. The function $G_q{}'$ incorporates the effects of symmetry so as to permit only Raman active pairs to contribute to the sum. It also accounts for the dependence of scattering efficiency on q'. The latter effect is difficult, if not impossible to calculate, and most second order Raman spectra are interpreted using a simple ansatz for $G_q{}'^{lm}$, usually that it is independent of q'. Only if interbranch phonon pairs are ignored ($l = m$) and if the q' dependence of $G_q{}'$ is ignored does equation (1) produce a near replica of the one-phonon density of states, on a frequency scale which must, of course, be halved. When interbranch processes ($l \neq m$) are included the spectra are much more complicated and the extraction of one-phonon density of states from second order Raman spectra - even in the simplest crystalline cases - is nearly hopeless.

Now, for a glass or disordered solid, even though the Brillouin zone or well-defined phonon branches do not strictly exist, there are dynamical fluctuations of all wave lengths down to $\lambda_{min} \approx a$; where a is a characteristic interparticle spacing, so that something approximating phonon branches in the crystal will exist. In addition there are, in an amorphous material, *static* fluctuations in displacement or configuration over the same span of wave lengths. For a glass the amplitudes of these static fluctuations are typically those appropriate to thermal fluctuations near the glass transition temperature, rather than at ambient temperatures. The Raman spectrum of a glass is then typically dominated by disorder-induced, one-phonon scattering. It is a process wherein a phonon of wave vector $\bar{q}'$ and frequency ω_q' pairs with a static fluctuation of wave vector $\bar{q}-\bar{q}'$ and frequency zero to couple to the light while satisfying the momentum conservation condition: $\bar{k}_i-\bar{k}_s = \bar{q}$.

The spectrum is then obtained by summing over all such pairs; that is summing over q'. Formally this can be obtained from equation (1) by replacing $S_l(q',\omega_q')$ by $S_s(q')$, the static structure factor of the glass. From this point of view it is easily seen that $I(q,\omega)$ will only approximate the one phonon density of states if: (1) the q' dependence in $G_q'^{sm}$ is ignored and (2) $S_s(q')$ is a constant. The condition $S_s(q')$ = Const. is equivalent in real space to complete structural randomness, a condition never met in any glass. More realistically $S_s(q')$ will have structure indicative of the short, intermediate and long-range (if any) order in the glass. The extreme of complete long range order (LRO) occurs when $S_s(q') = A\delta(q'-q_c)$, in which case sharp features are recovered in the Raman spectrum at $\omega^1_{q_c}$. Thus the disorder-induced, one-phonon scattering in a glass at best reveals the one-phonon density of states as viewed through the "window" represented by $S_s(q')$. That window is uniformly open only for a completely random material. Otherwise the Raman spectrum reveals only a *weighted* phonon density of states.

In addition, of course, two phonon scattering is also present in a glass, and moreover we have ignored the fact that the kernel of Eq. (1) represents a pairwise-factored, four-point correlation function, which amounts to ignoring anharmonic effects. Nevertheless, Raman scattering can be useful in the study of both the statics and dynamics of glasses - as pointed out by Shuker and Gammon.[1] But one must be aware of the pitfalls.

P. A. Fleury
Bell Laboratories
Murray Hill, New Jersey 07974 USA

REFERENCE

1. R. Shuker and R. W. Gammon, Phys. Rev. Lett. *25*, 222 (1970).

EXPERIMENTAL STUDIES OF OPTICAL ENERGY TRANSFER IN GLASSES

W. M. Yen

Department of Physics
University of Wisconsin
Madison, Wisconsin 53706, USA

We review here experimental results obtained in the study of some rare earth (4f) activated insulating glasses using laser based site selective spectroscopic techniques such as fluorescence line narrowing (FLN). These studies have led to considerable progress in our understanding of the microscopic structure and the nature of certain excitations in the neighborhood of the probe ion. In addition, laser spectroscopy has allowed the investigation of intrinsic properties of optical transitions of ions at specific glass sites. Through these properties the relaxation behavior of individual ions has been determined and has become a problem of considerable current interest. We also present here a synopsis of studies of energy transfer in glasses. These processes arise because of inter-ionic interactions which lead to spatial and spectral diffusion of the energy. We again demonstrate that the advent of new experimental techniques has provided us with new information regarding these processes. We conclude by briefly discussing various prospects in this area of research.

I. INTRODUCTION

The spectral properties of optically active glasses have been investigated intensively in the past two decades so that a generaı outline of their behaviour has been fairly well established. Earlier investigations, all of which utilized conventional spectroscopic techniques, played an important role in establishing our framework of understanding based mainly on the behaviour of equivalent ions and centers in crystalline materials. Indeed, the majority of phenomena which influence excited states in crystals have been demonstrated to occur in the disordered phases. We note, however, that much of the work in this area has been motivated by the practical necessities of various technologies and thus at times has failed to address more fundamental aspects of these problems.

The basic difference between crystals and glasses is of course the inherent disorder that is intrinsic to the latter. A paramagnetic ion doped into an insulating glass, for example, experiences local environments which vary, sometimes drastically, from site to site. These variations affect not only the nature of the static local (crystal) field at a given site but also may alter the bonding which keeps the center in place. The net result is that the energy levels and transition strengths (radiative and non-radiative) of individual ions will show a variation from site to site. The spectra of active glasses exhibit, in optical parlance, very large inhomogeneous broadening because of this disorder. The width of a given transition is then a composite of individual ion transitions distributed among the ensemble of local environments.

Conventional spectroscopic techniques which utilize broad band sources do not differentiate between homogeneous or inhomogeneous spectra and hence sometimes the interpretation of the latter can become ambiguous.[1] The advent of narrow line tunable laser spectroscopy has remedied this situation to a very large extent and has allowed the extraction of intrinsic properties of transitions be they in the gaseous or condensed phases. Techniques have been developed, e.g. absorptive hole burning and fluorescence line narrowing (FLN), which effectively reduce or totally suppress the statistically varying inhomogeneous contributions to the spectra.[2] These techniques coupled with their pulsed time resolved variants have provided unique new tools with which to probe disorder effects on optically excited states. Many of the demonstrations of these techniques on glassy materials have occurred in rare earth (4f) doped insulating glasses of sundry compositions and it is on this genre of glasses that we will concentrate our attention here.

This article is organized as follows: In the next section we will review very briefly the nature of the spectra and the optical transitions of paramagnetic 4f ions in glasses and summarize some prototypical laser spectroscopic techniques which have been used in the study of these materials. The following sections will present summaries of investigations of the static and dynamic properties of 4f ions in glasses as they have been determined through laser based spectroscopic studies. We conclude with some general remarks regarding future prospects in this area of investigation.

II. PROPERTIES OF 4f IONS IN GLASS

The optical properties of trivalent 4f ions in crystals in the visible and near IR are well understood in terms of the weak crystal field approximation.[3,4] This is a consequence of the relatively good shielding provided by higher n shells which attenuate various lattice perturbations on the 4f active electrons. The spectra of these ions is composed of a set of sharp levels which can be directly traced to their free ion or Russell-Saunders origins. The majority of commonly encountered transitions are intraconfigurational, i.e. $4f \rightarrow 4f$, and hence are weak because of parity selection rules; it follows that the radiative lifetimes of fluorescences in these materials are generally relatively long and the intrinsic (Heisenberg) widths can be extremely sharp by optical standards. The weak ion-phonon or lattice coupling and the sharpness of the transitions make 4f ions very useful as probes of various excitations and interactions in the condensed phases. This in conjunction with newly developed spectroscopic techniques have been put to good use in investigating the disordered solids which are of interest here.

When 4f ions are placed into glasses they replace the network cation which forms the glass or they can act as network modifiers. Many simple and multicomponent glass matrices have been shown to accommodate activator ions such as the rare earths. The precise manner in which dopants enter into the glass structure depends on the relative sizes, valencies and bonding of the constituents involved.[5]

The structure of glasses is inherently disordered and lacks any specific symmetry or long range periodicity. Thus the activator ion experiences a random distribution of local fields. Additional disorder arises in compound glasses because of the mixing of anion coordinations in the vicinity of the dopant and on occasion phase separation is also known to occur in certain classes of these materials.[6] The net result is that large variations of random origin are produced in the sites occupied by the individual 4f ions and as a consequence the observed optical transitions show considerable inhomogeneous broadening. Fig. 1 illustrates the extent of this broadening in the $^4I_{9/2} \rightarrow {}^4F_{3/2}$ absorption of Nd^{3+} in glasses of several compositions; we note that the magnitude of the broadening as well as the splitting of the crystalline field components are sensitive to the composition of the glassine host.[7]

As we have already mentioned, various laser spectroscopic methods have

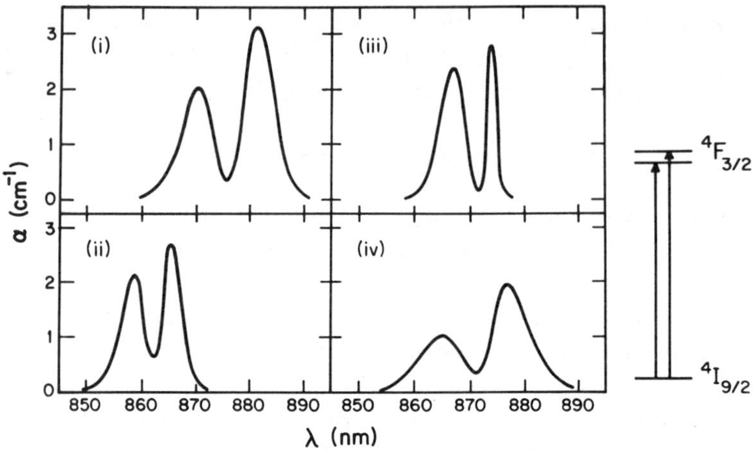

Fig. 1. Optical absorption coefficient, α, vs. wavelength for the $^4I_{9/2} \rightarrow {}^4F_{3/2}$ transition of Nd^{3+} in glasses of several compositions. Temperature is 4K. The compositions are as follows: (1) 67 GeO_2 · $15Na_2O$ · 18 BaO, (ii) 60 ZrF_4 · 34 BaF_2 · 6 NdF_3, (iii) 55 P_2O_5 · 30 Rb_2O · 10 CaO · $5Al_2O_3$, (iv) 75 S_2O_2 · 9 B_2O_3 · 9 Na_2O · 6 K_2O · 1 BaO. Doping levels varied. (From. Ref. 7).

provided us with a way to investigate properties which are intrinsic to the transition but which lie buried in the inhomogeneities of the spectra. Fluorescence line narrowing (FLN) has been the most commonly used technique in the study of 4f doped glasses. Historically, Denison and Kizel[8] first demonstrated line narrowing in an Eu^{3+} laser glass while Motegi and Shionoya[9] pioneered the use of tunable lasers in a FLN study of the Eu^{3+} doped into a pentaphosphate glass. The information derived through FLN studied is complementary to that obtained through the use of other techniques such as absorptive hole burning. Hole burning has been used extensively in the study of ions in crystals and was observed very early in activated glasses in the course of stimulated emission.[10] This technique however has not generally been pursued as a spectroscopic tool for the study of glasses.

In FLN, a tunable laser or narrowband source is used to excite a selected portion of an inhomogeneously broadened distribution.[2,5] In the limit of no interactions which might transfer the excitation to other portions of the whole distribution, the resulting fluorescence arises only from that subset of ions which is in resonance with the laser, the emission spectrum is then said to have been narrowed. In the case where the fluorescence and the excitation are in resonance, the FLN signal will have a spectral width which is a convolution of the laser width and the intrinsic or homogeneous width of the prototypical ion in the excited subset.[11,12] For transitions which are not in resonance with the excitation, additional inhomogeneous contributions leading to narrowed but not "laser" narrow signals arise. These intermediate state widths may be traced to the multiparameters which describe the local fields. It is known as the "accidental degeneracy effect" in FLN.[13] In addition, in cases where inhomogeneous broadening is as large as those encountered in glass, it is possible to simultaneously excite ions which occupy different components of the ground Stark manifold, (at finite temperatures), thus some care need be exercised in the interpretation of FLN spectra.

Glasses containing 4f ions are particularly attractive to study using FLN. Because of their relative insensitivity to lattice effects, the inhomogeneous broadening introduced by the glass disorder are large but not prohibitively so. A multitude of convenient fluorescing levels belonging to diverse ions of the series span the near IR and visible where, of course, tunable laser sources are readily available for resonant FLN studies. Generally, keeping in mind the complications due to degeneracies cited in the previous paragraph, the experimental requirements for line narrowing spectroscopy in 4f doped glasses are not very stringent and FLN can be easily demonstrated. The FLN signal of the simple case of Yb^{3+} in a silicate glass is shown in Fig. 2.

III. STATIC LASER SPECTROSCOPY OF 4f IONS IN GLASSES

FLN and related laser based techniques not only allow the suppression of certain inhomogeneous contributions but also permit us to selectively single out specific active sites for investigation. The importance of these properties to the study of inhomogeneously broadened systems cannot be overemphasized for they have provided us with a way to probe into the microscopic structure of glasses in the vicinity of the active center.

In their original paper, Motegi and Shionoya first demonstrated extreme laser induced site selectivity in a glass.[9,14] They excited various portions of the inhomogeneously broadened $^7F_o \rightarrow {}^5D_o$ transition of Eu^{3+} in a $Ca(PO_3)_2$ glass and observed that spectral changes occurred in the $^5D_o \rightarrow {}^7F_1$ emission as they tuned the laser across the absorption. These energy changes were interpreted by the authors as arising from site dependent Stark or "crystalline" field splittings. Subsequent studies by Sussman et al.[15] and by Brecher and Riseberg[16] in other glasses have confirmed this earlier assertion. For example, Fig. 3 illustrates the changes observed in the $^5D_o \rightarrow {}^7F_1, {}^7F_2$ emission as one excites across the ^5D_o absorption of Eu^{3+} in a silicate glass. Dramatic changes occur in the $^7F_1$ (triplet) Stark structure which are clearly connected with large variations of the local electrostatic field configurations at the Eu^{3+} sites.

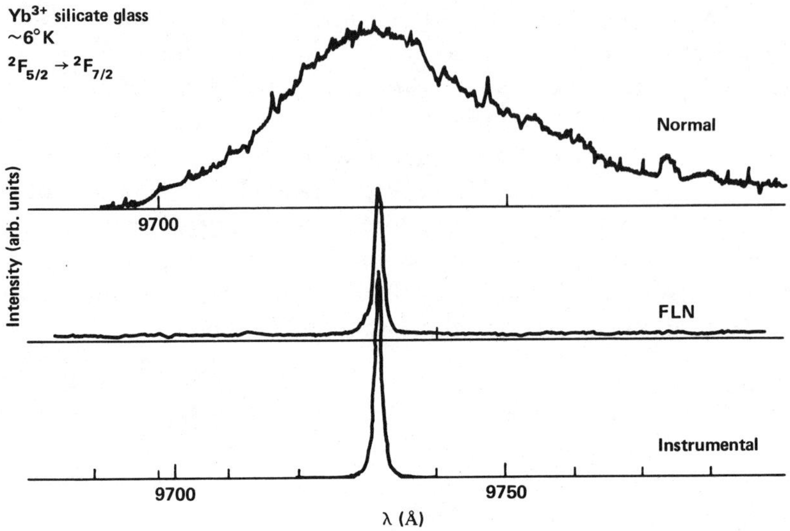

Fig. 2. FLN and normal fluorescence spectra of the ${}^2F_{5/2}{}^{(1)} \rightarrow {}^2F_{7/2}{}^{(1)}$ transition of Yb^{3+} in a silicate glass. Transition is between the lowest components of the excited and ground state Stark manifolds. The instrumental width is limited by the spectrometer resolution.

With the energy level diagrams obtained through these studies, it is possible to obtain certain insights into the nature of the crystalline fields responsible for the observed behaviour of the energy levels. Using the weak field approximation, for example, Brecher and Riseberg have been able to extract appropriate field parameters, B_q^k, in the expansion of the perturbing crystal field[17]

$$V = \sum_{k,q,i} B_q^k \left(C_{-q}^{(k)}\right)i \tag{1}$$

for Eu^{3+} in a number of glasses. In Eq. (1), $C_{-q}^{(k)}$ are tensor operators and the terms in the sum depend on the symmetry at the sites in question. Fig. 4 shows a representative result obtained by these workers in a fit of the ${}^7F_1$ and ${}^7F_2$ manifolds of Eu^{3+} in a silicate glass. Only even B_q^k play a role in the energy level determination. The variations in these parameters required to describe the sites in a glass encompass the whole range of values which have been encountered in crystals. In this figure, the dramatic changes of the ${}^7F_1(A_2)$ states are due principally to changes in the B_{20} parameter implying large variations in the axial symmetry of the different sites, for example.

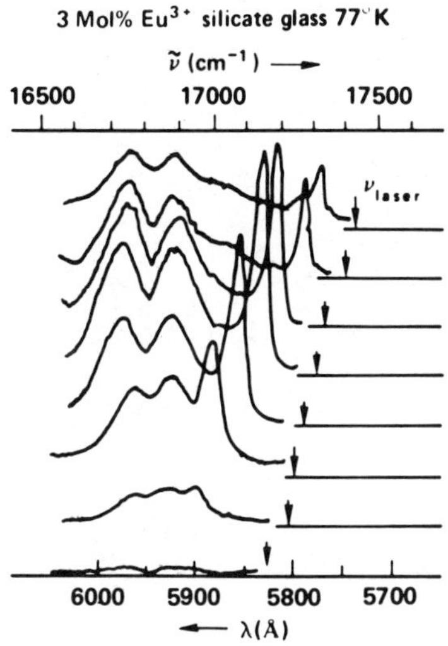

Fig. 3. Variation of the FLN spectra of Eu^{3+} in phosphate glass as a function of $^7F_o \rightarrow {}^5D_o$ excitation wavelength (denoted by arrows). Transition shown is the non resonant $^5D_o \rightarrow {}^7F_1$ fluorescence. The $^7F_1$ manifold entails three components which are clearly visible; note the large variation of the lowest component of this manifold.

The crystal field also affects the radiative transition probability by admixing odd parity into the 4f wavefunctions. It follows that fluorescence lifetimes will also vary from site to site in the inhomogeneous distribution.[18,19,20] The general non-exponential decay of fluorescence in activated glasses when excited with a broadband source may then be directly attributed to the different lifetimes summed over the whole distribution. Under narrowband laser excitation such as in FLN experiments, the lifetimes observed for selectively chosen subsets of ions become exponentials or nearly so. Additional parametric information may then be derived on the site symmetries through the study of these transition strengths. Lifetime variations by factors of 2 to 5 across inhomogeneous broadened profiles are commonly encountered in 4f glass systems. Fig. 5 illustrates the site-to-site lifetime changes of the $^2F_{5/2}$ state of Yb^{3+} in a silicate glass.

Crystal field parameters derived in the energy level and lifetime studies have been used by Brecher and Riseberg to develop geometrical models of the glass environment surrounding the 4f center.[16] Though only the effects of the first shell of neighbours are treated and the models do not yield absolute structural determinations, these studies do presage the beginning of a new

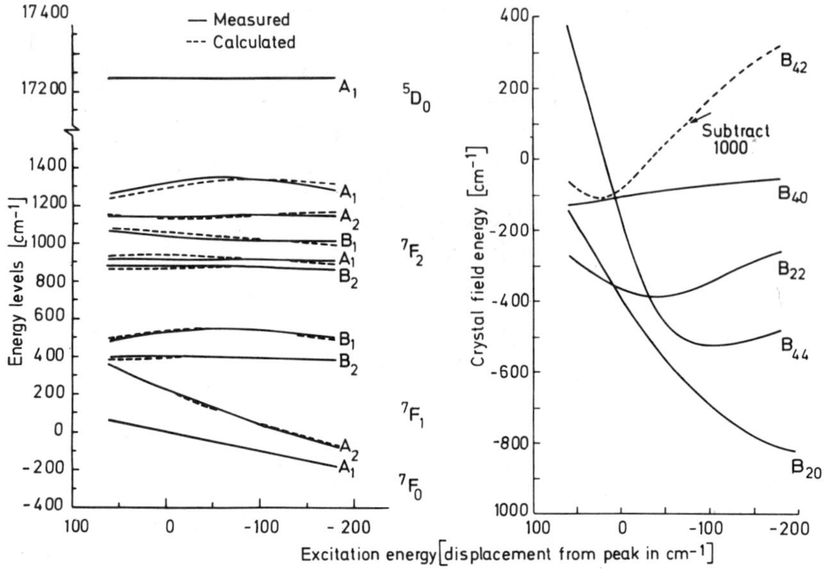

Fig. 4. Measured variations of the ^7F_o, $^7F_1$ and $^7F_2$ levels of Eu^{3+} in a silicate glass (left) and calculated parameters B_q^k (right) as a function of FLN excitation energy. The representations used on the left belong to C_{2v} site symmetry. (From Ref. 5 and 16).

stage in microscopic studies of disordered systems. For example, Brawer and Weber[21] have carried out an extensive computer based simulation of pure and 4f doped simple glasses (BeF_2). Glass configurations generated in this way lead to 4f site symmetries which are generally consistent with results deduced from optical studies.

FLN studies have also been used to investigate assisted transitions to determine the nature of various collective excitations which couple to electronic transitions. A recent example of such studies is the work of Hall et al.[22] These workers were able to observe and analyze the vibronic sidebands accompanying the $^6P_{7/2}$ to ground state emission of Gd^{3+} in a metaphosphate glass under FLN conditions (Fig. 6). Spectra obtained in this way again differs from those obtained via conventional spectroscopies (e.g. Raman) in that the vibronics obtained in FLN are localized at specific glass sites.

Finally, laser spectroscopy has been used to determine excited state structure of ions in glasses[23] and has been shown to be a potentially powerful tool to study structural changes in glasses such as those which might occur in phase separation.[24] A much more detailed description of the status of this field is to be found in some recent reviews by Weber.[5,25]

IV. RELAXATION AND HOMOGENEOUS LINEWIDTHS OF IONS IN GLASSES

The $T \simeq 0K$ homogeneous linewidths of 4f transitions in crystals comprise some of the narrowest widths measured to date by optical means. Results of such studies are discussed in greater detail elsewhere in this volume.[26] The situation for the same ions in glasses is not much different at low temperatures as linewidths ~20 MHz have been reported (Eu^{3+}, ${}^5D_0 \rightarrow {}^7F_0$ transition).[27] Basically, neglecting various hyperfine interactions which can be important, the intrinsic zero temperature widths of 4f transitions are determined by their radiative lifetimes; for fluorescing states because of selection rules these lifetimes are generally relatively long (μ-ms) hence the Heisenberg width is extremely (optically) narrow, of the order of M-kHz.

As the temperature is increased various interactions between collective excitations of the solid and the electronic levels come into play and limit the lifetime or the coherence of the excited state. It follows that transitions invariably broaden homogeneously as the temperature is increased. The thermal linewidth behaviour of 4f transitions in crystals is well established as arising from phonon-ion interactions.[28] Temperature dependent studies of the

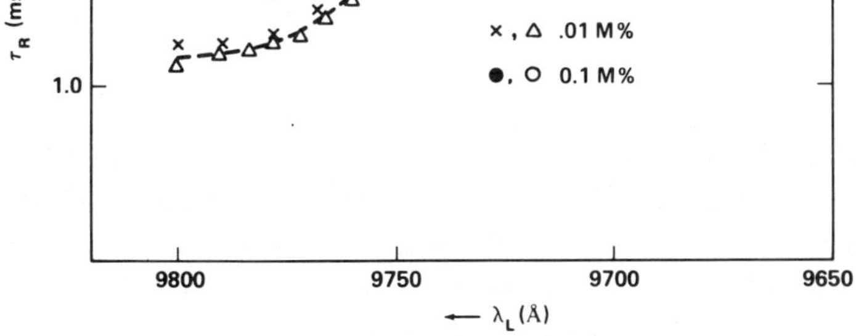

Fig. 5. Radiative lifetime changes as a function of laser pump frequency within the ${}^2F_{5/2}{}^{(1)} \rightarrow {}^2F_{7/2}{}^{(1)}$ transition of Yb in silicate glass, T = 77K. FLN decays are exponential throughout this pump range. Changes in the behavior at increased concentrations arise from ion-ion interactions and lead to a constant lifetime across the line when the Yb^{3+} concentration exceed 0.5 M%.

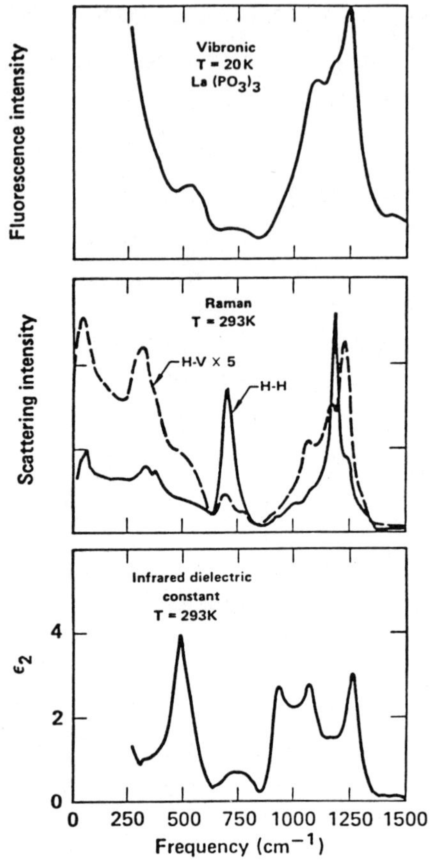

Fig. 6. Comparison of the vibronic, polarized Raman and infrared spectra of Gd in a La $(PO_3)_3$ glass. The vibronic sideband is obtained by inducing FLN in the $^6P_{7/2} \rightarrow {}^8S_{7/2}$ transition of Gd^{3+} in the near uv (312 nm). (From Ref. 22).

linewidth of 4f ions in glasses have been carried out using FLN techniques by a succession of workers.[19,20,29,30] As Huber has pointed out, there is presently some controversy as to the source of the interaction leading to the thermal broadening of optical transitions in glasses.[31] The experimentally observed facts are however well established.

Selzer et al.[19,29] conducted the first comprehensive measurement of the optical homogeneous widths as a function of temperature. Measurements were conducted on the $^5D_o \rightarrow {}^7F_o$ transition of Eu^{3+} in silicate and in BeF_2 glasses. They observed an anomalously large linewidth in the glass transition as compared to its crystal counterpart and obtained a near T^2

dependence of the linewidth in the ~5-100K region. They also reported that a smooth variation in the homogeneous linewidth occurs as one pumps across the inhomogeneous line. These properties have since been observed in a variety of 4f ions doped into glasses of different composition and very interestingly seem to imply universality in the thermal dependence of this homogeneous glass property. A measurement by Hegarty and Yen[30] in a Pr^{3+} glass which spans the 5-300K temperature range appears in Fig. 7. The dotted curve is the phonon induced linewidth of the same transition $\left({}^3P_o \rightarrow {}^3H_4\right)$ in crystalline LaF_3.[28] The low temperature anomaly is clear as is the smooth T^2 variation of the glass linewidths over the whole temperature region.

Additional features of this problem have been investigated recently by Pellegrino,[7] who conducted a comprehensive study of the linewidth behaviour of the ${}^4F_{3/2} \rightarrow {}^4I_{9/2}$ transition of Nd^{3+} as a function of glass composition in ten different glasses. Invariably the linewidth is found to

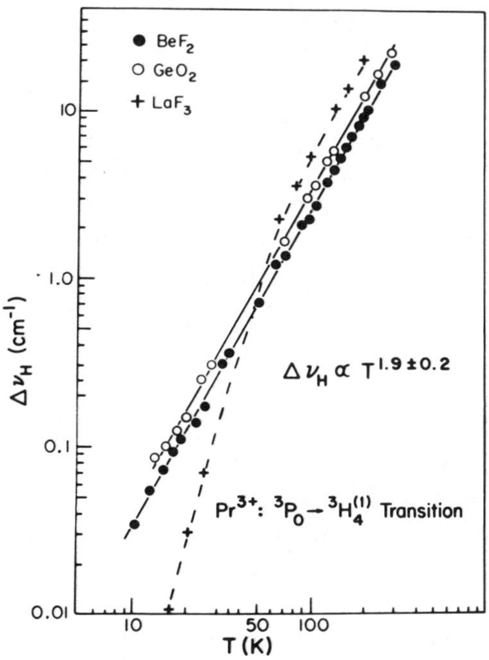

Fig. 7. Temperature dependence of the homogeneous linewidth of the ${}^3P_o \rightarrow {}^3H_4^{(1)}$ transition of Pr^{3+} in a beryllate and in a germanate glass. Dashed line represents the behaviour of the linewidth of the same transition in LaF_3 crystals. (From Ref. 30).

vary as T^2 and no correlation is found between the value of the homogeneous width and the extent of inhomogeneous broadening in the transition. This latter observation corroborates earlier observations and would indicate that the linewidth of transitions in glasses is independent of the Stark structure of the interacting states. The structure is known to dominate relaxation of paramagnetic ions in crystals at low temperature through one (direct) and two (Orbach) resonant phonon processes. Conversely, the implication is that these phonon-ion interactions are either suppressed or weakened in some way by other mechanisms.

Pellegrino et al. have also shown the homogeneous widths of the Nd^{3+} transition at ambient temperatures scales as

$$\Delta\nu_H \propto (\bar{v}_s)^{-2.6 \pm 0.2} \tag{2}$$

where $\bar{v}_s$ is the average sound velocity (over t and ℓ modes) in the specific glass. The same dependence on $\bar{v}$ is observed where the Raman process contributes dominantly to the transition width in crystals.[32] These observations, i.e. no correlation with the inhomogeneous width and the $\bar{v}$ dependence, disagree with the predictions of the earlier Lyo and Orbach[33] theory and would seem to implicate some Raman process as the source of the broadening. A similar study conducted by Morgan and coworkers on the ^5D_o transition of Eu^{3+} in glass are consistent with the Nd^{3+} results.[34]

The general difficulty in interpreting the linewidth behaviour lies in the absence of any cross over behaviour in the $\sim T^2$ observed. The notoriously poor thermal conductivity of insulating glasses has impeded the study of optical properties below $\sim$5K and indeed a comprehensive investigation of low temperature glass linewidth would seem to be in order. The prevalent suspicion remains that amorphous modes must play some role in the relaxation of ions; however in order to define this role, the nature of these excitations throughout the energy spectrum must be fully explored.[35]

We note in passing that the understanding of this specific problem in glasses is of considerable interest to laser technology. This is because the ratio $\Delta\nu_H/\Delta\nu_{IH}$ ultimately determines the energy which can be extracted from glass amplifiers, hence this ratio plays a crucial role in the performance efficiency of high power stimulated devices.[36]

V. ENERGY TRANSFER IN 4f DOPED GLASSES

As the concentration of active ions in a solid is increased, various dynamic effects originating in interactions amongst the ion begin to appear. These interactions result in the transferal of optical energy between the centers and a subsequent change in the excited state properties of the ions. A considerable number of phenomena, e.g. sensitization, fluorescence quenching, optical trapping, to name a few, are direct manifestations of energy transfer and a considerable volume of literature exists demonstrating their existence in ordered and disordered systems.[37]

Generally the transferal of energy entails the exchange of excitation from an excited ion (donor, D) to one that is in the ground or in an energetically lower state (acceptor, A). Several transfer processes are schematically illustrated in Fig. 8. There are several distinct aspects to the problem of energy transfer in solids. Firstly there is the determination of the microscopic interactions between D and A ions which leads to the transfer;

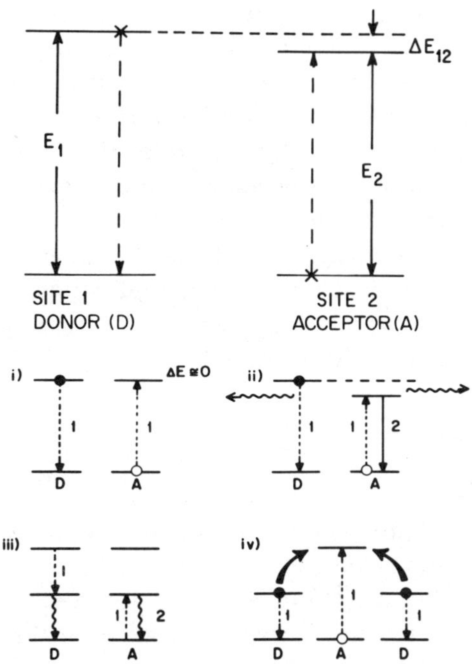

Fig. 8. Schematic representation of energy transfer between donor and acceptor ions. The donor (D) is initially in an excited state and transfers its energy non-radiatively to an acceptor in the ground state (broken arrows). The process may not conserve energy by a mismatch amount, ΔE_{12}, in which case phonon assistant need be invoked. Several process are illustrated in the following sequence: (i) Resonant D → A transfer, (ii) Phonon assisted sensitization, (iii) Cross relaxation and (iv) Upconversion. Wiggly arrows represent phonon emission. (Ref. 40).

this yields the appropriate parametric dependences of the ion-ion Hamiltonian. The second aspect deals with the translation of the microscopic parametric dependences to the macroscopic observables of the solid under consideration. The latter translation often involves ensemble averages over random variables with the averaging procedures being dependent on the nature of energy migration in the D system.

The theory of energy migration in ordered and disordered systems has been reviewed most recently by Holstein et al.[38] and by Huber in this volume and elsewhere.[31,39] We will again limit our discussion here to 4f ion activated glass where indeed the majority of the laser based studies have been conducted. A review of recent experimental studies of transfer in 4f crystalline systems has also been completed and many of the processes discussed there have direct analogies in disordered amorphous systems.[40]

In the past two decades, conventional spectroscopic studies have helped establish beyond doubt the phenomena of energy diffusion and transferal in 4f doped glasses.[41] Invariably in these studies various aspects of the ion-ion interactions are not measured directly but are inferred from the behaviour of a macroscopic observable. Thus, for example, the literature is replete with studies in which the nature of the microscopic interaction is extracted from concentration dependent studies of the decay time of some emission. Reliance of these dependences often leads to fallacious conclusions. The pitfalls arise because conventional broadband studies are not capable of measuring the dynamics which occur within like ion systems such as those leading to energy diffusion within the excited state of the D system.

A. INTRALINE TRANSFER

The time resolved version of FLN or TRFLN has provided with the methodology to probe into like ion dynamics by allowing us to study the dynamic behaviour of selectively chosen subsets of ions beneath an inhomogeneous distribution.[42] Fig. 9 illustrates TRFLN schematically and shows the behaviour of the line narrowed spectra of the $^2F_{5/2}$ state of Yb^{3+} in a silicate glass following pulsed excitation. The latter trace irrevocably demonstrates transfer between Yb^{3+} ions which are identical except for their relatively small inhomogeneous crystal field variations. Two additional features of the experimental results shown in Fig. 9 have proven crucial to theoretical developments. The time dependent changes in the FLN spectra indicate that transfer occurs from the originally excited subset to the whole inhomogeneous distribution simultaneously and that the linewidth of the FLN component does not broaden as a function of time. It is very probable that these properties are intrinsic to incoherent intra-ionic transfer in like ion

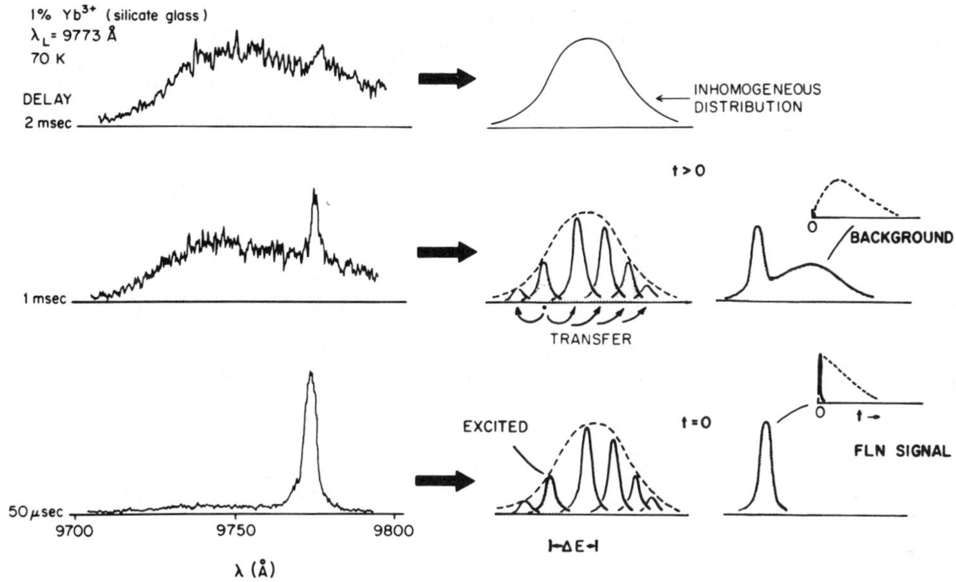

Fig. 9. Time evolution of the TRFLN spectra of the ${}^2F_{7/2}{}^{(1)} \rightarrow {}^2F_{5/2}{}^{(1)}$ transition of 1% Yb^{3+} in silicate glass at 70°K. Inserts represent a schematic representation of the TRFLN process.

systems as similar behaviour has been observed in all systems, crystals or glasses, studied to date except in a single isolated instance.[43,44]

Earlier theories of transfer properly identified the basic ion-ion interaction leading to energy diffusion as arising from electrostatic interactions between excited and ground state ions, i.e. multipolar, exchange etc. These theories however generally required some form of resonance to exist in the energies of the ions in question for the transfer to occur. For the TRFLN results cited above, it should be clear that additional non-resonant energy shifting mechanisms are required. More recent theories, notable among them the theory of Holstein and coworkers,[38] have remedied the situation by incorporating the role of phonons in a fundamental way and have led to a clearer understanding of the parametric dependences of the microscopic ion-ion interaction.

The role phonons play in the transfer process manifests itself not only in bridging the energy mismatch, ΔE_{12}, between the excited and unexcited ions but also through the temperature signature of the transfer process. Fig. 10 shows for example the temperature dependence of the TRFLN spectra of Yb^{3+} in silicate glass taken at a constant delay following the pulse excitation and attests to the increase in the transfer rate with increasing temperature. Also as in the case of crystals, a rate asymmetry is observed when the high and low energy wings of the inhomogeneous distribution are pumped under TRFLN conditions with the transfer rate from the high energy wings being faster. This asymmetry persists until $kT \simeq \Delta\nu_{IH}$ and implicates thermal population factors as its source.

A number of experimental studies have been carried out on the energy transfer properties within an inhomogeneously broadened 4f glass transition. In their original paper,[9] Motegi and Shionoya reported and analyzed time dependent spectral changes in the $^5D_o \rightarrow {}^7F_1$ transition in their Eu^{3+} glass; the spectra observed in this case suffers from the "accidental degeneracy" broadening since the transition is not resonant with the laser $\left({}^7F_o \rightarrow {}^5D_o\right)$. Studies of $^5F_o \rightarrow {}^7D_o$ TRFLN in Eu^{3+} glasses have subsequently been carried out by a number of workers.[45,46] Similar studies have been conducted on the $^2F_{5/2} \rightarrow {}^2F_{7/2}$ transition of Yb^{3+} in silicate glass[13,18,47] and in the $^4F_{3/2} \rightarrow {}^4I_{9/2}$ transition of Nd^{3+} in a glass laser host.[36]

As it has been pointed out,[31] the meaningful quantity to analyze in these experiments is R(t) which is defined as

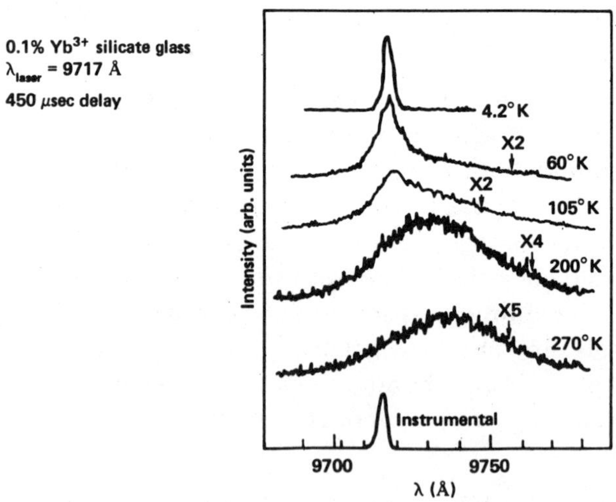

Fig. 10. TRFLN spectra of the $^2F_{7/2}{}^{(1)} \rightarrow {}^2F_{5/1}{}^{(1)}$ transition of Yb^{3+} in glass as a function of increasing temperature taken at a constant delay. At higher temperatures, the spectra shows no narrowing indicating that transfer has already taken place at this delay.

$$R(t) = \frac{I_{FLN}}{I_T} \tag{3}$$

where I_{FLN} is the intensity of the line narrowed component and I_T is the intensity of the total emission, i.e. narrowed plus broadened background component. The analysis of results in TRFLN have not been as detailed as those which have been conducted in crystals simply because many of the studies anteceded theoretical developments in this area.[48] By and large, the Inokuti-Hirayama model has been used to analyze TRFLN glass data and this model suffers from certain limitations.[49] In this model, which was principally devised for optical trapping, the excited ion energy is transferred to an acceptor and is subsequently radiated from there. Thus in the case of transfer within an inhomogeneously broadened state, the ions in the background behave as traps and any additional diffusion of the optical energy is precluded.

The Inokuti-Hirayama treatment over-estimates the rate of decay of R(t) in crystals where backtransfer is known to occur.[48] Regardless, with an average radiative lifetime normalized out of the equation, R(t) takes the generalized form of

$$R(t) = -\frac{4\pi N}{3} r_o^3 \left[W_o T\right]^{3/s} \Gamma(1-3/s) \tag{4}$$

with r_o being the distance of closest approach between ions, W_o being the transfer rate between ions separated by r_o and $\Gamma(x)$ being the gamma function. s depends on the nature of the interionic interaction and is the power in the spatial dependence of the latter Hamiltonian; for a dipole-dipole interaction, it follows that s=6.

Fig. 11 shows a plot of a modified version of R(t) as a function of t according to Eq. (4) for Yb^{3+} in silicate glass. Plots of this type allow the extraction of W_o and s. For Yb^{3+} and Eu^{3+} transitions studied to date, s=6 throughout indicating that electric dipole-dipole interactions between ions are responsible for the intraline transfer.[45,46,47]. The values of W_o measured for 4f ions in glass are also consistent with those measured in their crystalline counterparts.

The good agreement obtained in fitting R(t) with the Inokuti-Hirayama model may be due in part to the magnitude of the inhomogeneous width encountered in glasses. An initial large downwards transfer step, i.e. $-\Delta E_{12}$, would minimize any backtransfer especially at low temperatures, thus the intrinsic approximations of the model may in certain cases accurately

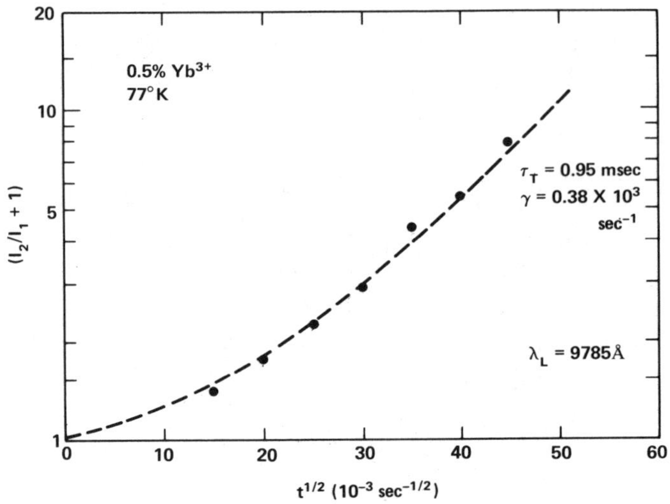

Fig. 11. Inokuti-Hirayama analysis of a quantity related to R(t) in text. The plot indicates a $t^{3/s}$, s=6 behavior commensurate with electric dipole-dipole transfer. Transition analyzed is as in Fig. 9 and 10, Yb^{3+} concentration was 0.5 M%.

represent the situation in glasses. Various additional details have not been incorporated in these analysis of R(t). For example, we have already noted that individual ion sites across the inhomogeneous distribution are characterized by different oscillator strengths and hence different radiative lifetimes. The incorporation of these variations into the transfer probability W and into R(t) poses a very complex and difficult problem and hence their influence on the results has not been fully ascertained. It may be however that the gross averaging procedure inherent to this model may smooth over these variations in a fortuitous way.

Detailed studies of the precise nature of the microscopic interaction leading to transfer in glasses are also less prevalent than those in crystals. However, we may arrive at a number of conclusions based on the comparison of TRFLN results in glass with those obtained in ordered systems. For example, the narrowed component transfer to the whole inhomogeneous background simultaneously implies that W_{12} is independently or weakly dependent on the transfer energy mismatch, ΔE_{12}. Among the phonon assisted transfer mechanisms proposed by Holstein et al., the only term in their expansion which shows this property is the so called "one phonon, second order" process acting at each of the coupled sites. Not surprisingly this has been shown to be the most prevalent mechanism governing intraline transfer in crystals.[50] Additionally, theory predicts a $\sim T^3$ increase in the rate of transfer as a function of temperature; such a dependence has indeed been reported by Avouris et al. in their study of the $^5D_0$ state of Eu in glass. Though the evidence is not complete, it is once again very probable that the same phonon mediated multipolar interactions that are effective in crystals are also responsible for energy diffusion in glasses.

The early time behaviour of R(t) in glasses has not been studied because of experimental problems connected with laser resonant TRFLN studies. Nor has a detailed study been carried out of the R(t) dependence on temperature over an extended range. These are areas which clearly merit additional theoretical and experimental consideration.

B. INTERLINE TRANSFER

The majority of the literature in optical energy transfer is addressed to cases in which the donor ion transfers its energy to an acceptor ion which is distinct from the donor or else the transfer is effected across an unlike electronic level in an identical ion. The former leads to processes such as sensitized luminescence in the acceptor system (Fig. 8b) while the latter are responsible for cross relaxation fluorescence quenching (Fig. 8c), for example. Annihilation of two excited ions resulting in fluorescence upconversion are also considered to fall in this general class of transfer phenomena (Fig. 8d). Regardless, the end result is the irreversible loss of energy from the excited donor state. We will classify such transfer as being interline.

A multitude of interline transfer processes involving many ions and ion combinations has been demonstrated experimentally in crystals and in glasses. For our purposes here, the property of interest becomes the decay characteristics of the total inhomogeneous fluorescence of the donor state following pulsed excitation. As discussed in a previous chapter,[31] the solution of the decay poses a formidable if not impossible theoretical problem. Various approximations have been made in order to make the problem tractable. We need only reiterate some the salient features of these models.

The characteristic of the temporal changes of the donor population as measured by the fluorescence decay depends on the relative magnitudes of the D—D and the D—A transfer rates. In the limit of D—D rates being very slow compared to D—A rates, the decay is describable in terms of the Inokuti-Hirayama relation, Eq. (4), and will exhibit a characteristic $t^{3/s}$ dependence.[49] In the opposite limiting case, (fast diffusion) the D—D rate is assumed to be extremely fast and the donor population decay follows a simple exponential decay.[51] When the D—D and D—A rates are comparable the Yokota and Tamimoto diffusion equations[52] or the hopping model of Burshtein are often employed.[53] In these intermediate regions the decay of the donor population is initially non-exponential changing gradually to an exponential at long decay times.

In the experimental studies, which have generally been done using conventional sources, the decay of the fluorescence is analyzed in terms of some appropriate model and the D—A rate is extracted from the analysis. These procedures need assume *a priori* the relative rate of the D—D transfer compared to the D—A rate. TRFLN provides us with a way to measure intraline transfer directly which is equivalent to a determination of the D—D rate. Thus measurements of both intraline and interline donor transfer properties are required in order to obtain a complete description of the situation.

In a series of recent studies, Hegarty and co-workers[54] have investigated the intraline (FLN) and interline (trapping, quenching) of Pr^{3+} in a crystalline system. By relying on the rapid T^3 dependence of the intraline (TRFLN) or D—D transfer within the inhomogeneously broadened donor state, they were able to change this rate continuously from slow to very fast compared to the D—A rate. From the behaviour of the donor decay, they were then able to establish quantitative limits to the validity of the various models cited above.

Though a number of comprehensive studies of interline transfer in glasses exist,[55,56,57] no studies which measure D—D and D—A dynamics simultaneously on the same system have appeared. The situation in glasses is of course expected to be more complex owing once again to the inhomogeneous variation within the donor and acceptor systems and the incorporation of this additional randomness to the various models remains to be done. However, allowing for the latter additional modifications, the general expectation is that models which have been substantiated in crystals again will be applicable to 4f doped glasses. For example, Weber first demonstrated the applicability of the Yokota-Tamimoto relations in a Cr^{3+} doped Eu glass and obtained a decay rate proportional to the energy diffusion coefficient $D^{3/4}$ as predicted by the theory.[55] A number of more recent studies of Nd^{3+} in laser glasses have also appeared. In these, various quenching processes are demonstrated and analyzed in terms of the models above.[57]

VI. DISCUSSION AND CONCLUSIONS

We have summarized developments in the study of the optical properties of activated glasses since the advent of tunable laser spectroscopy. The contribution of these new technique to structural studies of glasses has already been considerable. We expect that rapid advances will continue to occur in this area as these unique spectroscopic tools find a more general acceptance in the study of amorphous materials.

A number of pressing problems remain. The principal one deals with the origin of linewidths (and relaxation) of transitions in glass. As we have seen, many experimental facts have been established and are shown to be more or less universal to the linewidth behaviour. Extremely low temperature measurements (0-2K) are required to help establish the role of TLS in these processes, while consideration should be given to the nature of high energy TLS in disordered systems.

TRFLN has given us an additional tool with which to gather a more fundamental and direct understanding of energy migration in glasses. A number of these investigations are currently in progress and results in the experimental investigations should provide additional impetus to theoretical advances. We note that though the focus here has been on optical energy transfer in 4f glasses, the processes involved are prototypical to transfer of excitation in any inhomogeneous, amorphous system.

Finally, though not explicitly stated we have dealt here with systems in which the ion-ion coupling is extremely weak by any standards. Consequently, the question of coherence or extended states never seriously enters into our considerations and the excitation can be and is treated as localized on individual ions. It will be of some interest to seek and consider systems in which the interionic couple can be increased so that it becomes comparable to the inhomogeneities of the system. Coherence in transfer could then likely be studied in these systems.

Acknowledgements: A great deal of interest and involvement in the area of glasses has resulted from continuing discussions with and guidance from Dr. M. J. Weber of LLNL. Interactions with Prof. D. L. Huber, Dr. J. Hegarty and Dr. S. A. Brawer are also acknowledged with thanks. This work has been supported by the National Science Foundation, by the Army Research Office, by the University of Wisconsin Graduate Research Committee, and by NATO. Their support is noted with appreciation.

REFERENCES

1. A survey of conventional spectroscopic studies in glasses is to be found in J. Wong and C. A. Angell, *Glass Structure by Spectroscopy*. (Dekker, New York, 1976).
2. P. M. Selzer in *Laser Spectroscopy of Solids, Topics in Applied Physics*, Vol. 49, W. M. Yen and P. M. Selzer, eds. (Springer Verlag, Berlin, 1981) Ch. 4.
3. G. H. Dieke, *Spectra and Energy Levels of Rare Earth Ions in Crystals*, (Wiley-Interscience, New York, 1968).
4. S. Hüfner, *Optical Spectra of Transparent Rare Earth Compounds*, (Academic Press, New York, 1978).

5. M. J. Weber, in Ref. 2, Ch. 6.
6. D. R. Uhlmann and A. G. Kolbeck, Phys. Chem. Glasses **17**, 146 (1976).
7. J. M. Pellegrino, W. M. Yen and M. J. Weber, J. Appl. Phys. **51**, 6332 (1980).
8. Yu V. Denisov and V. A. Kizel, Opt. Spectrosc. **23**, 251 (1967).
9. N. Motegi and S. Shionoya, J. Lumin. **8**, 1 (1973).
10. E. Snitzer and C. G. Young; *Lasers*, Vol. 2, A. K. Levine, ed. (Dekker, New York, 1966).
11. T. Kushida and E. Takushi, Phys. Rev. **B12**, 824 (1975).
12. J. Hegarty, R. T. Brundage and W. M. Yen, Appl. Opt. **19**, 1889 (1980).
13. M. J. Weber, J. A. Paisner, S. S. Sussman, W. M. Yen, L. A. Riseberg, C. Brecher, J. Lumin. **12/13**, 729 (1976).
14. Mono frequency laser induced FLN was demonstrated first in a Nd^{3+} glass. L. A. Riseberg, Phys. Rev. Lett. **28**, 789 (1972).
15. S. S. Sussman, J. A. Paisner, W. M. Yen and M. J. Weber, Bull. Am. Phys. Soc. **20**, 44 (1975).
16. C. Brecher and L. A. Riseberg, Phys. Rev. **B21**, 2607 (1980).
17. B. G. Wybourne, *Spectroscopic Properties of Rare Earths* (Wiley-Interscience, New York, 1965).
18. J. A. Paisner, S. S. Sussman, W. M. Yen and M. J. Weber, Bull. Am. Phys. Soc. **20**, 447 (1975).
19. P. M. Selzer, D. L. Huber, D. S. Hamilton, W. M. Yen and M. J. Weber: in *Structure and Excitations in Amorphous Solids*, AIP Conf. Proc. 31, 328 (1976).
20. P. Avouris, A. Campion and M. A. El-Sayed, J. Chem. Phys. **67**, 3397 (1977).
21. S. A. Brawer and M. J. Weber, Phys. Rev. Lett. **45**, 460 (1980).
22. D. W. Hall, S. A. Brawer and M. J. Weber, Phys. Rev. **B25**, 2828 (1981).
23. J. Hegarty, W. M. Yen and M. J. Weber, Phys. Rev. **B18**, 5816 (1978).
24. M. J. Weber, J. Hegarty and D. H. Blackburn, in *Borate Glasses*, L. D. Pye, V. D. Frechette and N. J. Kreidl, eds. (Plenum Press, New York 1978) p. 215.
25. M. J. Weber, J. Non-Cryst. Sol. **47**, 117 (1982); in *Amorphous and Liquid Semiconductors*, W. E. Spear, ed. (Univ. of Edinburgh, Edinburgh, 1978) p. 645.
26. R. G. Brewer, this volume.
27. R. M. Macfarlane and R. M. Shelby, unpublished. See also R. M. Macfarlane, this volume.
28. W. M. Yen, W. C. Scott and A. L. Schawlow, Phys. Rev. **136**, A271 (1964).
29. P. M. Selzer, D. L. Huber, D. S. Hamilton, W. M. Yen and M. J. Weber, Phys. Rev. Lett. **36**, 813 (1976).
30. J. Hegarty and W. M. Yen, Phys. Rev. Lett. **43**, 1126 (1979).
31. D. L. Huber, this volume.
32. J. M. Pellegrino and W. M. Yen, Phys. Rev. **B24**, 6789 (1981).
33. S. K. Lyo and R. Orbach, Phys. Rev. **B22**, 4223 (1980).
34. J. R. Morgan, E. P. Chock, W. D. Hopewell, M. A. El-Sayed and R. Orbach, J. Phys. Chem. **85**, 747 (1981).
35. S. K. Lyo, Phys. Rev. Lett. **48**, 688 (1982).
36. S. A. Brawer and M. J. Weber, App. Phys. Lett. **35**, 31 (1979). See also W. M. Yen and M. J. Weber, Optical linewidths in glass and their relation to hole burning, Lawrence Livermore Laboratory, Report ELR 79-107, (1979), unpublished.
37. See for example, J. C. Wright in *Radiationless Processes in Molecules and Condensed Phases*, F. K. Fong, ed. (Springer Verlag, Berlin, 1976) Ch. 4.
38. T. Holstein, S. K. Lyo and R. Orbach, in Ref. 2, Ch. 2.
39. D. L. Huber, in Ref. 2, Ch. 3.
40. W. M. Yen in *Spectroscopy of Rare Earth Ions in Crystals*, R. M. Macfarlane and A. A. Kaplyanskii, eds. (North Holland, Amsterdam, to be published).
41. See for example, R. Reisfeld, Structure and Bonding **30**, 65 (1976).
42. W. M. Yen, J. Lumin. **18/19**, 639 (1979).
43. W. M. Yen and P. M. Selzer, in Ref. 2, Ch. 5.
44. M. Harig, R. Charneau and H. Dubost, J. Lumin. **24/25**, 643 (1981).
45. W. M. Yen, J. A. Paisner, S. S. Sussman and M. J. Weber, Lawrence Livermore Laboratory, Rpt. UCRL-76481 (1975).
46. P. Avouris, A. Campion and M. A. El-Sayed, Chem. Phys. Lett. **50**, 9 (1977).

47. R. T. Brundage, M. Shulavitch and W. M. Yen, to be published.
48. D. L. Huber, D. S. Hamilton and B. B. Barnett, Phys. Rev. **B16**, 4642 (1977).
49. M. Inokuti and H. Hirayama, J. Chem. Phys. **43**, 1978 (1965).
50. R. Flach, D. S. Hamilton, P. M. Selzer and W. M. Yen, Phys. Rev. Lett. **35**, 1034 (1975).
51. D. L. Huber, Phys. Rev. **B20**, 2307 (1979).
52. M. Yokota and I. Tamimoto, J. Phys. Soc. Jpn. **22**, 779 (1967).
53. A. I. Burshtein, Zh. Eksp. Teo. Fiz. **62**, 1695 (1972) [Sov. Phys.-JETP **35**, 882 (1972)].
54. J. Hegarty, D. L. Huber and W. M. Yen, Phys. Rev. **B23**, 6271 (1981); Phys. Rev. **B25**, 5638 (1982).
55. M. J. Weber, Phys. Rev. **B4**, 2932 (1971).
56. A. R. Speed, G. F. J. Garlich and W. E. Hagston, phys. stat. sol. (a)**27**, 477 (1975).
57. See for example: A. G. Avanesov, T. T. Basiev, Yu. K. Voron'ko, B. I. Denker, A. Ya. Karasik, G. V. Maksimova, V. V. Osiko, V. F. Pisarenko and A. M. Prokhorov, Zh. Eksp. Teo. Fiz. **77**, 1771 (1979) [Sov. Phys. *JEPT 50*, 886 (1980)] and references therein.

DISCUSSION

Experiments on Optical Energy Transfer

It was emphasized that while FLN (fluorescence line narrowing) data can be used to model the local environment, they do not provide an absolute structural determination. This method was compared with the more conventional neutron diffraction techniques.

The role of phonons in inter-ion energy transfer was then discussed. It was noted that the temperature dependence of the time resolved fluorescent line narrowing spectra of Yb^{3+} in silicate glasses (Yen, Figure 10) indicates that the transfer rate increases with increasing thermal phonon population. However, it was felt that the low temperature data are insufficient to substantiate the claim that the entire inhomogeneous band appears simultaneously following the excitation pulse. While characteristic excitation diffusion lengths are thought to be of order 20Å, there is little direct evidence for this value. Indeed transient grating (or degenerate four-wave mixing) experiments in crystals have placed an upper limit of ~500Å.

It was also pointed out that nonradiative processes have not been considered important in analyzing FLN experiments. In practice only sharp emission lines with high quantum efficiency have been used in these experiments. Nonradiative processes are considered energy independent (at least over $\Delta\nu_{INH}$) and describable by a single lifetime. The net effect is to reduce the radiative quantum efficiency uniformly across the inhomogeneous line.

J. Ryan, Chairman
J. Hegarty, Discussion Leader

COMMENTS

Correlation Functions and Rare Earth Environments

Rare-earth-containing glasses normally involve several constituents which means that the total correlation function T(r) obtained from a single diffraction experiment (cf. Wright's section) is fairly complicated. The rare earth ions are usually a minor constituent and hence the interpretation of such a correlation function to give reliable information on the environment of the rare earth ions is extremely difficult. An alternative approach is to study selectively the rare earth ions using one of the special techniques to vary their neutron scattering length while keeping the scattering lengths of all the other constituent elements constant. The difference correlation function $\Delta T(r)$ obtained from such an experiment is analogous to the result of an EXAFS measurement at the absorption edge of the rare earth.

In particular the anomalous dispersion technique has been employed to investigate the environment of Sm in vitreous $Sm_2O_3-Al_2O_3-GeO_2$ and isotopic substitution to study Dy in $NaF-DyF_3-BeF_2$.[1] In the latter each Dy atom is on average surrounded by 7.3 ± 0.5 fluorine atoms at a distance of 2.301 ± 0.005Å with an r.m.s. bond length variation of 0.086 ± 0.005Å. The distribution of oxygen atoms around Sm in vitreous $Sm_2O_3-Al_2O_3-GeO_2$ is much broader and at $Q_{max} = 8.64Å^{-1}$ the first Sm-0 peak is not fully resolved. Fitting a Gaussian distribution the low r side at the peak yields a coordination number of 7.5 ± 1. The mean bond length is 2.42 ± 0.01Å and the r.m.s. bond length variation 0.19 ± 0.01Å. In both the glasses studied the rare earth environment is similar to that found in related crystalline materials.

A. C. Wright
Department of Physics
University of Reading
Reading, England

REFERENCES

1. A. C. Wright, G. Etherington, J. A. E. Desa, and R. N. Sinclair, J. de Physique (in press).

LASER SPECTROSCOPY OF SOLIDS*

Richard G. Brewer and Ralph G. DeVoe

IBM Research Laboratory
San Jose, California 95193, USA

Remarkably narrow optical homogeneous linewidths of the order of 1 kHz have now been observed in low temperature zero-phonon transitions of dilute impurity ion crystals, such as Pr^{3+} in LaF_3. Novel nonlinear optical resonance techniques have been devised for this purpose using ultrastable phase locked cw dye lasers where the measurements are performed either in the frequency domain (hole burning) or in the time domain (coherent optical transients). These studies effectively bring the Mossbauer effect into the optical region. Hence, the observed linewidths are no longer limited by inhomogeneous strain broadening (~5 GHz) or even by static local fields due to neighboring spins (~100 kHz). However, weak magnetic field fluctuations from local spins are readily detected. As an example, spin decoupling and line narrowing, which are well known in NMR, are observed in an optical transition of Pr^{3+}:LaF_3 at 2°K where the $^{19}F-^{19}F$ dipolar interaction is quenched and the optical linewidth drops from 10 to 2 kHz, clearly demonstrating the spin broadening mechanism. Results will be discussed in terms of a Monte Carlo line broadening theory.

---

* This work is supported in part by the U.S. Office of Naval Research.

I. INTRODUCTION

Solids have long played an important role in the development of laser physics. Indeed, the first laser was a solid state device, namely, Maiman's ruby laser.[1] Shortly thereafter the field of nonlinear optics commenced when Franken et al.[2] transformed a ruby laser beam into its second harmonic by passing it through a quartz crystal. And the first coherent optical transient effect, the photon echo, was detected later in ruby by Hartmann et al.[3]

Laser spectroscopy,[4] on the other hand, has been concerned largely with atomic and molecular systems in the gas phase, and it is only recently that the new techniques of laser spectroscopy have been applied to solids. Within the last three years, for example, it has become clear that the optical homogeneous linewidths of certain rare earth impurity ion crystals can be exceedingly narrow, the order of 1 kilohertz or less.[5–7] These zero phonon transitions are the optical analogs of the Mossbauer effect. Such narrow linewidths suggest a number of new precision measurements, either in the time or frequency domain. In the time domain, coherent transients can be examined to reveal new aspects of the dynamic interactions occurring in solids. In the frequency domain, the structural details of solids can be probed as in nuclear magnetic resonance. Also, the possibility of developing solid state optical clocks exists and with it the potential for testing fundamental theories, such as the general theory of relativity.

In this article, I will restrict the discussion to recent dynamic studies of the optically active impurity ion Pr^{3+} in a host crystal of LaF_3 at $\sim 2°K$. This study tests our understanding of the Pr^{3+} optical linebroadening and dephasing mechanisms, which are largely magnetic in origin, through such experimental optical techniques as free induction decay[5] and magic angle line narrowing.[6] This work has advanced recently to a new level of precision due to the development of a highly stable tunable ring dye laser that possesses a linewidth of about 300 Hertz.[8] On the theoretical side, our success with Monte Carlo calculations of linebroadening in $Pr^{3+}:LaF_3$ will be reviewed.[9] Even more recently, we have been encouraged by an analytic treatment which explains the main features of linebroadening in this impurity ion solid and reveals as well limitations in the use of the optical Bloch equations for the case of solids.[10]

II. FREE INDUCTION DECAY THEORY

The two most important methods for measuring optical dephasing times have been the photon echo[3] and free induction decay (FID),[11] which we now consider. The simplest model is that of a collection of two-level atoms which are resonantly excited (prepared) by a coherent light wave. The atoms thereby are transformed from an initial stationary state to a superposition or mixed state which displays a time-dependent behavior both during the preparation stage and afterwards. Once the excitation is removed, the system freely radiates a coherent beam of light in the forward direction-the free induction decay effect. The atoms are perturbed of course by various time-dependent interactions which get them out of phase and produce a damped emission. An example of FID in Pr^{3+}:LaF_3 is shown in Figure 1.

III. BLOCH EQUATIONS

To interpret the observed decay rate, various theoretical models can be applied. We begin with the Schrödinger equation of motion in density matrix form

$$i\hbar\dot{\rho} = [H,\rho] + \text{relaxation terms} \tag{1}$$

which will be used to derive Bloch equation solutions[12] for a two-level atomic system before proceeding to more advanced cases. The Hamiltonian

$$\mathbf{H} = H_0 + H_1$$

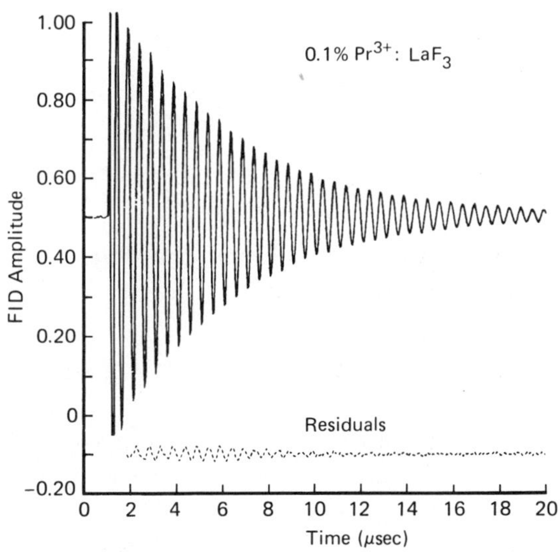

Fig. 1. A computer plot of 400 points of optical FID of 0.1 at.% Pr^{3+}:LaF_3 at 1.6°K. The experimental data are overlaid on a damped cosine, and the residuals indicate that the dephasing time of 5.10 μsec has an uncertainty of less than 1%. The signal is power broadened.

contains the free atom part H_0 with eigenenergies

$$E_i = \hbar\omega_i \ (i = 1,2)$$

where $\omega_{21} = \omega_2 - \omega_1$ and the index 2 labels the upper state and 1 the lower state. In the presence of a light wave

$$E(z,t) = E_0 \cos(\Omega t - kz) ,$$

the atom undergoes a transition $1 \leftrightarrow 2$ due to the atom-field interaction

$$H_1 = -\mu \cdot E(z,t)$$

where the electric dipole matrix element is

$$\mu_{12} = \langle 1|\mu|2\rangle .$$

The equations of motion for the slowing varying terms become

$$\dot{\tilde{\rho}}_{12} = (-1/T_2 + i\Delta)\tilde{\rho}_{12} + 1/2\, i\chi(\rho_{22} - \rho_{11}) \tag{2a}$$

$$\dot{\rho}_{22} - \dot{\rho}_{11} = -(\rho_{22} - \rho_{11})/T_1 + (\rho_{22}^{\circ} - \rho_{11}^{\circ})/T_1 + i\chi(\tilde{\rho}_{12} - \tilde{\rho}_{21}) \tag{2b}$$

with the definition

$$\rho_{12} = \tilde{\rho}_{12} e^{i(\Omega t - kz)}$$

and neglecting nonresonant terms. Here, the Rabi frequency χ and the tuning parameter Δ are defined by

$$\chi = \mu_{12}E_0/\hbar \text{ and } \Delta = -\Omega + \alpha + \omega_{21}$$

where α is a shift in the transition frequency ω_{21} due to an inhomogeneity in the local environment, i.e., static magnetic or crystalline Stark fields in the case of Pr^{3+}:LaF_3.

In the optical Bloch model, the decay behavior is introduced in (2) by phenomenological population and dipole decay times T_1 and T_2 as in NMR. Thus, with the definitions

$$u = \tilde{\rho}_{12} + \tilde{\rho}_{21} ,\ v = i(\tilde{\rho}_{21} - \tilde{\rho}_{12}), \text{ and } w = \rho_{22} - \rho_{11} ,$$

we cast (2) into the Bloch equation[13]

$$\frac{dB}{dt} = \beta \times B$$

which describes a precessional motion of the Bloch vector B about an effective field β with components

$$B = iu + jv + kw \, ,$$

$$\beta = i\chi + k\Delta \, .$$

This is mathematically equivalent to a spin precessing in a magnetic field.[14]

Bloch equation solutions are derived from (2) using a Laplace transform technique[15] and yield for the steady-state preparation

$$\tilde{\rho}_{12}(0) = \frac{i\chi(-i\Delta + 1/T_2)(\rho_{22}^{o} - \rho_{11}^{o})/2}{\Delta_2 + 1/T_2^2 + \chi^2 T_1/T_2} \tag{3}$$

At time t=0, the excitation ends and the FID begins because the laser frequency is switched suddenly to a new value ($\Omega \rightarrow \Omega'$). The FID solution follows from (2a) as

$$\tilde{\rho}_{12}(t) = \tilde{\rho}_{12}(0)e^{(-1/T_2 + i\Delta)t} \, , \; t > 0 \, . \tag{4}$$

The FID expressed as a field amplitude

$$E_{12}(z,t) = E_{12}(z,t)e^{i(\Omega t - kz)} + c.c.$$

obeys Maxwell's wave equation

$$\frac{\partial E_{12}}{\partial z} = -2\pi i k N \mu_{12} \langle \tilde{\rho}_{12}(t) \rangle \tag{5}$$

where the bracket

$$\langle \tilde{\rho}_{12}(t) \rangle = \frac{1}{\sqrt{\pi}\sigma} \int_{-\infty}^{\infty} g(\Delta)\rho_{12}(\Delta,t)d\Delta \tag{6}$$

denotes an average over a Gaussian inhomogeneous lineshape $g(\Delta) = e^{-(\Delta/\sigma)^2}$. The observed FID signal appears as a heterodyne beat

$$F(t) = \tfrac{1}{2}E_0E_{12}e^{i(\Omega-\Omega')t} + \text{c.c.} \tag{7}$$

due to the laser frequency switch $\Omega \rightarrow \Omega'$ at $t=0$, a process which terminates the excitation of the initially prepared packet and allows sensitive detection of the free precession signal with low noise. Omitting trivial factors, the resulting FID Bloch solution is of the form

$$F(t) \sim \chi^2 \left[1 - \frac{1}{\sqrt{1+\chi^2T_1T_2}}\right] e^{-(t/T_2)(1+\sqrt{1+\chi^2T_1T_2})} \,. \tag{8}$$

The preexponential factor displays a nonlinear intensity dependence in contrast to NMR where FID is a first order process due to the small inhomogeneous broadening. Similarly, the damping term in (8) exhibits power broadening through the term $\chi^2T_1T_2$ where $T_1 >> T_2$ in Pr^{3+}:LaF_3. In the limit $\chi^2T_1T_2 << 1$, the decay time becomes $\tfrac{1}{2}T_2$.

While the Bloch theory has played an important role in NMR and quantum optics, it also is limited in that it ignores the details of the basic dipolar interactions such as Pr^{3+}—F and F—F which broaden the Pr^{3+} optical transition. It also ignores the presence of a frozen core of fluorine nuclei surrounding each Pr^{3+} ion, the LaF_3 crystal structure, and the dependence of linewidth on the Pr^{3+} ($I = 5/2$) magnetic substrate. Before considering the Monte Carlo calculation which overcomes these difficulties, let us consider the current status of the experiments.

IV. FREE INDUCTION DECAY EXPERIMENTS

The optical Pr^{3+} transition of interest is ${}^3H_4 \rightarrow {}^1D_2$ which lies conveniently in the yellow region at 5925 Å. Due to the low electric field site symmetry (C_2), all electronic degeneracy is removed. Thus, for the ground electronic state ${}^3H_4$, there are $2J + 1 = 9$ Stark split *singlet* states with splittings of order 50 cm^{-1}. Consequently, as noted by Bleaney[16] and by Teplov,[17] all first order magnetic hyperfine interactions vanish in the absence of an external magnetic field. The nuclear quadrupole interaction and the second order magnetic dipole hyperfine interaction generate three doubly degenerate hyperfine states for each Stark split singlet ($I_z = \pm5/2$, $\pm3/2$ and $\pm1/2$ since $I = 5/2$) with splittings of the order of 10 MHz. Only the lowest crystal field Stark split states of ${}^3H_4 \rightarrow {}^1D_2$, the zero phonon line, are examined. Hence, three equally intense optical transitions occur, $I_z'' \leftrightarrow I_z' = \pm5/2 \leftrightarrow \pm5/2$, $\pm3/2 \leftrightarrow \pm3/2$, and $\pm1/2 \leftrightarrow \pm1/2$, and overlap because of the large inhomogeneous strain broadening of ~5 GHz.

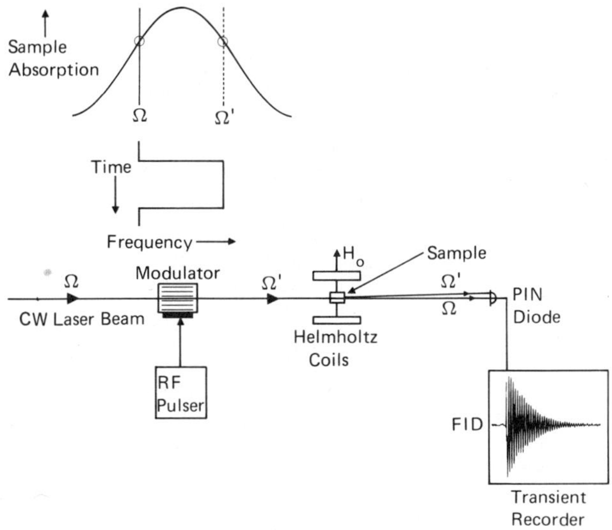

Fig. 2. Acousto-optic modulator laser frequency switching technique for observing coherent optical transients such as FID and photon echoes.

The technique for observing optical FID relies on laser frequency switching[18] as shown in Figure 2.[5] The external beam of a phase-locked cw ring dye laser of high frequency stability passes through an acousto-optic modulator prior to exciting a 0.1 at.% Pr^{3+}:LaF_3 crystal which is immersed in liquid helium at ~1.6°K. A single packet of Pr^{3+} ions within the inhomogeneous lineshape is coherently prepared by a laser beam (~5 mW at the crystal) when the modulator is driven by a 110 MHz rf source which is gated on for a 400 μsec period. FID follows when the rf frequency is switched suddenly from 110 to 108 MHz and is detected by a photodiode as a 2 MHz heterodyne beat signal in transmission.

V. LASER PHASE LOCKING

To detect ultraslow dephasing times by FID, the laser frequency must remain fixed within the Pr^{3+} homogeneous linewidth for the preparation interval ~T_2, otherwise the observed decay merely reflects laser frequency jitter. Figure 3 shows a phase modulation technique for phase/frequency locking a ring dye laser to a reference cavity. The method, which was proposed by Drever, has yielded laser linewidths as narrow as ~100 Hertz.[19,8] A part of the external beam of the laser is phase modulated by passing it

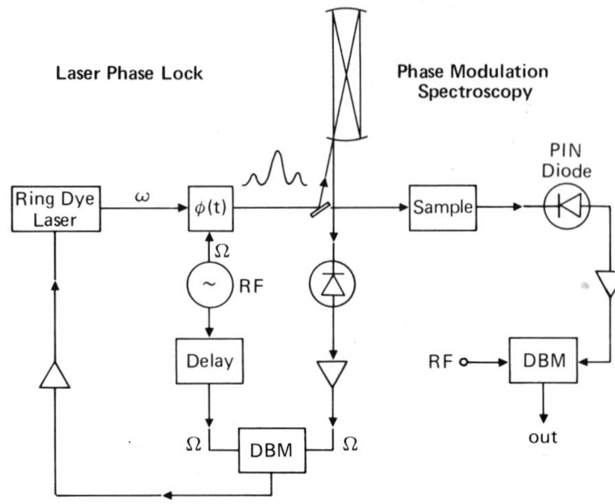

Fig. 3. Schematic of apparatus for phase-locking a cw ring dye laser. The sample and detection circuit to the right show how narrow hole burning signals can be detected by phase modulation spectroscopy with basically the same apparatus.

through an electro-optic crystal that is driven continuously by a 40 MHz rf generator, and the beam then strikes an acoustically isolated 50 cm confocal reference cavity having a 1.5 MHz bandwidth. The laser field now contains pairs of sidebands located symmetrically about the center frequency. When the center frequency coincides with a mode of the reference cavity, the central component will be stored in the cavity for its ringing time while the sidebands will be reflected at the end mirror. For this condition, the reflected light and some of the stored light which leaks out of the cavity will produce at a photodiode two heterodyne beats of opposite phase which just cancel. If, however, the laser phase or frequency fluctuates, this balanced condition will be upset because the stored light retains memory of the laser's frequency at a previous instant while the reflected light monitors the instantaneous laser frequency. In this circumstance, the two heterodyne beat signals no longer cancel but produce an error signal in a fast high gain servo loop that drives the laser frequency to a stable operating point.

Important advantages of this method are (1) Laser amplitude noise can be reduced to the shot noise limit because the beat frequency can be selected in a region where the laser noise is low and because the balancing technique automatically eliminates laser noise not associated with the error signal. (2) The response time of the optical cavity is fast in the reflection mode allowing a large servo loop bandwidth and thus a narrow laser linewidth. (3) The heterodyne beat error signal can be monitored in a dispersion mode (Figure 4) which offers a sharp discriminant and locates the lock point at zero amplitude, independent of the size of the error signal. In Figure 4, the phase-locked laser beam is monitored with a second 50 cm confocal cavity and shows that the laser linewidth is no larger than 300 Hertz rms.

The phase modulation technique can be used also for detecting low

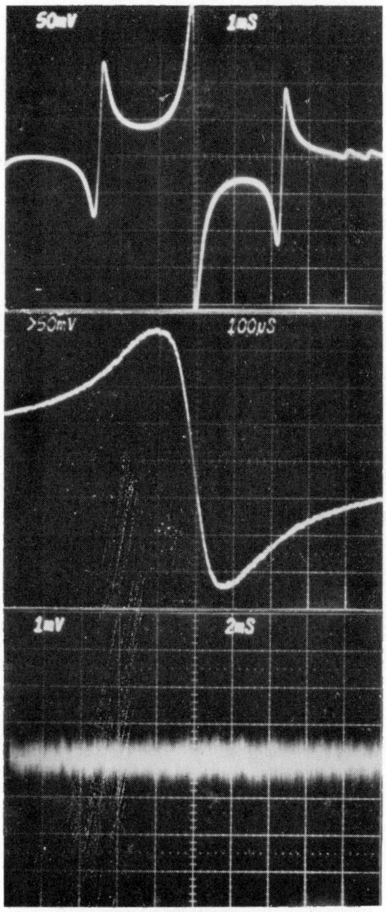

Fig. 4. Top two traces: Experimental dispersive lineshape of a phase-locked laser beam (40 MHz modulation frequency) as seen by a frequency swept 50 cm confocal cavity in reflection. The laser beam is phase locked to a second 50 cm cavity and exhibits in the third trace, where the first cavity is not swept, a 300 Hz rms laser linewidth.

amplitude noise lineshapes (hole-burning) under steady-state conditions as demonstrated initially by Bjorklund.[20,21] In combination with a laser of high frequency stability, extraordinarily high resolution optical studies can now be performed in solids.

VI. DATA ACQUISITION

The realization of tunable lasers of high frequency stability, assures that FID or other coherent transients with long optical dephasing times can be monitored reproducibly with high precision. Figure 1 shows the excellent fit of a damped cosine function which is overlaid on 400 experimental points of the Pr^{3+}:LaF_3 data. In fact, the two curves cannot be distinguished and the residuals indicate that the decay time can be determined to an uncertainty of less than 1%. The experimental points were acquired from an analog signal using a transient digitizer, and the data was stored in an IBM Personal Computer and then transferred to an IBM 3033 computer for least squares analysis, graphics or other data handling.

VII. MAGIC ANGLE LINE NARROWING

The FID technique offers a way of measuring the optical homogeneous linewidths of a solid such as Pr^{3+}:LaF_3 without the influence of inhomogeneous broadening. Thus, in Figure 1 the linewidth is about 10 kHz HWHM, i.e., about the same magnitude as that encountered in NMR. What is the broadening mechanism? The $^1D_2$ radiative lifetime of 0.5 msec sets a limiting value of 160 Hertz HWHM; phonon-ion interactions at 1.6°K are negligible at this point in time as are $Pr^{3+}-Pr^{3+}$ interactions in the dilute samples studied. The dominant effect is the fluctuating part of the Pr$-$F magnetic dipole interaction.

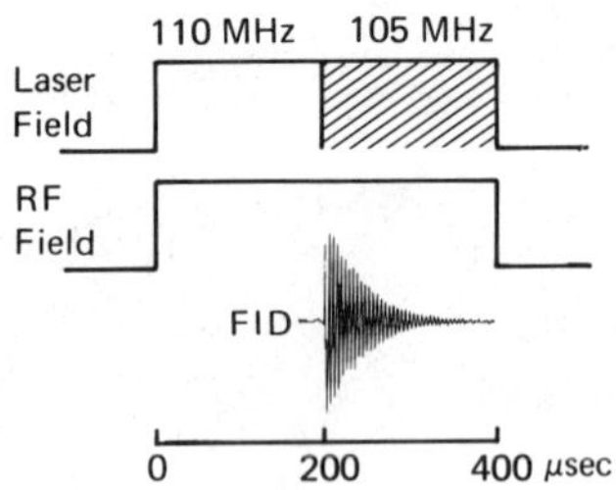

Fig. 5. Magic angle pulse sequence showing the laser field frequency shift and the F spin decoupling radio frequency field with time. The Pr^{3+} ions are coherently prepared by the laser field in the initial 200 μsec interval and then exhibit optical FID when the laser frequency is suddenly switched 2 MHz at t=200 μsec.

The dipolar mechanism has been demonstrated convincingly by an optical magic angle line narrowing experiment[6] which we now discuss. The pulse sequence is shown in Figure 5 and involves not only laser frequency switching for producing FID but the simultaneous application of an rf pulse in near resonance with ^{19}F nuclei. The basic idea is that pairs of F nuclei throughout the crystal undergo mutual spin flips which produce a fluctuating magnetic field at each Pr^{3+} site. The $^1D_2$ and $^3H_4$ states of Pr^{3+} fluctuate in energy correspondingly and thus the optical transition broadens. With a suitable rf field applied, the F precessional motion about its effective field tends to average out the F−F dipolar interaction and as a result the Pr−F dipolar interaction is quenched (see Figure 6). This effect can be viewed as a kind of motional narrowing. A detailed theory, which is given elsewhere,[6] predicts that the Pr^{3+} linewidth is given by

$$\Delta\nu(\beta) = \Delta\nu(0) \cos\beta \cdot \tfrac{1}{2}(3\cos^2\beta - 1) \tag{9}$$

where $\beta = \tan^{-1}(\gamma_F B_x/\Delta_F)$ is the angle that the effective field of the F nucleus makes with the z axis in its rotating frame, B_x being the rf field amplitude.

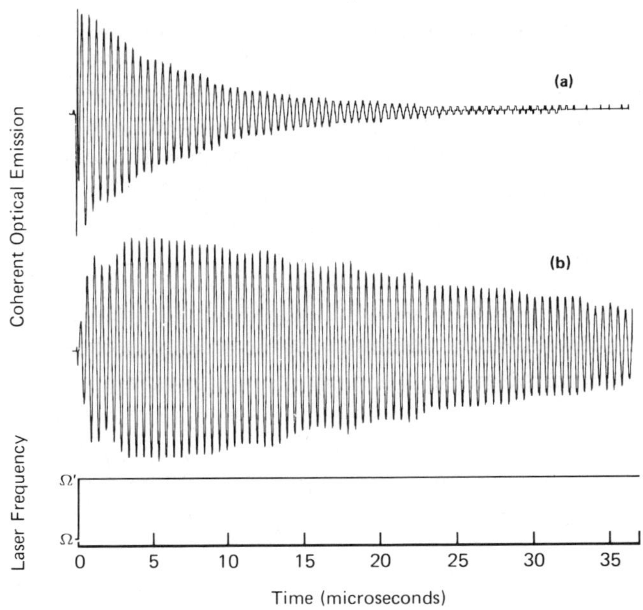

Fig. 6. Optical FID in 0.1% Pr^{3+}:LaF_3 at 1.8°K in the presence of a static magnetic field B_0=130G ⊥ c axis: (a) with no rf field where T_2=15.6 μsec (10.2 kHz) and (b) under magic angle conditions with an rf field B_x=25G where T_2=66 μsec (2.4 kHz).

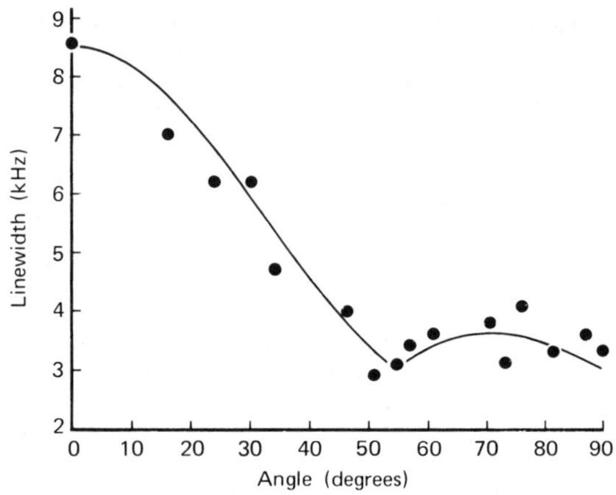

Fig. 7. Pr^{3+} optical linewidth versus angle β expressed in degrees. Solid circles: experimental points for the case $B_0 = 130G \parallel$ c axis and $B_x = 25G$. Solid curve: Eq. (9) with frequency offset of 3 kHz included for residual broadening.

From (9), we see that the Pr^{3+} linewidth should drop to zero when $\beta = \pi/2$, the F resonance condition, and $\beta = \cos^{-1}\left(\frac{\sqrt{3}}{3}\right)$, the magic angle condition. In practice, the linewidth drops from 10 to 2 kHz, but the measurements should be repeated using the improved phase locked laser to determine the origin of the residual width. Figure 7 shows that the experimental results agree rather well with Eq. (9) and thus unequivocally verifies the magnetic dipolar broadening mechanism.

VIII. MONTE CARLO THEORY[9]

We now require a more detailed theory of linebroadening than the simplistic T_1, T_2 description of the Bloch equations. We first replace Eq. (2a) by

$$\dot{\tilde{\rho}}_{12} = (-\gamma + i(\Delta + \delta\omega(t))\tilde{\rho}_{12} + 1/2\, i\chi(\rho_{22} - \rho_{11}) \tag{10}$$

where the Pr^{3+} transition frequency fluctuates as

$$\delta\omega(t) = -\left(\gamma_I' - \gamma_I''\right)\gamma_s\hbar \sum_k \frac{3\cos^2\theta_k - 1}{r_k^3} I_k S_{kz}(t) \tag{11}$$

due to random flipping of the fluorine spin $S_z(t)$ arising from F–F mutual spin flips which affect the secular part of the Pr−F dipolar interaction. Here, the F spin is labeled S and the Pr spin I, the population decay term $\gamma = 1/2(\gamma_1 + \gamma_2)$, and the gyromagnetic ratio of the $^1D_2$ and $^3H_4$ states are γ_I' and γ_I''. The FID solution of (10) now assumes the form

$$<\tilde{\rho}_{12}(t)> = <\tilde{\rho}_{12}(0) \exp \left[(-\gamma + i\Delta)t + i \int_0^t \delta\omega(t')dt'\right]> . \tag{12}$$

The quantity $\tilde{\rho}_{12}(t)$ involves the phase history of a single Pr^{3+} ion and therefore must be averaged over the distribution of frequency fluctuations occuring at different Pr^{3+} sites both during the preparative period $t \lesssim 0$ and afterward $t \gtrsim 0$. In addition, averages are to be performed over the local inhomogeneous static magnetic and crystalline Stark fields.

The following assumptions enter into the calculation. (1) The sudden jump approximation of the kth fluorine spin $S_{kz}(t)$ assumes that it can have only two values +1/2 and −1/2 and that it jumps instantaneously between these two values at random times and positions in the lattice at an average

TABLE I

Linewidths of ^{141}Pr in LaF_3 Due to Magnetic Broadening

Transition	Method	Linewidth FWHM (kHz)
		Inhomogeneous
rf($\Delta I_z'' = \pm 1$)	Monte Carlo theory	82
	Van Vleck second moment	84.5
	cw rf-optical double resonance[22]	
	$I_z'' = 1/2 \rightarrow 3/2$	$180 \pm 10^a (\sim 100)^b$
	$I_z'' = 3/2 \rightarrow 5/2$	$200 \pm 10^a (\sim 100)^b$
	Optically detected rf transients[23]	
	$I_z'' = 3/2 \rightarrow 5/2$	230 ± 25^a
		Homogeneous
optical ($^3H_4 \leftrightarrow {}^1D_2$)	Monte Carlo theory	16.8
($\Delta I_z = 0$)	FID experiment[5]	20.2^b

[a]Earth's magnetic field
[b]Static external field ⩾16G

rate W. (2) The LaF_3 crystal structure $\left(P3Cl-D_{3d}^{3}\right)$ is assumed for the nearest 125 unit cells (2250 fluorines) surrounding a Pr site. (3) The c axis of the LaF_3 crystal is parallel to an external magnetic field. (4) The number of fluorine spin flips per unit time follows a Poisson distribution.

The resulting computer program is not only capable of generating Pr^{3+} optical homogeneous linewidth but the static magnetic inhomogeneous linewidth as well. Table 1 summarizes these calculations and the current experimental results. For an rf transition of the $^3H_4$ Pr^{3+} ground state, a value of 82 kHz is obtained for the inhomogeneous magnetic broadening which compares favorably to a Van Vleck second moment calculation and to optically detected rf measurements for external fields in excess of the local field of 16G.

Notice in Figure 8 that the optical dephasing time is rather insensitive to the assumed fluorine mean flip time T. Actually, the parameter T need not be assumed, but rather it can be calculated from the methods of moments[24] to be $T \sim 10T_2 = 170$ μsec which yields a 19 μsec (16.8 kHz) dephasing time compared to the experimental value 15.8 μsec (20.1 kHz). The agreement is unusually good by the standards of previous theories considering that the discrepancy is only 15%.

Other conclusions emerge from these studies largely due to the fact that only a few fluorine nuclei, those close to the Pr^{3+} site, are found to contribute to optical dephasing. For example, our lattice size of 2250 fluorines is about

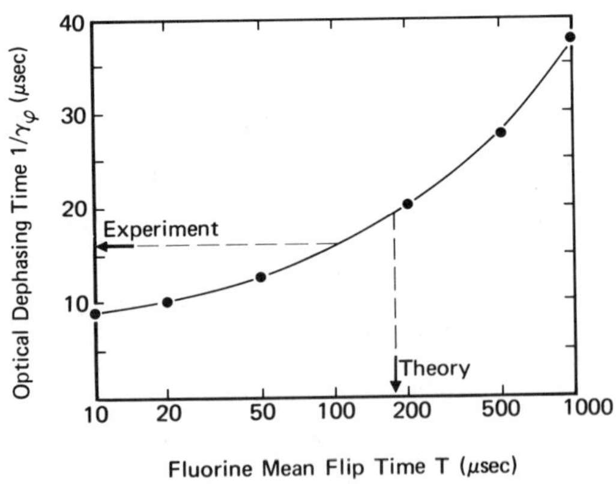

Fig. 8. The Pr^{3+} optical dephasing time versus the fluorine mean flip time T. The optical FID result is 15.8 μsec and the theoretical fluorine spin flip time $T=170$ μsec.

two orders of magnitude larger than necessary. Secondly, the small number of fluorines required implies that correlations between nearby fluorines do not strongly affect the optical linewidth. Thirdly, the static Pr^{3+} magnetic moment can polarize the nearest neighbor fluorines and detune them from the bulk fluorines so that they are incapable of undergoing spin flips. This frozen core effect, which is well known in ESR, reduces the Pr^{3+} linewidth from the value it would have if the nearest neighbors were also flipping. Inclusion of the frozen core model in the Monte Carlo calculation shows, contrary to intuition, that the optical linewidth varies slowly with the Pr^{3+} magnetic moment $\mu_z^{Pr} = \gamma_I I_z \hbar$ or magnetic substate I_z.

IX. GAUSSIAN MODULATION MODEL

Very recently, we have attempted to obtain analytic solutions for the optical dephasing problem using a Gaussian modulation model. This approach, which will be reported elsewhere,[10] has enjoyed some success and should be a useful method in the future for describing the nature of optical dephasing in these many body systems.

REFERENCES

1. T. H Maiman, Nature **187**, 493 (1960).
2. P. A. Franken, A. E. Hill, C. W. Peters and G. Weinrich, Phys. Rev. Lett. **7**, 118 (1961).
3. N. A. Kurnit, I. D. Abella and S. R. Hartmann, Phys. Rev. Lett. **13**, 567 (1964).
4. See for example *Laser Spectroscopy V*, edited by A. R. W. McKeller, T. Oka and B. P. Stoicheff (Springer-Verlag, New York, 1981).
5. R. G. DeVoe, A. Szabo, S. C. Rand and R. G. Brewer, Phys. Rev. Lett. **42**, 1560 (1979).
6. S. C. Rand, A. Wokaun, R. G. DeVoe and R. G. Brewer, Phys. Rev. Lett. **43**, 1868 (1979).
7. R. M. Macfarlane, R. M. Shelby and R. L. Shoemaker, Phys. Rev. Lett. **43**, 1726 (1979).
8. R. G. DeVoe and R. G. Brewer (unpublished).
9. R. G. DeVoe, A. Wokaun, S. C. Rand and R. G. Brewer, Phys. Rev. B**23**, 3125 (1981).
10. E. Hanamura, R. G. DeVoe and R. G. Brewer (to be published).
11. R. G. Brewer and R. L. Shoemaker, Phys. Rev. **A6**, 2001 (1972).
12. R. G. Brewer in *Frontiers in Laser Spectroscopy*, Les Houches Lectures (North-Holland, New York, 1977), p. 341.
13. F. Bloch, Phys. Rev. **70**, 460 (1946).
14. R. P. Feynman, F. L. Vernon and R. W. Hellwarth, J. Appl. Phys. **28**, 49 (1957).
15. A. Schenzle and R. G. Brewer, Phys. Rev. A**14**, 1756 (1976).
16. B. Bleaney, Physica (Utrecht) **69**, 317 (1973).
17. M. A. Teplov, Zh. Eksp. Teor. Fiz. **53**, 1510 (1967) [Sov. Phys. JETP **26**, 872 (1968)].
18. A. Z. Genack and R. G. Brewer, Phys. Rev. A**17**, 1463 (1978); R. G. Brewer and A. Z. Genack, Phys. Rev. Lett. **36**, 959 (1976).
19. R. W. P. Drever, J. L. Hall, F. V. Kowalski, J. Haugh, G. M. Ford and A. Munley (unpublished).
20. G. C. Bjorklund, Opt. Lett. **5**, 15 (1980); G. C. Bjorklund and M. D. Levenson, Phys. Rev. A**24**, 166 (1981).
21. The theory of phase modulation laser spectroscopy is described in A. Schenzle, R. G. DeVoe and R. G. Brewer, Phys. Rev. A**25**, 2606 (1982).

22. L. E. Erickson, Phys. Rev. **B16**, 4731 (1977).
23. R. M. Shelby, C. S. Yannoni and R. M. Macfarlane, Phys. Rev. Lett. **41**, 1739 (1978).
24. N. Bloembergen, Physica (Utrecht) **15**, 386 (1949); I. J. Lowe and S. Gade, Phys. Rev. **156**, 187 (1967).

DISCUSSION

Laser Spectroscopy

The initial discussion centered on the shape of the free induction decay (FID) signal. The observed decay was exponential at very long time and did not show the Gaussian component at short time expected from theory. On the timescale of the measurement this component would occur at extremely short times, too short to be observed with the present detection system. It was pointed out that these Gaussian components have been observed in similar measurements in gaseous systems. For instance in the D line of sodium the component was found to extend out to 300-400 psec. The FID decay in the present instance showed a small falling off from exponential behavior at the longest times but this may have been artifactual. Such a dropping off is, on the other hand, consistent with some analytical theories.

The question of the applicability of this coherent technique to rare earth impurities in glasses was raised. As yet it has not been applied though no intrinsic impediments could be foreseen. Based on previous discussions on the anomalously fast dephasing times observed in glasses at low temperature, a suggestion was made that the dephasing times may be too short for FID application. If inhomogeneous broadening dominates then FID is of limited value since it can then provide no information on the homogeneous linewidth.

The other interesting parameter in these measurements was the temperature. All of the experiments were performed at 2K and to date no information was available at other temperatures. The dependence on temperature would be interesting as it might provide insight on the nature of the interaction of the ions with the lattice phonons or with other excitations. Since the observed homogeneous linewidth decreased on the application of a magnetic field a question as to the effect of larger magnetic fields was raised. The dependence of the linewidths on magnetic field was weak, however, so that using higher fields would not be expected to make a sizeable difference.

It was pointed out that the narrowest linewidths observed in these experiments are the narrowest optical lines seen to date and rival the Mossbauer widths. The possibility of using such narrow lines as frequency standards was alluded to.

Since FID is one of many coherent transient techniques capable of measuring dephasing times and homogeneous linewidths, a comparison between the FID technique and the photon echo technique (see the paper by Macfarlane) was asked for. It was pointed out that the FID technique requires that the laser linewidth be narrower than the homogeneous linewidth, otherwise the T_2' inhomogeneous dephasing term dominates. While this entails complex laser stabilization and feedback techniques the fact is that extremely narrow cw laser widths on the order of 1 kHz or less are now possible. At such linewidths only a very narrow energy range of optical sites are sampled from within the inhomogeneous line and site-to-site variations in dephasing times can be investigated in a controlled way.

In the photon echo technique, on the other hand, the laser linewidth can be much broader than the homogeneous width, thereby relaxing the most stringent requirement of FID. The T_2' dephasing term due to the spread in energy packets excited is canceled out by the rephasing π pulse. In the situation where the homogeneous linewidth is narrower than even the linewidth achievable with cw lasers ($\sim$1 kHz) then the echo technique can still operate. The experimental apparatus is naturally considerably cheaper and simpler than the FID apparatus. As a disadvantage of the broader laser linewidth used in echo experiments, one necessarily averages over many types of sites thereby losing some site-to-site sensitivity. It was agreed in general that these techniques are complementary to an extent and act as tests for each other. The actual experimental conditions usually dictate the most feasible technique.

J. Hegarty, Chairman
J. Ryan, Discussion Leader

MEASUREMENT OF OPTICAL DEPHASING BY SPECTRAL HOLEBURNING IN RARE EARTH DOPED INORGANIC GLASSES

R. M. Macfarlane and R. M. Shelby

IBM Research Laboratory
San Jose, California 95193

Spectral hole burning has been observed in rare earth doped silicate glasses, and used to measure homogeneous linewidths (Γ_h). For Eu^{3+}, holes are burned by optical pumping of nuclear quadrupole levels, and our low temperature measurement of Γ_h shows that the previously determined $T^{1.8}$ temperature dependence holds down to at least 1.6K. On the other hand, we find a linear temperature dependence for the $^1D_2 \leftrightarrow {}^3H_4$ transition of Pr^{3+} glass from 1.6K-20K, a result which was confirmed by picosecond accumulated photon echoes. Finally, the $^4G_{5/2}$ level of Nd^{3+} shows hole widths limited by fast ($T_1 = 55$ psec) nonradiative relaxation. Current theories do not predict these results.

I. INTRODUCTION

Recently there has been considerable experimental[1–7] and theoretical[8–12] activity on the question of the temperature dependence of the homogeneous linewidths (Γ_h) of optical transitions of ions and molecules in glasses. Qualitatively it is already clear that these widths are typically much greater ($\sim 10^2 - 10^3$) than the corresponding widths in crystals, and the temperature dependence is weaker, i.e., ~T to T^2 rather than exponential or T^7. Rare earth doped inorganic glasses have been studied using fluorescence line narrowing (FLN) and in all cases investigated so far, i.e., Eu^{3+} in silicate,[1,2] phosphate,[5] and borate[5] glasses, and Pr^{3+} in BeF_2 and GeO_2 glasses,[3] the temperature dependence of Γ_h over the range 10K to 300K is T^γ with γ between 1.8 and 2.3. On the other hand, in hole burning studies of molecules in organic glasses both linear and quadratic[7] temperature dependences have been reported over more restricted temperature ranges. Several authors have proposed theories of line broadening based on a coupling of "two level systems" (TLS) or disorder modes to the impurity together with a coupling of the TLS to a phonon bath. Most of the theories[8–12] have addressed the question of the $T^{1.8}$ dependence of Γ_h found in rare earth doped inorganic glasses and since these theories are couched in rather general terms it is of some interest to determine how universal this temperature behavior is.

We have recently observed that long lived spectral hole burning occurs in rare earth doped inorganic glasses, and this promises to be a significant new tool for high resolution spectroscopic studies of these materials as it has been for crystals and organic glasses. In hole burning experiments a narrow bandwidth laser bleaches a hole in an inhomogeneously broadened spectral line. If the hole is shallow ($\leqslant$10%), its width is simply twice the homogeneous linewidth. Since the resolution is limited by the laser width, hole burning is particularly useful at low temperatures and for narrow homogeneous linewidths. We have used it to study several systems and report some of the results here. Perhaps the principal result is that for the $^1D_2 \leftrightarrow {}^3H_4$ transition of Pr^{3+} in silicate glass, Γ_h does not have the near quadratic temperature dependence observed in other rare earth doped glasses, but instead shows a linear dependence between 1.6K and 20K.

II. EXPERIMENTAL

A. HOLE BURNING

Holes were burned by irradiating samples with ~1 W/cm^2 of cw dye laser light having a spectral width of ~2 MHz. Exposure times varied from ~10 sec at low temperatures to ~500 sec at high temperatures. For detection of the holes the laser was attenuated by a factor of 30-1000 and the sample fluorescence was monitored as the laser frequency was scanned; i.e., an excitation spectrum of the hole was measured. The burning time and intensity were varied to check for laser heating and saturation effects. If care is not taken, these can affect the measured width by up to 50% so short exposure times and shallow (<10%) holes were used for the measurement of

Γ_h. This is more difficult at the higher temperatures where longer exposures are needed. Below 4.2K the sample was immersed in liquid helium and above this temperature an exchange gas cryostat was used.

B. PICOSECOND ACCUMULATED PHOTON ECHOES

Since the time scale of hole burning experiments is rather long (~ secs) we were concerned about the possibility of slow changes in the environment affecting the value of the homogeneous linewidth obtained in this way. We therefore used another technique to measure Γ_h — picosecond accumulated echoes — which makes the measurement on the time scale set by the population decay from a storage level, here the fluorescence decay time of $^1D_2$ which is 128 μsec. This technique was recently demonstrated by Hesselink and Wiersma[13] in measurements on pentacene in naphthalene, and has also been given a theoretical analysis by them.[14]

In these experiments the sample is irradiated by train of pulse pairs from a synchronously pumped mode locked R6G dye laser. The separation between pulses of a pair is a variable τ and the repetition rate was 246 MHz (i.e., 3.8 nsec between pairs of pulses). One pair of pulses produces a population grating in frequency space with a periodicity $1/\tau$ and one pulse from the next pair stimulates an echo.[13] An important feature of the technique is that the signal accumulates over the lifetime of a storage level which determines the time scale of the homogeneous width measurement. In order to form this accumulated grating a well-defined phase relationship is necessary between the optical fields of the two pulses. This is in contrast to the usual two-pulse photon echo. The amplitude of the grating depends on the loss of optical phase coherence of the sample in the time 2τ and the echo decays as τ increases like $\exp(-2\tau/T_2)$ where $(\pi T_2)^{-1}$ is the homogeneous linewidth. Further details of the experimental technique are given in Refs. 13-15.

III. RESULTS

A. Eu^{3+}: SILICATE GLASS

This sample had a composition 74.75% SiO_2, 15% Na_2O, 5% BaO, 5% ZnO and was doped with 0.25% Eu_2O_3. The glass composition is essentially the same as used by Selzer *et al.*[1] but with a factor of ten less Eu^{3+} doping. The Eu^{3+} ion is an attractive probe for these measurements because its $^7D_0 \leftrightarrow {}^5D_0$ transition at 5800 Å is between two rather well isolated J=0 levels. The inhomogeneous broadening of this band is ~100 cm^{-1}.

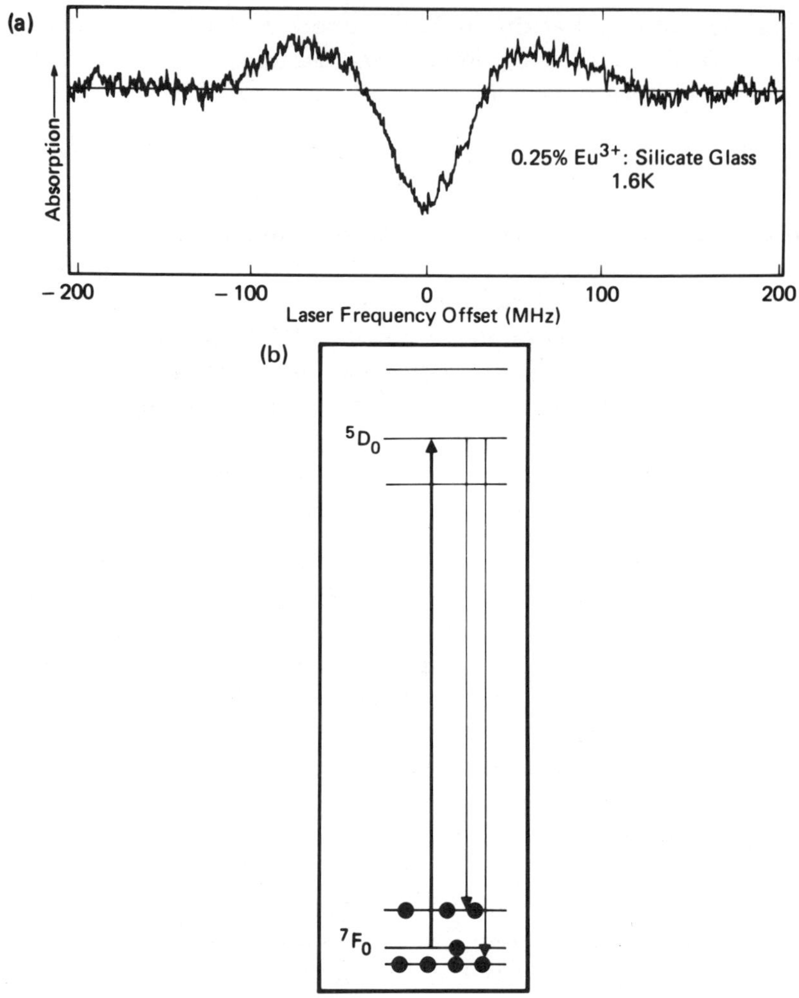

Fig. 1. a) Hole burned at 5800 Å in the $^5D_0 \leftrightarrow {}^7F_0$ absorption band of 0.25% Eu^{3+} in silicate glass at 1.6K. Note the increased absorption or antiholes adjacent to the hole. b) Energy level diagram showing redistribution of population in nuclear quadrupole levels of Eu^{3+} by optical pumping, leading to a hole at the laser frequency.

Homogeneous linewidths were first studied by Selzer *et al.*[1] who showed from fluorescence line narrowing (FLN) measurements that $\Gamma_h \propto T^{1.8\pm0.2}$ in the range 8-90K, and also by Avouris *et al.*[2] who found $T^{2.1 \pm 0.1}$ from 200K to 350K. We have used the greater resolution of hole burning to obtain a low temperature measurement of Γ_h at 1.6K. In this system hole burning is due to a mechanism which has been demonstrated in crystals[16,17] but not previously seen in glasses, i.e., optical pumping of nuclear hyperfine levels. This assignment was based on the fast hole recovery time (20 secs) and the presence of adjacent "antiholes" or increased absorption from nuclear quadrupole levels which have additional population due to optical pumping (see Fig. 1). This parallels the behavior observed in crystals[16,17] but the holes are broader and the spin lattice relaxation much faster. The data of Selzer *et al.*[1] is well fit by the expression $\Gamma_h = 13\ T^{1.8}$ MHz which would extrapolate to 30 MHz at 1.6K. Our value of 24 ± 3 MHz is in good agreement with this, showing that the $T^{1.8}$ dependence probably holds down to very low temperatures. Extrapolation gives a linewidth of 11 kHz at 20 mK, at which level nuclear spin contributions to the homogeneous linewidth are expected to become significant.[17] This mechanism of optical pumping of the nuclear quadrupole levels operates only when Γ_h is less than the quadrupole splittings Δ_Q. The position of the antiholes, and knowledge of typical values[17] of Δ_Q shows that at 1.6K $\Gamma_h \sim \Delta_Q$. At temperatures above ~3K we find that hole burning due to this mechanism no longer occurs, and because of the very weak absorption no permanent holes were observed.

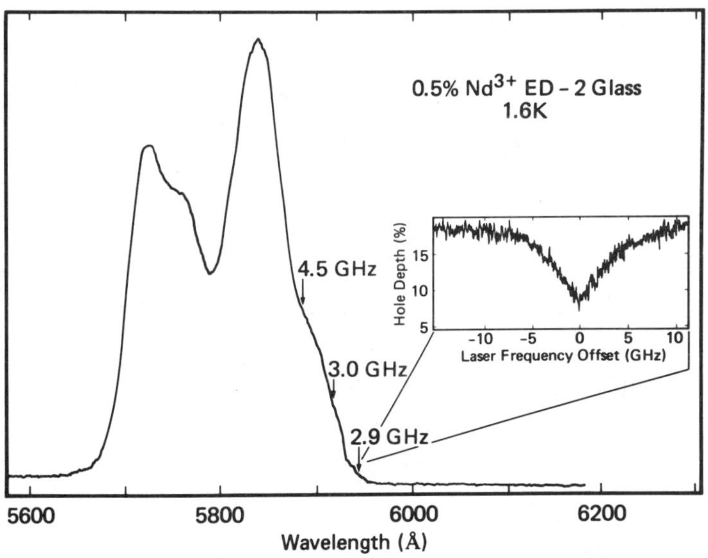

Fig. 2. Absorption band of the inhomogeneously broadened $^4I_{9/2} \rightarrow {}^4G_{5/2}, {}^2G_{7/2}$ transitions in Nd^{3+}:ED-2 glass. The homogeneous linewidths obtained from hole burning at different frequencies within the band are shown. The inset shows a hole burned at the long wavelength edge, i.e., in the lowest "crystal field" component.

B. Nd^{3+}: SILICATE GLASS

This sample was an Owens Illinois ED-2 glass containing 0.5% Nd^{3+}. Holes were burned in the ${}^4I_{9/2} \rightarrow {}^4G_{5/2}, {}^2G_{7/2}$ band which peaks around 5800 Å. The absorption spectrum at 1.6K is shown in Fig. 2. The splitting of the free ion J levels is comparable to the inhomogeneous broadening resulting in a band of overlapping "crystal field" levels with a total width of 600 cm^{-1}. The inset to Fig. 2 shows a hole burned at the low energy edge of the band at 5942 Å. At this wavelength the lowest energy crystal field level is preferentially irradiated. The hole width of 5.8 GHz corresponds to a homogeneous linewidth $\Gamma_h = 2.9$ GHz and a dephasing time $T_2 = (\pi\Gamma_h)^{-1} = 110$ psec. Holes burned at 4.2K, and also in a sample of 2% Nd^{3+}, gave the same result, apparently eliminating thermally induced dephasing or $Nd^{3+} - Nd^{3+}$ interactions as the source of broadening. Instead we assign the width to T_1 processes, i.e., nonradiative relaxation to the next lowest electronic level which is 1200 cm^{-1} away. This is close to the maximum internal vibration frequency in ED-2 glass of $\sim$1100 cm^{-1}.[18] Typically, lifetimes in this glass are about 100 times faster[18] than for the same level in a crystal such as $YAlO_3$[19] where T_1 for ${}^4G_{5/2}$ is 30 nsec. The low quantum yield of fluorescence from this level in our ED-2 sample is consistent with lifetime limited dephasing with $T_1 = 1/2\,T_2 = 55$ psec. These results are in agreement with accumulated echo data which is reported elsewhere.[16]

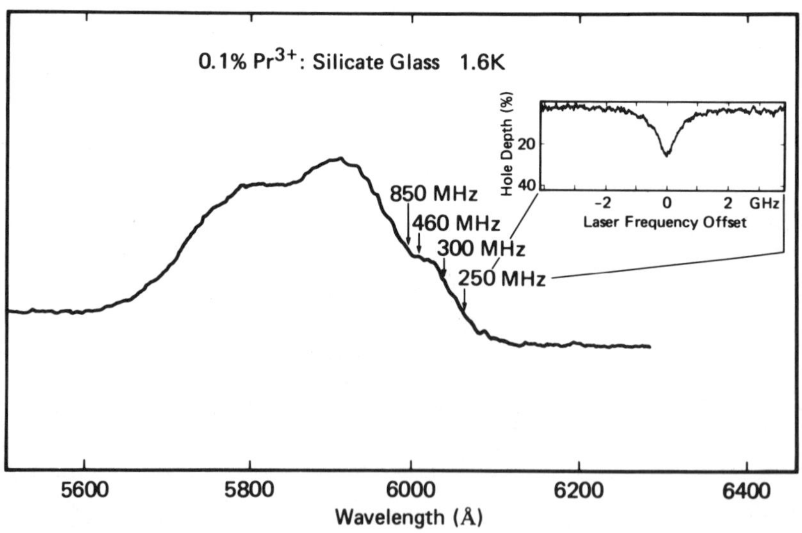

Fig. 3. ${}^3H_4 \leftrightarrow {}^1D_2$ absorption band of Pr^{3+} in silicate glass at 1.6K. The arrows show the homogeneous linewidths measured at different positions within the band. Inset shows a hole burned at the longest wavelength (6060 Å).

As the laser was tuned to higher frequencies the holes became noticeably non-Lorentzian and broader (Fig. 2). This is probably due to excitation of overlapping higher lying crystal field levels which are even faster relaxing than the lowest one. The hole widths give a measure of the energy dependence of T_1 in this case.

C. Pr^{3+}: SILICATE GLASS

The low temperature absorption bandshape of the $^1D_2 \leftrightarrow {}^3H_4$ transition around 5880 Å is shown in Fig. 3. There are at least four poorly resolved peaks corresponding to inhomogeneously broadened crystal field components of $^1D_2$. As in the case of Nd^{3+} the hole width is a function of frequency within the band. At the low energy edge of the band which consists predominantly of the lowest crystal field component, 500 MHz wide holes were burnt yielding Γ_h = 250 MHz or T_2 = 1.3 nsec. The fluorescence decay time was T_1 = 128 μsec so this contributes a negligible 1.2 kHz to the width. The temperature dependence of Γ_h was studied up to 20K (where the burning became very slow) and in contrast to the other cases of rare earth doped glasses[1–5] the temperature dependence is not close to T^2 but rather was linear (see Fig. 4). Because of this unexpected linear dependence we measured the homogeneous linewidth by picosecond accumulated photon echoes[13–15] since, as mentioned earlier, the time scale of this measurement is faster ($\sim T_1$ = 128 μsec). We found that the accumulated echo data confirmed the linear temperature dependence observed by hole burning (see Fig. 4), i.e., that no significant environmental motion was occurring on the timescale of 100 μsec-100 sec.

The observation of a linear temperature dependence in Pr^{3+} doped glass is interesting because it shows that the approximately T^2 behavior in not universal in rare earth doped glasses. Several theories[8–11] have been proposed to explain the near quadratic temperature dependence usually observed. In these theories the rare earth ions are coupled to a distribution of low frequency modes or "two level systems" (TLS) with varying barrier heights. These represent disorder modes in the glass. The bath consists of phonons which induce scattering between the wells of the TLS. Recently, Reinecker and Morawitz[10] have pointed out that the results of such theories are very sensitive to the parameters describing the TLS and that a detailed numerical solution predicts a linear temperature dependence for Γ_h down to approximately one-tenth of the Debye temperature, and then a cross over to a quadratic dependence at very low temperatures. Our experiments are probably still in the low temperature regime however where a quadratic dependence is theoretically predicted. Hayes *et al.*[6] obtain a linear

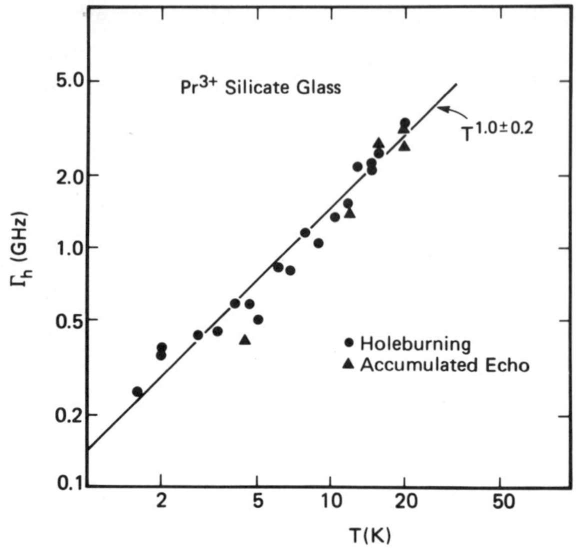

Fig. 4. Temperature dependence of the homogeneous width of the lowest ${}^1D_2 \leftrightarrow {}^3H_4$ transition of Pr^{3+}: silicate glass obtained from hole burning and picosecond accumulated echo measurements at 6060 Å.

temperature dependence for Γ_h by assuming a very low frequency cut-off (~2K) for the density of states of the TLS in organic glasses. There appears to be no evidence that such a model is applicable to the present case. There is clearly need for further work - both theoretical and experimental.

IV. CONCLUSION

We report the observation of long lived spectral hole burning in inorganic glasses containing rare earth ions and use it to measure homogeneous linewidths in Eu^{3+}, Nd^{3+} and Pr^{3+} doped silicate glasses. In the case of Eu^{3+}, holes burn due to optical pumping of the nuclear quadrupole levels and low temperature measurements show that the $T^{1.8}$ dependence of Γ_h reported by Selzer *et al.*[1] holds down to 1.6K. In Nd^{3+} glass T_1 limited dephasing is observed due to rapid nonradiative relaxation. The case of the ${}^1D_2 \leftrightarrow {}^3H_4$ of Pr^{3+} is interesting because a linear temperature dependence is found to Γ_h between 1.6 and 20K. This result is different from other rare earth glasses, and like them, is not explained by current theories.

ACKNOWLEDGEMENTS

We thank Dr. M. J. Weber for providing all of the doped glass samples, and Prof. A. J. Sievers for stimulating discussions in the early stages of this work.

REFERENCES

1. P. J. Selzer, D. L. Huber, D. S. Hamilton, W. M. Yen and M. J. Weber, *Phys. Rev. Lett.* **36,** 813 (1976).
2. P. Avouris, A. Campion and M. A. El-Sayed, *J. Chem. Phys.* **67,** 3397 (11977).
3. J. Hegarty and W. M. Yen, *Phys. Rev. Lett.* **43,** 1126 (1979).
4. J. M. Pellegrino, W. M. Yen and M. J. Weber, *J. Appl. Phys.* **51,** 6332 (1980).
5. J. R. Morgan and M. A. El-Sayed, *Chem. Phys. Lett.* **84,** 215 (1981).
6. J. M. Hayes, R. P. Stout and G. J. Small, *J. Chem. Phys.* **73,** 4129 (1980); J. M. Hayes, R. P. Stout and G. J. Small, *J. Chem. Phys.* **74,** 4266 (1981).
7. J. Friedrich and D. Haarer, *App. Phys.* **B28,** 262 (1982).
8. T. L. Reinecke, *Solid State Commun.* **32,** 1103 (1979).
9. S. K. Lyo and R. Orbach, *Phys. Rev.* **B22,** 4223 (1980).
10. R. Reinecker and H. Morawitz, *Chem. Phys. Lett.* **86,** 359, (1982).
11. S. K. Lyo, *Phys. Rev. Lett.* **48,** 688 (1982).
12. D. L. Huber, J. Non. Cryst. Sol. **51,** 241 (1982).
13. W. H. Hesselink and D. A. Wiersma, *Phys. Rev. Lett.* **43,** 1991 (1978).
14. W. H. Hesselink and D. A. Wiersma, *J. Chem. Phys.* **75,** 4192 (1981).
15. R. M. Shelby, *Opt. Lett.* **8,** 88 (1983).
16. L. E. Erickson, *Phys. Rev.* **B16,** 4731 (1977).
17. R. M. Macfarlane, R. M. Shelby A. Z. Genack and D. A. Weitz, *Opt. Lett.* **5,** 462 (1980); R. M. Shelby and R. M. Macfarlane, *Phys. Rev. Lett.* **45,** 1098 (1980).
18. C. B. Layne, W. H. Lowdermilk and M. J. Weber, *Phys. Rev.* **B16,** 10 (1977).
19. M. J. Weber, *Phys. Rev.* **B8,** 54 (1973).

DISCUSSION

Spectral Holeburning

It was suggested that since these holeburning experiments and the fluorescence line narrowing experiments reported earlier addressed the same problem of anomalously large dephasing rates relative to rare earth doped crystals at low temperatures, both techniques should be used on the same system such that they overlap in temperature.

Two of the rare earth ions used in the experiment (Pr and Nd) have rather complicated Stark structure both in the ground and excited state which could not be resolved in either absorption or luminescence. In the holeburning experiment as the laser was tuned from the extreme low energy side of the broad absorption line, very fast dephasing times became evident corresponding to relaxation of the upper Stark components directly excited. Consequently it was not clear whether the laser could ever select only one component in the ground and excited states. If not, then complicated dephasing times corresponding to a mixture of times would result. It was pointed out that at the extreme low energy side of the line the contribution from upper stark components, if indeed present, would be very small.

The nature of the holeburning was next addressed. The effects of a "high" temperature anneal which, in other situations restored the system to its original state have not been studied yet in these systems. It would be of interest to see if, after cycling the temperature, the systems were completely restored or whether the features of a new hole were different in some way from the original one. Experimentally it was also observed that the hole could be filled by pumping the system at different optical frequencies within the inhomogeneous absorption lines but no understanding of this is available at present.

J. Hegarty, Chairman
J. Ryan, Discussion Leader

NONLINEAR PROPAGATION EFFECTS IN GLASS FIBERS

R. H. Stolen

Bell Telephone Laboratories
Holmdel, New Jersey 07733

Nonlinearities in optical fibers constitute a rich and diverse field of interest for devices, for the limits imposed on fiber transmission, and for the study of nonlinear optics. The basic properties of fiber nonlinear optics are the exchange of fiber length for optical power, the application of simple plane-wave theory and the importance of group-velocity dispersion and fiber polarization. The primary effects are stimulated Raman and Brillouin scattering, Kerr effects, parametric four-photon mixing, and self-phase modulation. Self-phase modulation in combination with group-velocity dispersion leads to solitons and related effects.

I. INTRODUCTION

Optical fibers have undergone intensive development as passive transmission media. However, these fibers are also extremely elegant nonlinear optical media. The most striking property of fiber nonlinearity is the exchange of length for power which permits the observation of nonlinear processes, such as stimulated Raman scattering, at powers orders of magnitude lower than in bulk materials.[1] One might expect that fiber waveguide modes, particularly in multimode fibers, would add tremendous complexity. In fact, many nonlinear effects have been studied in fibers and in all cases can be treated with a simple plane-wave formalism[1,2] as opposed to bulk nonlinear optics where transverse intensity variations impose additional difficulties. Optical nonlinearities in bulk materials were well studied in the 1960's.[3] Typically, the various processes were obscured by self-focusing and relaxation of the nonlinearity, neither of which takes place in glass fibers. Fiber and bulk nonlinear optics share the problem of the simultaneous presence of different nonlinear processes so that any attempt to study a given process requires techniques to reduce the importance of others. There are two important features of fiber nonlinearity which did not play much of a role in traditional bulk nonlinear optics. The first is group-velocity dispersion which has a most dramatic consequence in soliton effects; the second is the role of polarization.

The present treatment first deals with some of the properties of fiber nonlinearity which are common to all processes. In the next section, the basic nonlinear processes are briefly discussed. These processes are stimulated Raman and Brillouin scattering, Kerr effects, parametric four-photon mixing, and self-phase modulation. Finally, it is shown how self-phase modulation and group-velocity dispersion combine to produce solitons and related effects.

II. SPECIAL CHARACTERISTICS OF FIBER NONLINEAR OPTICS

The various nonlinear interactions discussed here arise from the complex third order susceptibility χ_3 and lead to effects of the form:

$$\exp\left[(g_R+ig_I)IL\right] \tag{1}$$

where I is the intensity and L is length. Processes with real gains g_R are Raman and Brillouin amplification which come from the imaginary part of χ_3 and parametric amplification which comes from the real part of χ_3. The imaginary gain g_I is a nonlinear phase shift which comes from the real part of χ_3 and is usually treated as a nonlinear refractive index.

The effect of interaction length is illustrated by Figure 1 where the high intensity region of a focused beam is compared with light coupled into a fiber. A focused beam quickly diverges so that focusing harder to raise the intensity causes a corresponding decrease in interaction length. Light coupled into a fiber does not diverge so that the high intensity of the focus continues until the light is lost by absorption. The core diameters of typical single-mode fibers range between 3 and 9 μm and losses are between 0.2 and 20 dB/km so enhancements can be around 10^8.

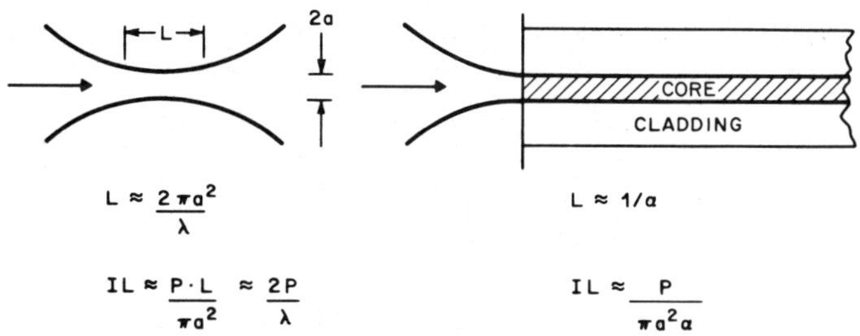

Fig. 1. Comparison of the interaction length of a focused Gaussian beam with focused light coupled into an optical fiber.

The linear fiber loss is usually handled by replacing the actual fiber length ℓ by an effective length L.[1]

$$L = \frac{1-e^{-\alpha\ell}}{\alpha} \tag{2}$$

For a long fiber, L is the absorption length $1/\alpha$ while for short fibers L is the actual length ℓ.

The intensity I becomes P/A where P is the power at the fiber input end. A is an effective core area and is derived by expanding the electric field in the wave equation in terms of fiber waveguides modes Ψ_m.[2]

$$\left(\frac{\partial^2}{\partial z^2} + \nabla_t^2 - \frac{n^2}{c^2}\frac{\partial^2}{\partial t^2}\right)\vec{E} = \frac{4\pi}{c^2}\frac{\partial^2 \vec{P}}{\partial t^2}$$

$$\vec{P} = 4\left(\chi_3' - \iota\chi_3''\right)\vec{E}^3 \tag{3}$$

$$\vec{E} = \sum_{\ell,m} \vec{A}_{\ell m}(z)\Psi_m(r,\theta)\cos(k_{\ell m}z - \omega_\ell t + \phi_{\ell m})$$

where ℓ refers to the frequency, m the mode number, $k_{\ell m}$ the wavevector, and $\phi_{\ell m}$ the phase. Contributions to the polarization P from the second order susceptibility χ_2 are zero because of inversion symmetry. The solutions of the transverse part of the wave equation are the waveguide modes[4] and it is assumed that these are unaffected by the nonlinearity. With the slowly varying envelope approximation $\partial^2 A_{\ell m}/\partial z^2 << 2\iota k \partial A_{\ell m}/\partial z$

By integrating over the mode fields, the effect of the waveguide modes is reduced to a constant factor which has the dimensions of area^{-1}.

$$\frac{1}{A} = \frac{\int_0^{2\pi}\int_0^{\infty} \Psi_1\Psi_2\Psi_3\Psi_4 \, r dr d\theta}{D_1^{1/2} D_2^{1/2} D_3^{1/2} D_4^{1/2}} \tag{4}$$

$$D_m = \int_0^{2\pi}\int_0^{\infty} \Psi_m^2 \, r dr d\theta$$

The magnitude of this effective area is fairly close to the fiber core area so that in many cases a convenient approximation is to simply use the core area with no further calculations. For Raman gain in a single-mode fiber all the Ψ_m are the same while for parametric four-photon mixing all the Ψ_m can be different as will be discussed later. The important point is that, with the inclusion of this constant factor, the wave equations reduce to coupled wave equations of the same form as for uniform plane waves. Thus, the plane wave formalism for nonlinear optics can be said to apply to fiber nonlinearities exactly while it is only a first approximation for bulk nonlinear optics.

Interest in group-velocity dispersion in fibers parallels recent work in resonant systems such as sodium vapor where dispersions are enormous. In fibers, the magnitude of the group-velócity dispersion is small but because of the long lengths employed the net effect can be significant. Figure 2 shows the group-velocity dispersion of a typical weakly-doped silica-core fiber along with a curve showing the loss obtained in very good fibers.[5,6] One has the option of working in a regime of either positive or negative group-velocity dispersion or to utilize the zero dispersion point near 1.3 μm. The zero dispersion wavelength can also be shifted between 1.25 and 1.55 μm by suitable doping of the fiber.[6]

Most fibers do not preserve the state of polarization along the fiber. Such a fiber can be viewed as a collection of wave plates of various orders in combination with polarization rotators.[7] There are now several different types of highly birefringent single-mode fibers which maintain linear polarization over hundreds of meters.[8] Polarization-preserving fibers are especially useful for nonlinear optics and in fact, the first such fiber was made for use in a fiber Raman laser.[9] Linear polarization can be preserved in multimode fibers but only over relatively short lengths.[10] A birefringent fiber excited at 45° to the principal axes makes an approximate model for propagation in a non-

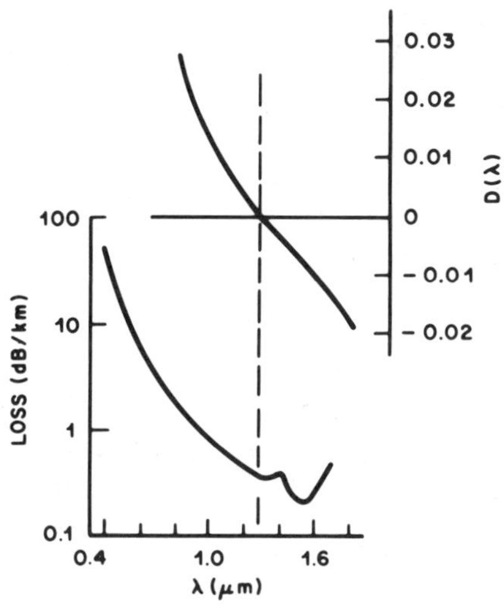

Fig. 2. Group-velocity dispersion and loss in high-quality, weakly-doped silica-core fiber. The scale for dispersion is in dimensionless units where $D(\lambda) = \lambda^2 d^2n/d\lambda^2$. D in ps/nm-km $= D(\lambda)/(C\lambda)$.

polarization preserving fiber as pictured in Figure 3. The state of polarization varies from linear to elliptical to circular as the light travels along the fiber. Different frequencies, will see the same birefringence but since the wavelengths differ the states of polarization will eventually get out of step after some polarization distance ℓ_p.[11]

$$\ell_p = \frac{1}{\delta n \delta \nu} \tag{5}$$

where δn is the birefringence expressed as a difference in refractive indices and $\delta \nu$ is the frequency separation in cm^{-1}.

III. BASIC NONLINEAR EFFECTS

A. STIMULATED RAMAN AND BRILLOUIN

Stimulated Raman scattering has been seen and studied in single and multi-mode liquid and glass-core fibers. Usually, many Stokes orders are generated with the pulse successively downshifted as it proceeds along the fiber.[1] In low-loss fibers we have seen 26 orders of Stokes output, each separated by about 450 cm^{-1} which is the peak Raman shift in silica glass. Feedback oscillators of various configurations have also been constructed with pump wavelengths ranging from the blue to the near IR. These

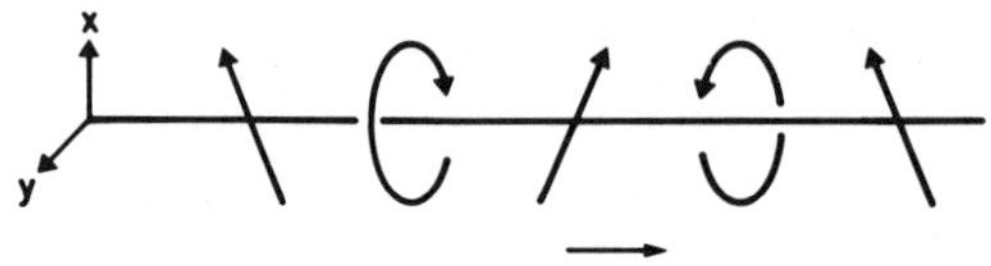

Fig. 3. Variation in the state of polarization along a birefringent single-mode fiber excited with linear polarization at 45° to the principal axes.

oscillators can be tuned over the broad (~300 cm^{-1}) Raman band of silica glass and it is possible to cascade the Stokes output to extend the tuning range many times.

Critical powers for single-pass stimulated emission and threshold powers for oscillators generally agree with simple theoretical predictions. A critical power for single-pass emission is defined as that pump power at the fiber input for which the Stokes and pump powers are equal coming out of the fiber.[12] This occurs for $\gamma = gPL/A$ where for a wide range of fiber parameters $\gamma = 16 \pm 2$. An approximate expression for critical power becomes:

$$P_t = 30\ d^2 \lambda \xi\ (mW) \tag{6}$$

Here the effective area is approximated by the core area, the core diameter d and pump wavelength λ are in μm, and ξ is the fiber loss in dB/km. The fiber is long enough that the effective length is the absorption length; for shorter fibers the threshold power is higher by the factor obtained from Eq. (2). It is assumed in Eq. (6) that the fiber preserves linear polarization. If polarization is not preserved, the Raman process averages over high and low gain regions as the states of pump and Stokes polarizations get in and out of step.[11] The result is to reduce the gain coefficient g by a factor of two. Threshold powers for oscillators are obtained from the usual conditions that gain equals loss. Thresholds are typically around 1W but have been as low as 200 mW.[11]

The effect of group-velocity dispersion on Raman amplification is illustrated by Fig. 4 which shows pump and Stokes pulses coupled into an optical fiber. Fig. 4a shows walk-off of pump and Stokes pulses due to group-velocity dispersion. This walk-off distance will depend on the pulse length, the frequency separation, and the magnitude of the dispersion. Walk-off can reduce the interaction length below the limit set by fiber absorption. Group velocity matching, illustrated in Fig. 4b, is possible by utilizing the differences in group velocity of different waveguide modes so

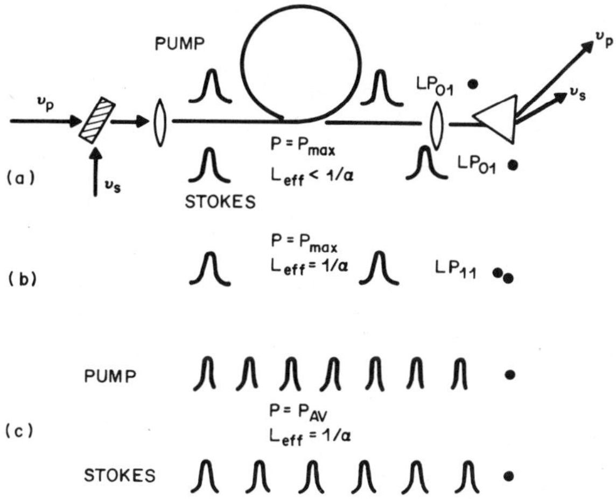

Fig. 4. Effect of group-velocity dispersion on Raman amplification. a. walk-off, b. group-velocity matching, c. effect of many closely-spaced pulses.

that pulses at pump and Stokes frequencies travel together. Group-velocity matching has been seen in both fiber Raman oscillators and in single-pass stimulated emission.[13] When the pump and Stokes pulses are spaced close together (Fig. 4c) so that each Stokes pulse passes through many pump pulses then the effective pump power is the average rather than the peak power. This same picture applies to amplification or stimulated scattering using typical nontransform limited optical pulses which contain strong sharp microstructure. The result is that one can usually use the average envelope of the pulse and knowledge of the details of the pulse structure is unimportant.

Brillouin scattering is similar to Raman scattering in that it arises from the imaginary part of the third order susceptibility although the interaction is with acoustic rather than optical modes of the glass. Because of wavevector matching considerations, the Brillouin shifted wave is amplified only in the backward direction while Raman gain is the same in either direction along the fiber. The peak Brillouin gain is more than two orders of magnitude larger than the Raman gain, the linewidths are sharp (~100 MHz) and the frequency shifts are small (~1 cm^{-1}). Because of these narrow linewidths typical lasers are inefficient pumps for Brillouin amplification and maximum Brillouin gain is obtained only with narrow line lasers. This dependence of Brillouin gain on pump linewidth provides a convenient method of favoring either stimulated Raman or Brillouin scattering.[14]

An approximate expression for the cw Brillouin critical power is:[15]

$$P_t = 2.2\, d^2\lambda^2\xi\Delta\nu_B \text{ (mW)} \tag{7}$$

The parameters are the same as for Eq. (6) and $\Delta\nu_B$ is the Brillouin linewidth in GHz. The pump linewidth is much narrower than the Brillouin linewidth; for a broad pump line of frequency width $\Delta\nu_P$ the critical power is increased by the factor $\Delta\nu_P/\Delta\nu_B$. It is also assumed in Eq. (7) that linear polarization is preserved in the fiber; if polarization is not preserved, the gain is reduced by a factor of two as for Raman emission. For a fiber of 0.2 dB/km loss at 1.5 μm and a core diameter of 5 μm the cw Brillouin critical power is 420 μW.

B. KERR EFFECTS

The optical Kerr effect refers to the nonlinear refractive index arising from the real part of the third order susceptibility χ_3. Its simplest form is the change in index of a linearly polarized wave on itself which is related to the power and fiber parameters by:

$$\begin{aligned} \delta n &= \frac{1}{2} n_2 E^2 \\ E^2 &= \frac{8\pi P}{ncA} \times 10^7 \\ \Delta\Phi &= \frac{2\pi L}{\lambda}\delta n \end{aligned} \tag{8}$$

P is pump power in Watts and n_2 is the well-known self-focusing coefficient[16] which leads to an intensity dependent phase shift $\Delta\Phi$. Any additional confinement caused by self-focusing is negligible.[17]

The induced index of a wave on a different wave with the same linear polarization is a factor of two larger than for a wave on itself. Different in this case means a different frequency, a different waveguide mode, or a wave propagating in the opposite direction. The difference in nonlinear refractive index between a wave on itself and an oppositely traveling wave of the same frequency turns out to be a significant cause of drift in fiber gyroscopes.[18]

The induced index of a linearly polarized wave on a wave of perpendicular polarization is two-thirds that of the wave on itself. The 1/3 comes from the difference in the parallel and perpendicular susceptibilities and the factor of two appears because the waves are different.

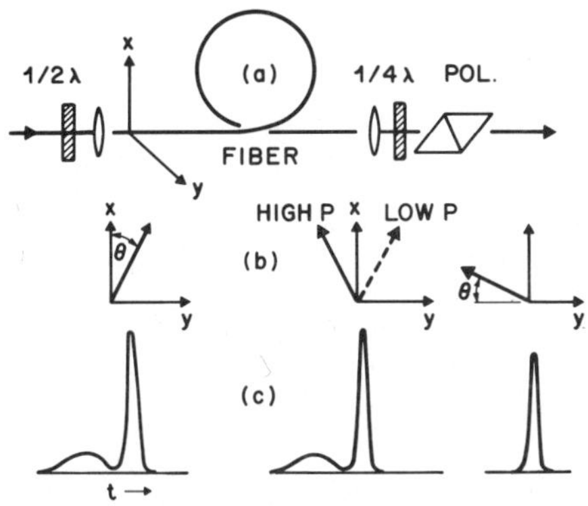

Fig. 5. Intensity discriminator for optical pulses utilizing an intensity dependent state of polarization out of the fiber.

The difference between parallel and perpendicular nonlinear indices can be exploited to make an intensity discriminator for optical pulses[19] as illustrated in Figure 5. The discriminator utilizes the intensity dependent state of output polarization that results when the light is not linearly polarized along one of the principal axes of the fiber. Rejecting the low power pulse with the polarizer also requires sacrificing some of the high power pulse except near input angles of 45° which then requires extremely high powers. Compromises are possible which produce satisfactory discrimination at reasonable powers although these powers are still high. This type of discriminator has found application in the separation of highly compressed optical pulses from a weaker uncompressed background as will be discussed in the section of soliton effects.

C. PARAMETRIC FOUR-PHOTON MIXING

Amplification of light can take place through the real part of the third order susceptibility by way of parametric four-photon interactions. Parametric amplification in a fiber absorbs two pump photons and creates one photon higher in frequency (anti-Stokes) and one photon lower in frequency (Stokes) than the pump frequency. This is illustrated in Figure 6.

Parametric interactions require phasematching since in general, the sum of wavevectors of the generated waves does not equal the sum of the pump wavevectors. Several different phasematching techniques are available in fibers which utilize various properties of waveguide modes.

The basic principle of phasematching with waveguide modes is illustrated

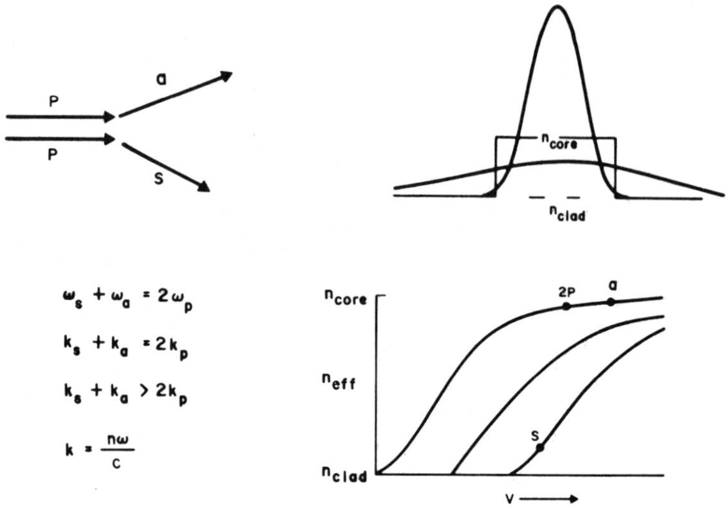

Fig. 6. a. Frequency and wavevector matching conditions for parametric four-photon mixing. Index dispersion leads to a wavevector mismatch. b. Illustration of increasing confinement of the fundamental mode with increasing frequency which leads to an increase of effective refractive index n_{eff} from n_{clad} to n_{core}. c. Plot of n_{eff} vs normalized frequency V illustrating a possible combination of modes to achieve phase matching.

in Figure 6. As frequency increases, the confinement of a given mode increases, and its effective refractive index becomes closer to the core index. A family of curves can thus be generated representing effective refractive index vs normalized frequency for the fiber waveguide modes.[4] Putting the pump, Stokes, and anti-Stokes waves into different modes can then correct for the wavevector mismatch from material dispersion.

It can be shown that the wavevector mismatch from material dispersion takes the simple form:[20]

$$\Delta k_m = 2\pi\lambda D(\lambda)\Omega^2 \tag{9}$$

where $D(\lambda)$ is the group-velocity dispersion $D(\lambda) = \lambda^2 d^2n/d\lambda^2$. Wavelength λ is in cm and frequency shift Ω is in cm^{-1}. Eq. (9) shows that at wavelengths close to the zero dispersion wavelength a much greater frequency shift is possible for a given waveguide mode correction. Also, as the sign of the dispersion changes beyond 1.3 μm, the order of the modes used for phasematching must be inverted. Equation (9) breaks down at $D(\lambda) \approx 0$ because of the importance of higher order terms but Δk_m does become very small and it is possible to achieve phasematching by the slope of the n_{eff} vs V curve of a single waveguide mode.[21] The birefringence in a polarization preserving fiber can also be used for phasematching.[22]

Phasematched parametric interactions have been observed both in low-power mixing experiments and in stimulated emission.[2,23] Parametric mixing leads to a decreasing linewidth with increasing length which eventually runs into limits because of fiber imperfections. Stimulated emission is only seen in relatively short fibers; stimulated Raman scattering dominates in longer fibers. The actual length involved varies with the pump wavelength and the particular phasematching processes and can vary from 10 cm to 50 meters.

The length for which stimulated parametric emission is favored over stimulated Raman scattering depends on the initial pump linewidth and the bandwidth of the parametric interaction. In like manner to the dependence of Brillouin amplification on linewidth the parametric gain decreases when the pump linewidth exceeds the gain bandwidth. In a simple picture, the parametric mixing bandwidth decreases as $1/\ell$ until the gain bandwidth equals the initial pump linewidth. Beyond this length the parametric gain decreases while the Raman gain is unaffected. Lengths calculated on this basis agree well with experience although the details of the actual process involve an intensity dependent gain bandwidth and broadening of the pump linewidth from self-phase modulation.[2]

If the frequency shifts involved are small the coherence length is often longer than the actual fiber length and no special effort is required for phasematching.[1] The coherence length is the distance in which polarization wave induced by the parametric interaction and the free running wave at the same frequency get out of step. For example, for a 0.5 μm pump wavelength and a frequency shift of 1 cm^{-1} the coherence length is about 2 km.

D. SELF-PHASE MODULATION

A common situation is to have many frequencies of comparable optical power rather than a single line which can be designated the pump. For example, the many axial modes of a typical cw laser or the different frequency components of a mode-locked laser. Now every line is both pump and signal and everything amplifies or depletes everything else. The result is an intensity dependent spectral broadening.

For optical pulses the problem of spectral broadening is most conveniently treated in the time domain where the interpretation is that of a phase shift of the center of the pulse due to the intensity dependent refractive index.[24] This is illustrated by Fig. 7a. The magnitude of the

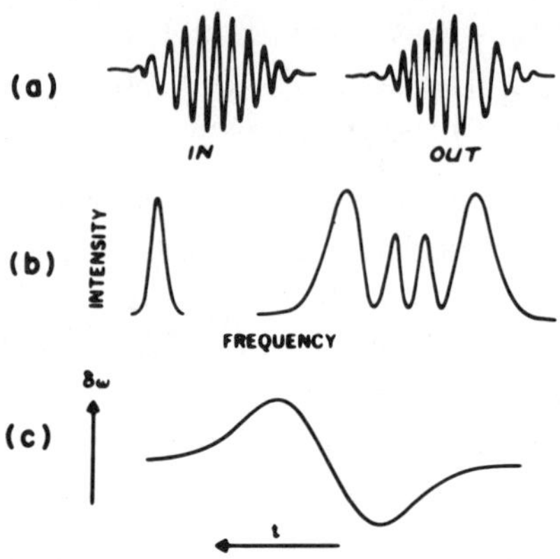

Fig. 7. a. Effect of the intensity dependent index on the phase of the electric field of an optical pulse. b. Typical spectra before and after self-phase modulation. c. Frequency chirp from self-phase modulation.

intensity induced index change is small but when added up over fiber lengths the phase shift at the peak of the pulse can be large.

Modulation of the phase results in a spectral broadening. In Fig. 7b are shown typical input and output spectra. The spectra calculated by a Fourier transform of the self-phase modulated pulse agree well with measured spectra.[25]

One of the most important features of the spectral broadening from self-phase modulation is that the instantaneous frequency varies along the pulse as shown in Fig. 7c. The leading part of the pulse is downshifted in frequency and the trailing part is upshifted with a large central region nearly linearly chirped. This linear chirp formed the basis for early proposals[26] to produce ultrashort optical pulses by first chirping a pulse by self-phase modulation and then compressing the pulse with a grating pair. Observation of the broadened spectrum out of a fiber also provides a relatively cheap and simple technique for peaking up a mode-locked laser as illustrated by Figure 8.

IV. SOLITON EFFECTS

Under the heading "soliton effects" we lump a growing class of effects which combine group-velocity dispersion and self-phase modulation from the nonlinear index. It has been shown that the wave equation including the nonlinear index and dispersion is the nonlinear Schrödinger equation of soliton theory.[27,28]

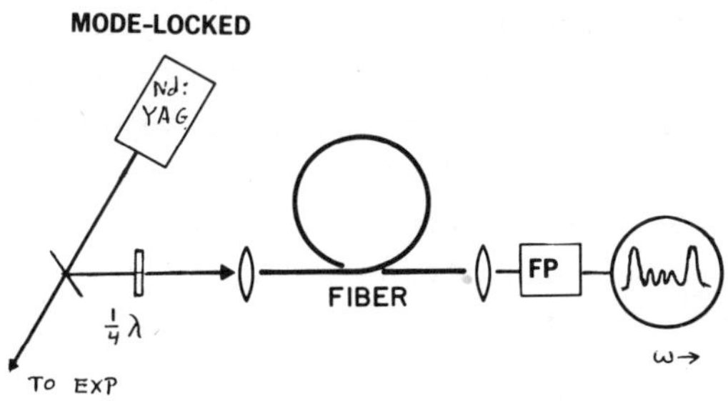

Fig. 8. Use of self-phase modulation in a fiber for peaking up a mode-locked laser. The broadest and cleanest spectrum corresponds to the shortest and cleanest mode-locked pulse.

$$i\left(\frac{\partial u}{\partial z}+k_1\frac{\partial u}{\partial t}\right)=k_2\frac{\partial^2 u}{\partial t^2}+k_3|u|^2u \tag{10}$$

where u(z,t) is the pulse envelope function, $k_1 = n/c$, $k_2 = \lambda D(\lambda)/(4\pi c^2)$, $k_3 = \pi n_2/(n\lambda)$, and λ is the vacuum wavelength. Depending on the wavelength as shown in Fig. 2, the dispersion $D(\lambda)$ can be either positive or negative.

Positive group-velocity dispersion combined with self-phase modulation increases pulse spreading. The resultant decrease in peak intensity thus limits the possible frequency chirp. This would appear to be a liability in using a fiber to chirp a pulse for subsequent grating compression.[26] In fact, it appears that the group-velocity dispersion acts to linearize the chirp so that the final compressed pulse is cleaner with smaller uncompressed side lobes than if there were no dispersion.[29] This type of fiber chirping combined with a grating pair has been used to compress 90 fs pulses at 6100 Å to a record 30 fs which is only 14 optical wavelengths long.[30]

At wavelengths of negative (or anomalous) group-velocity dispersion the fiber acts as its own compressor. In this way, 8.7 ps pulses from a mode-locked color center laser at 1.55 μm have been compressed to 250 fs.[31] The compressed pulse can be separated from the uncompressed background by utilizing intensity dependent birefringence.[19] At a much lower power, the compression due to the combination of nonlinearity and negative group-velocity dispersion just balances normal pulse spreading from group-velocity dispersion alone. At this power a pulse with hyperbolic secant amplitude will propagate unchanged in shape or amplitude-the so-called envelope soliton.[27]

At higher powers, pulses first narrow and then evolve in a complex manner which in general consists of a sequence of narrowings and splittings. At specific powers an initial sech pulse is restored periodically at a characteristic distance z_o. The distance z_o is the soliton period and is particularly useful as a normalized length and for comparing effects at different powers inasmuch as z_o depends only on the pulse length τ (FWHM) and the group-velocity dispersion as $D(\lambda)$ or D (see Fig. 2).

$$z_o = 0.322 \frac{\pi^2 c \tau^2}{D\lambda^2} = 0.322 \frac{\pi^2 c^2 \tau^2}{D(\lambda)\lambda} \quad (11)$$

The powers for restoration of a hyperbolic secant input pulse at z_o are 4,9,25....N^2 times the power for the fundamental soliton. As an example, the evolution of an $N = 3$ pulse is shown in Fig. 9; these pulse shapes are numerical solutions of the nonlinear Schrödinger equation.[32] At intermediate powers the pulse shape does not repeat at z_o.

The power required to form the fundamental soliton is:

$$P_1 = \frac{nc\lambda}{16\pi z_o n_2} A_{eff} \quad (12)$$

As for other nonlinear effects in fibers, the plane wave equations apply with the inclusion of the effective area derived from the integral over the mode fields. P_1 depends essentially on the soliton period z_o since practical values of λA_{eff} range only over about a factor of four while z_o can vary by many orders of magnitude depending on input pulse length and wavelength

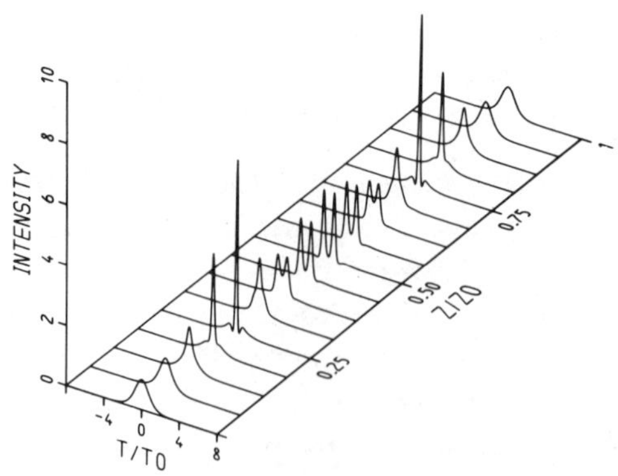

Fig. 9. Evolution of the pulse shape along a fiber for the $N = 3$ soliton.

deviation from the point of zero dispersion. For a typical core diameter of 9 μm at a wavelength of 1.5 μm, P_1 is about 1W for a z_o of 1 km. A z_o of 1 km corresponds to a 6 ps pulse length. A convenient rough approximation for P_1 is thus:

$$P_1 \approx [z_o]^{-1} \text{ watts} \tag{13}$$

where z_o is now measured in km.

Along with narrowing of the pulse in time comes a broadening in frequency. Fig. 10 shows calculated spectra for N = 3 at the fiber input, $z_o/2$ and z_o.[32,33] The periodic restoration of the pulse width along the fiber must be accompanied by a restoration in spectral width so the frequency spectrum broadens and then narrows again.

Soliton propagation in fibers was predicted long before the possibility of experimental observations.[27] Only recently have fibers of low loss and reasonably large negative group-velocity dispersion become available. It is fortuitous that the lowest fiber loss occurs in a regime of significant negative dispersion. Meaningful experiments require that the absorption length be long compared to the soliton period z_o. Bringing z_o down to 1 km requires the relatively large negative dispersion at 1.55 μm and a pulse length shorter than 7 ps. The other requirement was a laser providing high-quality picosecond pulses at 1.5 μm. These requirements were met by the NaCl F_2^+ mode-locked color center laser.[28]

The ideal investigation of pulse shape and frequency spectrum would sample various points along the fiber. Unfortunately at present, this requires destruction of the fiber and also places unrealistic demands on the long-term stability of the optical pulsed source. Experiments are more easily performed by using a fixed fiber length and varying the power by changing the input

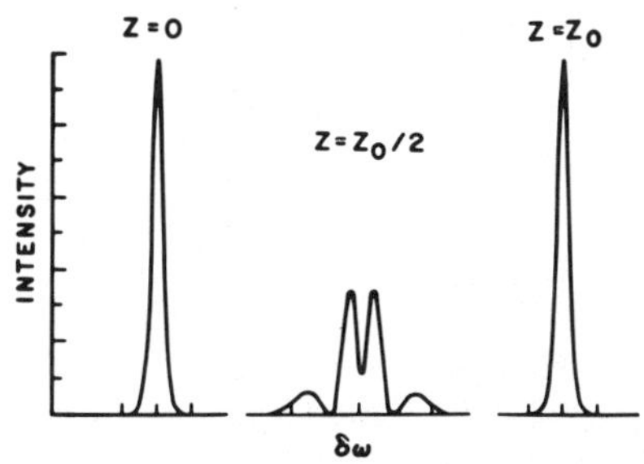

Fig. 10. Calculated spectra at three points along the fiber for the N = 3 soliton.

coupling. Some variation in effective length is possible by varying the pulse length but this is more useful for trimming purposes.

The first experiments were performed with 7 ps pulses at 1.55 μm in a 700 m fiber. The fiber was approximately half the soliton period in length and the compressions and splittings expected from theory were observed. Fig. 11 shows autocorrelation traces of the fiber output as a function of input power. At low power the pulse broadens as expected from group-velocity dispersion. At 1.2 W the output pulse matches the input pulse; this power is close to the calculated power for the fundamental soliton P_1. At four times P_1 ($N = 2$) pulse compression is observed. At $9P_1$ ($N = 3$) the pulse splits into two in agreement with the calculations of Fig. 9. Note that a three fold splitting in autocorrelation corresponds to two fold splitting of the pulse.

At the full soliton period z_0 the input pulse should be restored in both pulse length and spectrum. This has been observed in a 1.3 km fiber using 6.4 ps pulses at 1.55 μm.[33] In the experiment, pulse shape and spectrum were monitored as input power was increased. The pulse length was observed to alternately narrow and return to its input length while the frequency spectrum broadened and returned to the input width at the same powers. At higher powers, jitter in the laser pulse length and peak power became a significant source of discrepancy between theory and experiment.

The implications of the soliton effects have only begun to be explored. Problems of current interest are the effect of linear absorption on soliton

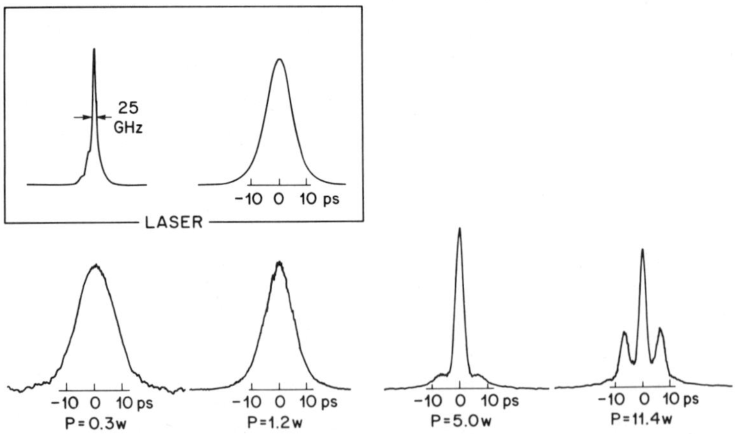

Fig. 11. a. Autocorrelation shapes and frequency spectrum of 1.55 μm laser pulses launched into a fiber whose length is one-half the soliton period. b. Autocorrelation traces of pulse output at low power and for input powers corresponding to the $N = 1$, 2 and 3 solitons.

properties[34]-early theory dealt only with lossless fibers, the limits of validity of the nonlinear Schrödinger equation-both in the soliton regime of negative group-velocity dispersion and at wavelengths of positive dispersion,[35] and the possibility of using soliton effects to increase the bit rate in fiber transmission.[36,37]

V. CONCLUSION

Fiber nonlinearities are of interest for such diverse reasons as their device potential, the limits they impose on single-mode fiber communication systems, and as an elegant medium for the study of nonlinear optics.

It is clear that the nonlinearities of a simple silica optical fiber constitute a rich and diverse field. The simultaneous presence of a variety of nonlinear processes can lead to incomprehensible and sometimes irreproducible spectra. Proper experimental conditions, however, can separate the various processes as was seen in the separation of stimulated Raman and Brillouin scattering and the separation of stimulated parametric mixing from stimulated Raman scattering. In some cases such as the intensity discrimination in soliton compression one takes advantage of different processes. Applications utilizing more than one nonlinear process may become prevalent as more is learned about fiber nonlinearities and more control is gained over the basic fiber structures.

REFERENCES

1. For reviews of fiber nonlinear optics see: E. P. Ippen, "Nonlinear Effects in Optical Fibers," in Laser Applications to Optics and Spectroscopy, S. F. Jacobs and M. O. Scully Eds. (Reading, Mass. Addison-Wesley, 1975) p. 213; R. H. Stolen, "Nonlinear Properties of Optical Fibers," in Optical Fiber Telecommunications, S. E. Miller and A. G. Chynoweth, Eds. New York: (Academic Press, 1979) p. 125; K. O. Hill, B. S. Kawasaki, D. C. Johnson, and Y. Fujii, "Nonlinear Effects in Optical Fibers," in Fiber Optics-Advances in Research and Development, B. Bendow and S. S. Mitra, Eds. (New York: Plenum, 1979) p. 211; R. H. Stolen, "Fiber Raman Lasers," in Fiber and Integrated Optics, D. B. Ostrowsky, Ed. (New York: Plenum, 1979) p. 157 and Fiber and Integ. Opt., 3, 21 (1980).
2. R. H. Stolen and J. E. Bjorkholm, "Parametric Amplification and Frequency Conversion in Optical Fibers," IEEE J. Quantum Electron. **QE-18**, 1062 (1982).
3. See for example: N. Bloembergen, "Nonlinear Optics," New York: W. A. Benjamin, 1965; R. W. Minck, R. W. Terhune, and C. C. Wang, "Nonlinear Optics," Appl. Opt. 5, 1595 (1966). F. Shimizu, "Numerical Calculation of Self-Focusing and Trapping of a Short Light Pulse in Kerr Liquids," IBM J. Res. Develop. **17**, 286 (1973).
4. D. Gloge, "Weakly Guiding Fibers," Appl. Opt. **10**, 2252, (1971).
5. H. Murata and N. Inagaki "Low-Loss Single-Mode Fiber Development and Splicing Research in Japan" IEEE J. Quantum Electron., **QE-17**, 835 (1981).
6. L. G. Cohen, P. Kaiser and Chinlon Lin "Experimental Techniques for Evaluation of Fiber Transmission Loss and Dispersion" Proc. IEEE, **68**,, 1203 (1980).
7. F. P. Kapron, N. F. Borelli, and D. B. Keck, "Birefringence in Dielectric Optical Waveguides," IEEE J. Quantum Electron. **QE-8**, 222 (1972); A. Simon and R. Ulrich,

"Evolution of Polarization along a Single-Mode Fiber," Appl. Phys. Lett. **31**, 517 (1977).

8. For reviews of polarization preserving fibers see: I. P. Kaminow, "Polarization in Optical Fibers" IEEE J. Quantum Electron. **QE-17**, 15 (1981); T. Okoshi, "Single-Polarization Single-Mode Optical Fibers" IEEE J. Quantum Electron.**QE-17** 879 (1981); D. N. Payne, A. J. Barlow and J. J. Ramskow-Hansen, "Development of Low- and High-Birefringence Optical Fibers," IEEE J. Quantum Electron. **QE-18**, 477 (1982).
9. R. H. Stolen, V. Ramaswamy, P. Kaiser, and W. Pleibel, "Linear Polarization in Birefringent Single-Mode Fibers," App. Phys. Lett. **33**, 699 (1978).
10. R. E. Wagner, R. H. Stolen, and W. Pleibel, "Polarization Preservation in Multimode Fibers," Electron. lett. **17**, 177, (1981); A. J. Snyder and W. R. Young, "Modes of Optical Waveguides," J. Opt. Soc. Am., **68**, 297 (1978).
11. R. H. Stolen, "Polarization Effects in Fiber Raman and Brillouin Lasers," IEEE J. Quantum Electron. **QE-15**, 1157 (1979).
12. R. G. Smith, "Optical Power Handling Capacity of Low Loss Optical Fibers as Determined by Stimulated Raman and Brillouin Scattering," Appl. Opt. **11**, 2489 (1972).
13. Chinlon Lin, R. H. Stolen, and R. K. Jain, "Group Velocity Matching in Optical Fibers," Opt. Lett. **1**, 205 (1977); Y. Ohmori, Y. Sasaki, M. Kawachi, and T. Edahiro, "Single-Pass Raman Generation Pumped by a Mode-Locked Laser," Electron Lett. **17**, 594 (1981).
14. E. P. Ippen and R. H. Stolen, "Stimulated Brillouin Scattering in Optical Fibers," Appl. Phys. Lett. **21**, 539 (1972); N. Uesugi, M. Ikada and Y. Sasaki, "Maximum Single-Frequency Input Power in a Long Optical Fiber Determined by Stimulated Brillouin Scattering," Electron. Lett. **17**, 379 (1981).
15. R. H. Stolen, "Nonlinearity in Fiber Transmission," IEEE J. Quantum Electron. **QE-68**, 1232 (1980).
16. D. Milan and M. J. Weber, "Measurement of Nonlinear Refractive Index Coefficients Using Time-Resolved Interferometry", J. Appl. Phys. **47**, 2497 (1976).
17. P. L. Kelly "Self-Focusing of Optical Beams" Phys. Rev. Lett. **15**, 1005 (1965).
18. S. Ezekiel, J. L. Davis, and R. Hellwarth, "Intensity Dependent Nonreciprocal Phase Shift in Fiber Gyros" in Proceedings of the International Conference on Fiberoptic Rotation Sensors and Related Technologies, (Springer-Verlag, New York, 1982); R. A. Bergh, H. C. LeFevre, and H. J. Shaw, "Compensation of the Optical Kerr Effect in Fiber-Optic Gyroscopes," Opt. Lett. **7**, 282 (1982).
19. R. H. Stolen, J. Botineau, and A. Ashkin, "Intensity Discrimination of Optical Pulses With Birefringent Fibers," Opt. Lett., **7**, 512 (1982).
20. K. O. Hill, D. C. Johnson, B. S. Kawasaki, and R. I. McDonald, "CW Three-Wave Mixing in Single-Mode Optical Fibers," J. Appl. Phys., **49**, 5098 (1978).
21. K. Washio, K. Inoue, and T. Tanigawa, "Efficient Generation of Near IR Stimulated Light Scattering in Optical Fibers Pumped in Low-Dispersion Region at 1.3 μm, Electron." Lett., **16**, 331 (1980); Chinlon Lin, W. A. Reed, A. D. Pearson, Hen-Tai Shang, and P. F. Glodis, "Designing Single-Mode Fibers for Near-IR (1.1 - 1.7 μm) Frequency Generation by Phase-Matched Four-Photon Mixing in the Minimum Chromatic Dispersion Region," Electron. Lett., **18**, 87 (1982).
22. R. H. Stolen, M. A. Bösch, and Chinlon Lin, "Phase Matching in Birefringent Fibers," Opt. Lett., **6**, 213 (1981).
23. R. H. Stolen, J. E. Bjorkholm, and A. Ashkin, "Phase-Matched Three-Wave Mixing in Silica Fiber Optical Waveguides," Appl. Phys. Lett., **24**, 308 (1974); R. H. Stolen, "Phase-Matched-Stimulated Four-Photon Mixing in Silica-Fiber Waveguides," IEEE J. Quantum Electron. **QE-11**, 100 (1975).
24. F. Shimizu "Frequency Broadening in Liquids by a Short Light Pulse," Phys. Rev. Lett. **19**, 1097 (1967).
25. R. H. Stolen and Chinlon Lin, "Self-Phase-Modulation in Silica Optical Fibers," Phys. Rev. A. **17**, 1448 (1978).
26. R. A. Fisher, P. L. Kelley and T. K. Gustafson, "Subpicosecond Pulse Generation Using the Optical Kerr Effect," Appl. Phys. Lett., **14**, 140 (1969).

27. A. Hasegawa and F. Tappert, "Transmission of Stationary Nonlinear Optical Pulses in Dispersive Dielectric Fibers, I. Anomalous Dispersion," Appl. Phys. Lett. **23**, 142 (1973).
28. L. F. Mollenauer, R. H. Stolen, and J. P. Gordon, "Experimental Observation of Picosecond Pulse Narrowing and Solitons in Optical Fibers," Phys. Rev. Lett. **15**, 1095 (1980).
29. H. Nakatsuka, D. Grischkowsky, and A. C. Balant, "Nonlinear Picosecond Pulse Propagation Through Optical Fibers with Positive Group Velocity Dispersion," Phys. Rev. Lett., **47**, 1910 (1981); D. Grischkowsky, and A. C. Balant, "Optical Pulse Compression Based on Enhanced Frequency Chirping," Appl. Phys. Lett., **41**, 1 (1982).
30. C. V. Shank, R. L. Fork, R. Yen, R. H. Stolen, and W. J. Tomlinson, "Compression of Femtosecond Optical Pulses," Appl. Phys. Lett., **40**, 761 (1982).
31. L. F. Mollenauer and R. H. Stolen "Solitons in Optical Fibers," Laser Focus **18** (April, 1982).
32. Calculations from W. J. Tomlinson.
33. R. H. Stolen, L. F. Mollenauer, and W. T. Tomlinson, "Observation of Pulse Restoration at the Soliton Period in Optical Fibers," Opt. Lett., **8**, 186 (1983).
34. A. Hasegawa and Y. Kodama "Signal Transmission by Optical Soliton in Monomode Fiber," Proc. IEEE **69**, 1145 (1981).
35. D. Anderson and M. Lisak, "Nonlinear Asymmetric Pulse Distortion in Long Optical Fibers," Opt. Lett. **7**, 394 (1982).
36. Y. Kodama and A. Hasegawa, "Amplification and Reshaping of Optical Solitons in Glass Fiber-II," Opt. Lett. **7**, 394 (1982).
37. K. J. Blow and N. J. Doran, "High Bit Rate Communication Systems Using Non-Linear Effects," (to be published).

DISCUSSION

Nonlinear Effects

The question of the comparison between fiber solitons and self-induced transparency (SIT) solitons was raised. Fiber solitons depend upon the nonlinearity of the refractive index whereas SIT solitons depend upon the nonlinearity of the absorption. The absorption is linear in the fiber case, so the attenuation is not anomalously low as in the SIT case. In both cases the pulse can slowly lengthen to compensate for attenuation by a linear loss.

SIT can be thought of as a coherent precession of all the dipoles, no matter how far off resonance, driven by the sech electric field envelope, or as coherent absorption and coherent reemission. Inhomogeneous broadening is in no way harmful if all of the oscillators have the same value of the dipole moment. If the inhomogeneous line consists of several overlapping inhomogeneous transitions with different dipole moments, the electric field pulse cannot have an area of 2π for all of them and complete transparency will be impaired. Features such as pulse breakup and peaking are particularly impaired by the presence of more than one dipole. It is likely that this could explain their apparent absence in phonon SIT in glasses. The pulse velocity in SIT is approximately $2/\alpha\tau$, where α is the absorption coefficient and τ is the characteristic time duration of the electric field E.

For a 2π soliton $\frac{2p}{\hbar}\int \epsilon(t)dt = 2\pi$, or roughly $\epsilon_m \sim \frac{\pi\hbar}{p\tau}$, where ϵ_m is the maximum value of ϵ. Then $V \sim \alpha p\epsilon/2\pi\hbar$, i.e. the more intense pulses travel faster. In the case of fiber solitons the negative linear dispersion $\left(\frac{\partial^2\omega}{\partial k^2}\right)$ is being compensated, on the average, by a nonlinear refractive index. The pulses may change shape but only in a repetitive manner. They travel at approximately c/n_o independent of intensity (assuming $n_2 I \ll n_o$.) Hasegawa has recently included a $\frac{\partial^3\omega}{\partial k^3}$ term but found that all of the features persist and a small renormalization gives results indistinguishable from those without it.

Fiber solitons might conceivably increase the usable bandwidth of fibers. They might also make it unnecessary to use complicated optical repeaters consisting of detectors and regenerators. Instead an in-phase cw beam could be copropagated in the repeater; the pulse would narrow and grow more intense to preserve the soliton condition much the reverse of its lengthening to compensate for a linear loss. The prospects of optical fiber transmission using solitons was deemed to deserve additional study.

P. A. Fleury, Chairman
H. Gibbs, Discussion Leader

COMMENTS

Light Scattering as a Probe of Low Frequency Excitations in Glasses

As detailed in some of the earlier chapters in this proceedings, the energy density of states of the tunneling levels responsible for many of the anomalous low temperature properties in amorphous materials has so far defied direct measurement. In principle inelastic light scattering should be able to probe this question. The intriguing possibility that tunneling levels may have contributed to low frequency Raman spectra in fused silica was raised by Winterling[1] in his 1975 temperature-dependent studies. There he pointed out the existence of an excess scattering at low temperatures and low frequencies (below about 10 cm^{-1}). This excess was beyond that expected from the simple model of disorder-induced scattering which should mirror the acoustic phonon density of states which should in the simplest case vanish with a frequency-squared dependence. Further, his spectra showed a suggestive upturn at low frequencies which might have arisen from some spectral central peak, possibly induced by tunneling level scattering. However, the resolution and contrast employed in Winterling's experiments and in many of the subsequent experiments on a–SiO_2 were insufficient to definitively determine stray light contributions, on the one hand, and high frequency tails to the inelastic Brillouin scattering, on the other.

Quite recently Lyons and coworkers[2] have studied, with very high contrast and high resolution both the Raman and Brillouin spectra of fused silica in the form of long optical fibers. These more detailed experiments have indeed verified the presence of an excess low frequency scattering evident below 4 cm^{-1}. However, as schematically shown in Figure 1, this excess does not continue to rise in intensity on approach to zero frequency (in particular below the frequency of the Brillouin peak, ~1 cm^{-1}). The observed nonzero intensity at nearly zero frequency, ($\Delta\nu \ll 1\ cm^{-1}$) is fully and quantitatively consistent with the tail of the lifetime- and instrumental-broadened Brillouin component. These results therefore are consistent with a

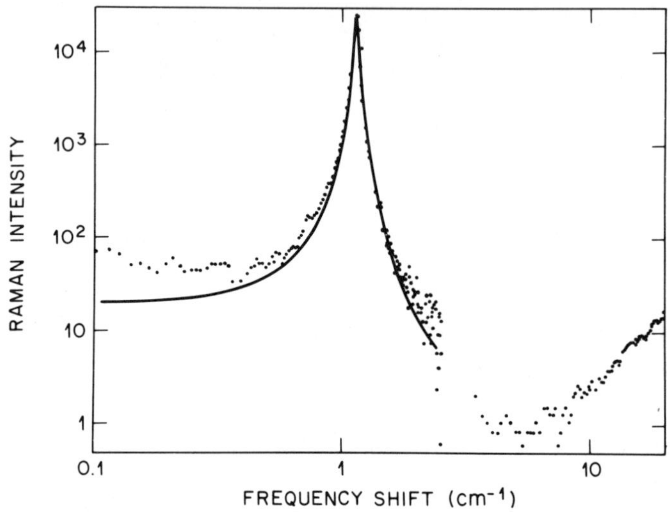

Fig. 1. Depolarized light scattering spectrum at 1.5K in a fused silica glass fiber (dots). The low frequency (<1 cm^{-1}) scattering is predominantly due to the tail of the lorentzian shaped inelastic Brillouin scattering from the acoustic phonon at 1 cm^{-1}. The slight excess near 0.1 cm^{-1} is accounted for by secondary forward Brillouin scattering of elastic Rayleigh scattering.

frequency independent excess scattering whose intensity is approximately a thousand times less than the peak Brillouin intensity. Furthermore, at low temperatures the dependence of this excess scattering upon temperature does not agree with the predictions made a few years earlier[3] on the temperature dependence expected for scattering from tunneling states.

There have been, however, several observations of quasielastic central peak features in the light scattering spectra of glasses. Fleury and Lyons in 1976 first reported[4] the observation of anomalous low frequency scattering in a variety of heavy metal oxide-containing glasses. Using a resonant iodine filter reabsorption technique, spectra like those obtained in Figure 1 were observed in a number of such glasses. The temperature dependence of this quasielastic feature varied depending upon the nature of the glass but generally both the linewidth and intensity decreased upon decreasing temperature. Furthermore, with the experimental sensitivity then obtainable the anomalous scattering could not be followed down in temperature below about 50K. Thus the identification of this feature as due to direct scattering from tunneling levels in these glasses is not supported and indeed was not made.

The experiments on fused silica fibers mentioned above attempted to observe such direct scattering in a simple, thoroughly studied single component glass in which it has been documented fully that tunneling levels do dominate the low temperature thermal and acoustic properties. The failure to observe tunneling levels in the Raman spectra of fused silica in combination with the observation of anomalous quasielastic central peaks in a variety of more complicated glasses containing more highly polarizable molecular groups constitutes an unsolved puzzle. This puzzle can be summarized briefly as follows.

In those glasses such as fused silica where tunneling levels have been definitively observed by acoustic and microwave resonance experiments, the most sensitive light scattering experiments to date have not revealed their presence. On the other hand, in those materials for which the anomalous light scattering central peaks have been observed, to date there have been no similarly-definitive studies of the behavior associated with the tunneling levels. Clearly some of the metal oxide glasses already studied in light scattering should be explored using the techniques of thermal conductivity, acoustic absorption, and phonon echoes, etc.

P. A. Fleury
Bell Laboratories
Murray Hill, New Jersey 07974 USA

REFERENCES

1. G. Winterling, Phys. Rev. *B*12, 2432 (1975).
2. K. B. Lyons, P. A. Fleury, R. H. Stolen and M. A. Bosch, Phys. Rev. *B*26, 7123 (1982).
3. N. Theodorakopoulos and J. Jackle, Phys. Rev. *B*14, 2637 (1976).
4. P. A. Fleury and K. B. Lyons, Phys. Rev. Lett. *36*, 1188 (1976).

OPTICAL LOGIC DEVICES, OPTICAL BISTABILITY AND GIANT NON-LINEAR EFFECTS IN SEMICONDUCTORS WITH POSSIBLE APPLICATIONS USING AMORPHOUS MATERIALS

S. D. Smith

Physics Department
Heriot-Watt University
Edinburgh, U.K.

We describe the series of all-optical circuit elements that have arisen by exploitation of very large third order non-linearity which has been discovered near the band gaps of III-V semiconductors. The biggest effects are found in InSb and $Cd_xHg_{1-x}Te$. Electron excitation from a broadened band tail is held to be responsible in some cases and may therefore be expected in the case of amorphous materials. Numerous device possibilities exist.

I. INTRODUCTION

The possibility of replacing electrical currents by optical beams in information processing devices has been greatly enhanced during the last year by the application of very large nonlinearities in semiconductors to optically bistable and derived devices. The most important of these is the transphasor or optical transistor in which one optical beam can control the amplification of another. The effects described in this paper are based upon the nonlinear Fabry Perot resonator in which the refractive index of the spacer material, and hence the optical thickness, is intensity dependent. This gives rise to dispersive optical bistability - an effect that may be understood by considering the input-output characteristic of the resonator. In Fig. 1, normally linear relationship between output and input (dotted lines) is drastically changed in the nonlinear case as the resonator tunes or detunes on to the laser frequency. As the resonator frequency approaches the laser frequency a greater proportion of the incident intensity circulates inside the resonator thus creating a larger nonlinear effect. Thus the characteristic bends from the minimum transmission line towards the maximum, i.e. becoming nonlinear. The process is catastrophic so that a *differential gain* region (output changing more than input) is reached followed by a region of negative slope. At this point the device switches to an upper state of transmission near the maximum allowed through the resonator. On now reducing the intensity, the internal field remains high until a lower level of incident intensity when switch-down occurs. The device thus displays hysteresis and the characteristics of a bistable element in which its state is automatically indicated by its transmission. In addition a differential gain mode can always be achieved by appropriate detuning. When the phase thickness of the resonator is changed by the addition of a second beam we achieve "the transphasor," the analogue of the transistor for electrical

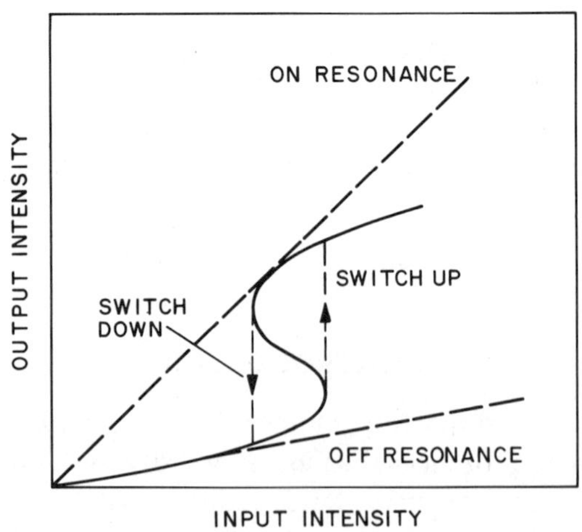

Fig 1. Non-linear relationship between input and output intensities for a bistable device.

currents. These devices in which light is controlled by light in principle constitute "all-optical circuit elements." The possibilities for fast processing will be discussed.

The first observations of optical bistability were made by Gibbs and McCall[1] in 1975 using sodium vapour. Other nonlinear materials such as Kerr liquids[2] and ruby[3] as well as hybrid devices[4] have since been used. The first use of semiconductors followed from the discovery of self-defocusing at modest laser powers in InSb[5] and saturable effects due to excitons in GaAs[6]. Bistable resonators in InSb and GaAs were reported[7,8] in 1979. Semiconducting materials such as those mentioned have been known to exhibit a comparatively large passive non-resonant $\chi^{(3)}$[9] (around $10^{-8} - 10^{-11}$ e.s.u.) but the discovery of strong nonlinear refraction in the region just below the optical bandgap in both materials gives susceptibilities many orders of magnitude higher with $\chi^{(3)}$ in the range $10^{-2} - 1$ e.s.u. In both cases the giant effect can be explained by saturation mechanisms although the details vary, the nonlinearity in GaAs relying on the existence of excitons whilst that in InSb is ascribed to interband transitions. The advantages of using semiconductors in optically bistable devices are that

(i) the nonlinear refraction is sufficiently high for small power densities (as low as 10 W/cm^2 or 8 mW incident) to be used;

(ii) small resonator thicknesses (4 – 100 μm) - important because one limit on switching time is related to the cavity field build up time and so to the cavity length.

These factors when combined suggest that small low power, low energy, fast switching devices may ultimately be possible with semiconductors and be consistent with an integrated optic approach to the system architecture.

II. EXPERIMENTAL

Optical bistability was observed using the coincidence of the 60 or 70 CO laser lines from 1930 cm^{-1} to 1660 cm^{-1} with the bandgap of InSb at 5K (around 1900 cm^{-1}) and 77K (around 1840 cm^{-1}). The laser radiation was obtained from an Edinburgh Instruments PL3 CO laser and controlled with an attenuator and spatial filter[10] to give a near perfect Gaussian profile over several orders of magnitude. Optically polished parallel crystals of thickness 560 μm and 130 μm were used as the resonators with the natural reflectivity R = 0.36. The entire beam was observed at the detector and the output-input characteristic is shown in Fig. (2a). It shows steps corresponding to successive changes of optical thickness of $\lambda/2$ with 5 such steps being observed at 5K. Optical bistability is clearly seen in fourth and fifth orders in both transmission and reflection. At 77K with a thinner sample and with a

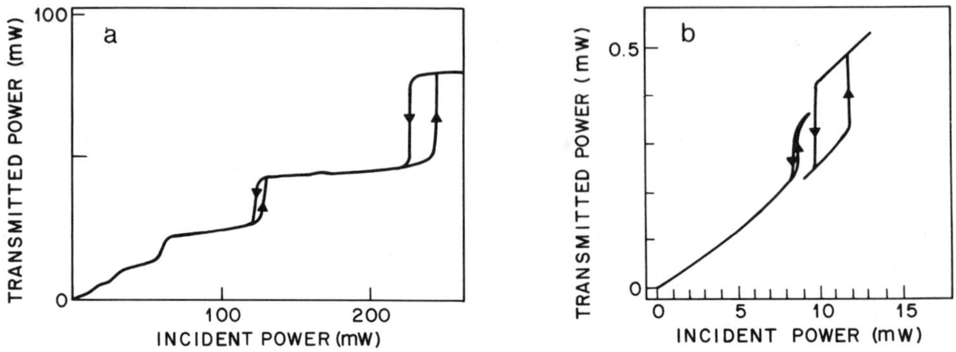

Fig. 2. a) Optical bistability in InSb at 5K. b) Optical bistability at 77K in InSb at an input power of 8 mW.

crystal coated to a reflectivity of 0.7, bistability is observed with an input power as low as 8 mW (Fig. 2b). This latter curve was also studied dynamically by slowly sweeping the input power with an electro-optic modulator and then observing the fast switching. The combined time constants of detector and electronics was not shown to be faster than 500 ns; but switching was observed at least as quickly as this limit.

The early work on the nonlinear refraction indicated an electronic origin.[11,12] It was therefore possible to predict that the use of an additional beam would also change the refractive index. Using the configuration shown in Fig. 5 we have been able to optically control the nonlinear Fabry Perot resonator and demonstrate differential signal gain. Since this occurs by the transference of phase thickness from one beam to the other we term the device the 'transphasor' in analogy with the transistor. In any practical processing device it is likely that switching and control will be achieved in such a manner. Signal gains of up to 10 are demonstrated.

III. MECHANISM OF NONLINEAR REFRACTION IN SEMICONDUCTORS

Nonlinear effects in both refraction and absorption have been reported in InSb just below the bandgap energy.[11] Absorption in the band tail is shown to be saturable at extremely low intensities (less than 1 W/cm^2). It is possible to dissociate this from the nonlinear refraction occurring at around 10 W/cm^2 but a substantially linear band tail absorption remains at this power density. The band tail seems not to be well explained in the literature.

Figure 2a gives us further information in that the intensity increment between each "step" increases with intensity. This implies that the nonlinearity is decreasing with increasing intensity, i.e. is itself saturating. Following the prediction of Javan and Kelley[13] we therefore examine the effect of saturation on refractive index. In atomic systems which are pumped below an absorbing transition nonlinear refraction arises due to the saturation of 'anomalous' dispersion. This gives a negative n_2 and self-defocusing is observed. If we therefore postulate that the band tail absorption can excite a system of oscillators at higher frequencies a change of refractive index

$$\Delta n = -\frac{2\pi}{\hbar n} \frac{|\mu|^2}{(\omega_o - \omega)} \tag{1}$$

is obtained for each transition which is blocked (i.e. saturated). A detailed theory for a semiconductor requires knowledge of the density of interband energy differences ω_o and the distribution of the excited population. First approximations to such a theory are advanced in this paper with some simple assumptions about scattering mechanisms and lifetimes for each step. This approach explains both the qualitative features and the overall size of the effect. One of the difficulties of the theory is to explain how for $\hbar\omega < E_G$ an excitation mechanism exists which can block higher frequency (interband) transitions. We return to this point later. Given that excitation does take place for the power densities involved a carrier density between 1×10^{14} cm^{-3} and 6×10^{15} cm^{-3} may be generated for the range of intensities used

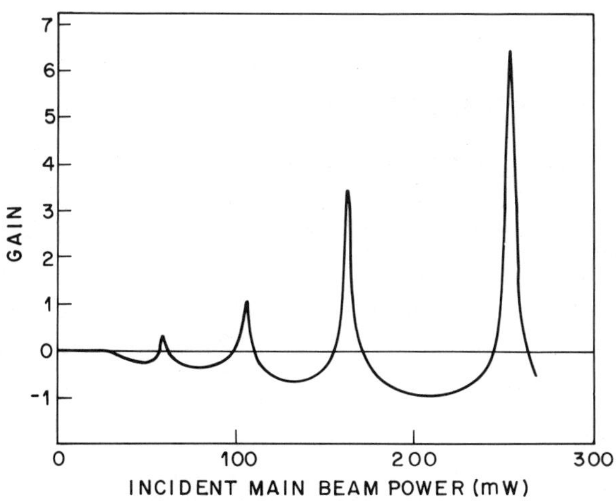

Fig. 3. Demonstration of differential gain by use of an additional beam to control refractive index of non-linear element.

and assuming a carrier recombination time of the order of 100's of ns. In general such optical generation of carriers can have the following effects giving rise to nonlinear refraction:

A. EXCITON SCREENING

In a material such as GaAs where there is a prominent exciton feature near the bandgap the electron-hole interaction can be screened out at certain carrier densities. For InSb this is around 1×10^{14} cm^{-3} close to the unexcited carrier density and therefore excitonic effects are not significant. In GaAs a density of 2×10^{16} cm^{-3} is required and this origin for nonlinearity has been demonstrated by Gibbs et al.[8] An n_2 of around 10^{-4} cm^2/KW has been observed in the presence of an absorption coefficient $\alpha \sim 10^3$ cm^{-1}.

B. FREE CARRIER PLASMA

The optically generated carriers constitute a free carrier plasma which gives a refractive index change given by

$$n_2(P) = \frac{-2\pi\, e^2\, \alpha\, \tau_R}{n\, m^*\, \hbar\omega^3} \tag{2}$$

This process has been invoked for the case of silicon[14] and favours long wavelengths. For the case of InSb we have conducted two-beam pump and probe experiments[15] at different frequencies which are not consistent with this mechanism. The experiments indicate that our effect is bandgap resonant.

C. DYNAMIC BURSTEIN-MOSS SHIFT

If we assume that a given number of carriers reach the bottom of the conduction band and that thermalization has occurred the lower states of the conduction band will be blocked. This is analogous to the Burstein-Moss shift induced by impurities but in which the carriers are excited by the laser through the band tail absorption. By integrating Eq. (1) over appropriate densities of states we can estimate the effect of blocking of these states on the refractive index and hence evaluate n_2. This yields[16]

$$n_2(B\text{–}M) = \frac{2\pi}{3n}\left[\frac{eP}{\hbar\omega}\right]^2 \frac{\alpha_{eff}(\omega)\tau_R}{\hbar(\omega_G-\omega)\hbar\omega} \tag{3}$$

where P is the interband momentum matrix element and $\alpha_{eff}(\omega)$ represents the frequency dependent carrier generation function here taken empirically. τ_R is the normal recombination time. The model overestimates the completeness of the filling by assuming that scattering processes are complete.

D. DIRECT SATURATION

This process is closely related to C but approaches the problem from a different limit. We assume that we can model the interband absorption with a set of two-level oscillators each of which is saturable and can be treated by standard nonlinear optical theory. The rate of pumping is controlled by the effective dipole moment (or empirically by $\alpha_{eff}(\omega)$) and the process by two time constants - the energy relaxation time T_1 within which population decays and the dephasing time T_2 which measures the uncertainty width of individual energy states. The calculation yields the result

$$n_2(D-S) = \frac{2\pi}{5n}\left[\frac{eP}{\hbar\omega}\right]^2 \frac{\alpha_{eff}(\omega)T_1}{\hbar(\omega_G-\omega)\hbar\omega} \tag{4}$$

Mechanisms C and D are related by comparing Eqs. (3) and 4 by

$$n_2(B-M) = \frac{5}{3}\, n_2\,(D-S)\cdot\frac{\tau_R}{T_1} \tag{5}$$

There are two possible limits for T_1:

(i) If T_1 corresponds to the interband recombination time τ_R and we assume no intraband scattering. This maximizes the contribution of C.

(ii) If we consider intraband scattering, one oscillator will be depopulated and reavailable for excitation as soon as an electron or hole moves out of a state k to a neighbouring state. The lowest limit of this would be if $T_1 = T_2$. Even taking this limit mechanism D gives a substantially large value for n_2 possibly within 100 of the measured value. This limit gives in fact the lowest estimate for n_2 and the high speed limit.

E. A MECHANISM FOR EXCITATION WITH $\hbar\omega_G < E_G$: T_2-TAILING

We return to the problem of the excitation of carriers for laser photon energies less than the energy gap. Mechanism D gives us a first method of suggesting a mechanism. Standard nonlinear optical theory for a two-level oscillator including power broadening or saturation is given by

$$\alpha(\omega,I) \sim \frac{\mu^2 T_2}{1 + (\omega_o - \omega)^2 T_2^2 + I/I_S} \tag{6}$$

where I_S is that saturation intensity. The dephasing time T_2 is related to intraband scattering mechanisms which for InSb give times of the order of 1 ps or less. The consequent line broadening

$$\Delta\omega = 1/c\pi T_2 \text{ cm}^{-1} \tag{7}$$

is therefore of the order of 30 cm^{-1}. Summing the number of oscillators representing the interband transitions and including this broadening mechanism we have the analogy of the simple models for free carrier absorption (a broadened zero frequency transition) and this gives a band tail capable of causing the excitation of the interband oscillators. The laser "pumping" is "off-resonance" and with many oscillators corresponds to the case well-known in atomic vapours of inhomogeneous broadening and one would expect a modified form of "hole burning". Including these population effects by means of a density matrix treatment with a constant T_2 we obtain

$$\alpha(\omega) = \frac{1}{3nc}\left[\frac{eP}{\hbar\omega}\right]^2 \left[\frac{2m_r}{\hbar}\right]^{3/2} \frac{\omega}{\hbar T_2}(\omega_G - \omega)^{-1/2} \tag{8}$$

whence by combination with Eq. (4) we find

$$n_2 \sim \frac{P^4}{\omega^4} \cdot \frac{T_1}{T_2}(\omega_G - \omega)^{-3/2} \tag{9}$$

The results of this evaluation are indicated in Fig. 4 and they show that reasonable order of magnitude agreement is obtained for plausible values of T_1 and T_2 although the frequency fit is not perfect.

Other possible excitation mechanisms may include impurity states and indeed we do observe impurity dependence of the effect. This is however rather weak and is directly related to the amount of absorption which itself would be sensitive to a change in the intraband scattering time T_2.

IV. SPEED OF RESPONSE

If an all-optical processing device is to be competitive with very large scale integration or Josephson junctions we would require the following features:

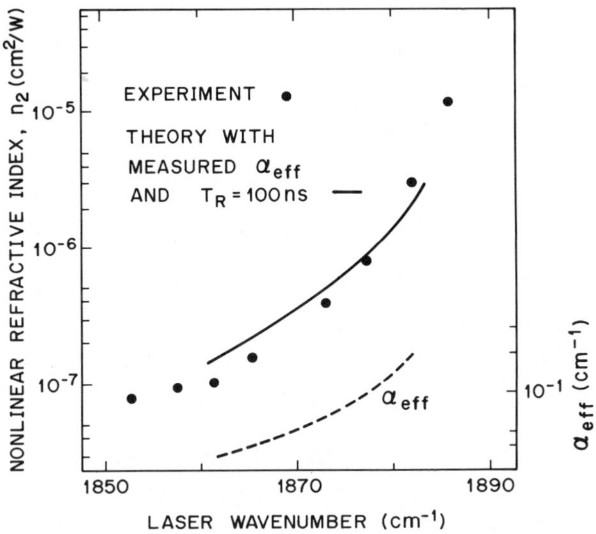

Fig. 4. In Fig. (4) we illustrate the bandgap resonant behaviour of the effect and show that with reasonable values, e.g. from the data of Miller[11] we extract $\alpha_{eff}(\omega)$ and with $\tau_R = 100$ ns, mechanisms (C) and (D) give good order of magnitude estimates for the size of this giant nonlinearity. It seems therefore that bandgap resonant saturation is basically responsible for the large nonlinear refraction in InSb.

(i) near picosecond switching time

(ii) micron dimension - this ensures cavity field build-up time $\sim$ ps

(iii) small holding power (mW)

(iv) small switching energy (pJ)

(v) fast response of nonlinearity.

For the case of GaAs Gibbs et al[6] have already shown that exciton features can be screened out in times of the order of ps but show a much longer recovery time and have demonstrated bistable switching for near visible wavelengths of the order of 10's of ns.

For InSb recent experiments have demonstrated an AND gate in which the bistable element is held near switching point with the holding power of the CO laser beam at about 20 mW and the device is then switched by the incidence of a 30 ps Nd:YAG laser pulse. Importantly a very definite threshold energy ($\sim$5 nJ in the first experiments) was observed. Simple

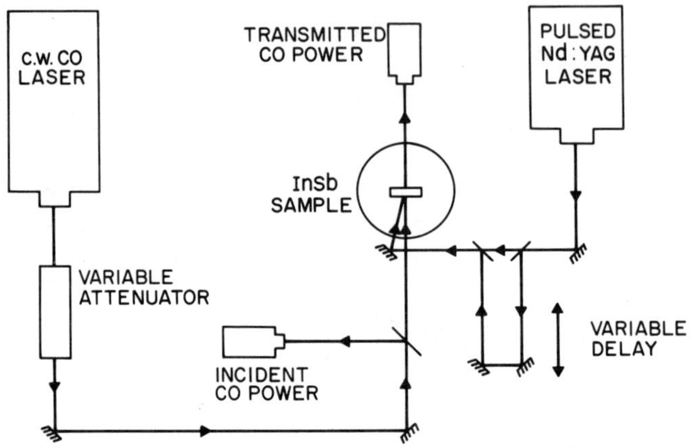

Fig. 5. Experimental arrangement for the fast AND gate experiment.

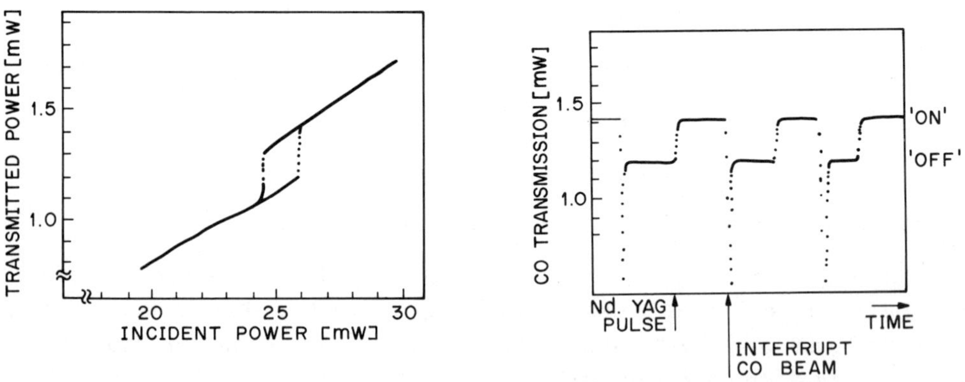

Fig. 6. Bistable switching induced by 5 nJ of Nd:YAG energy in a 30 ps pulse.

calculations reveal that sufficient electrons are excited by the Nd:YAG switching pulse to cause an adequate optical thickness change to be induced in less than 5 ps. The switch up time should therefore be limited by the build up time on the field in the InSb resonator. Since this is 200 μm thick the cavity build up time is of the order of 8 ps. It is therefore possible that we have demonstrated a bistable switching device with a switching speed of this order.

Switch down does not occur so quickly. On receiving the switching pulse the device remains switched up, thus acting as a single pulse detector until the holding power is reduced below the appropriate threshold. The response time of the non-linearity (of the order of the carrier lifetime) will then control how rapidly the device switches down. We have investigated this feature by use of a 'delayed AND gate'. In this case the switching pulse is divided into two with beam splitters. It was first checked that both pulses were required to cause switching this established AND gate operation. One of the pulses was then delayed up to 250 ns. As the carriers decay more and more pulse energy is required of the second switching pulse to induce switching. The clearly defined switching energy enables the device to be used to calibrate itself by use of attenuators in the experiment.[17] Figures 5, 6 and 7 illustrate the arrangement and the results in this experiment. A clear and definite exponential decay is indicated with a lifetime of around 90 ns.

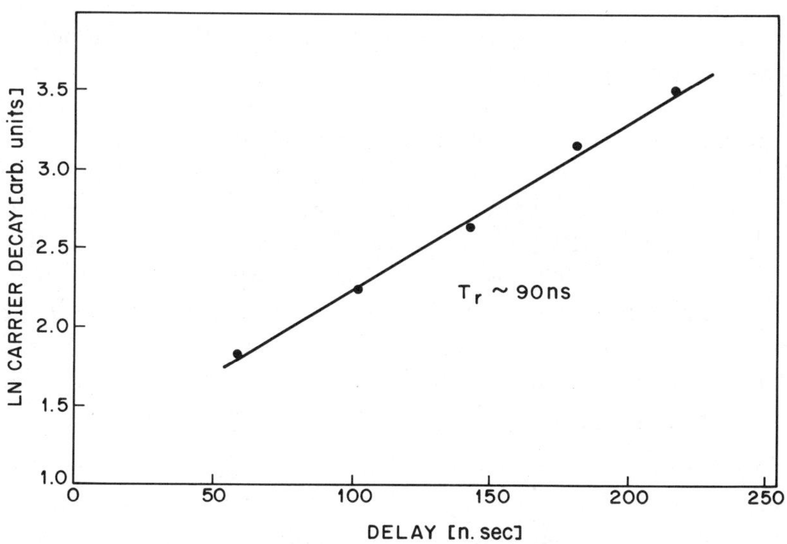

Fig. 7. Carrier recombination time determination using delayed AND gate.

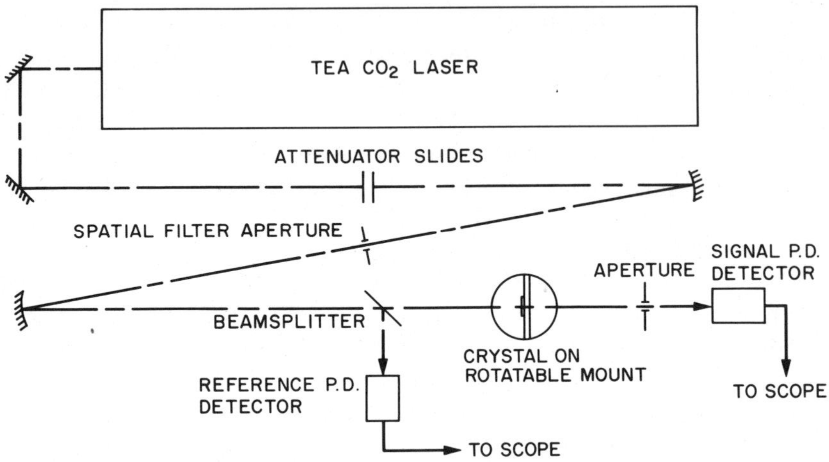

Fig.8. Equipment for observation of room temperature bistability in InSb at 9-11 μm using SLM TEA CO_2 injection locked laser.

Using different orders of bistability it is possible to investigate this phenomenon at different holding beam intensities. Reduced values of carrier lifetime upon laser intensity are indicated. This would be consistent with the conclusions of Hill, Miller and Parry[18] that the lifetime is at least partly limited by Auger processes.

The realization of bistable devices at room temperature is clearly of importance to practical application. Groups at Bell Laboratories (Miller, Chemla and P W Smith)[19] and at Arizona (Gibbs)[20] have demonstrated the phenomenon in multiple quantum well structures using GaAs. The non-linearities correspond to $\chi^{(3)} \sim 10^{-3}$ e.s.u. and typical intensities $\sim$50 kW/cm^2. We have recently demonstrated similar room temperature optical bistability in InSb at 10 μm using two-photon absorption as the carrier excitation mechanism.[21] The experiment and results are shown in Figures 8 and 9. Clear bistability is observed in an experiment using photon drag detectors and system time constants of the order of nanoseconds. Surprisingly the two-photon process is sufficiently sensitive that non-linearities corresponding to $\chi^{(3)} \sim 10^{-4}$ e.s.u. are demonstrated.

Recent work in Hungary by Hajto[22] has also shown a series of effects at room temperature and mW powers using $GeSe_2$ self-supporting films. Using mW powers from HeNe lasers a series of non-linear and bistable effects have been observed qualitatively rather different from those described above. The authors attribute them partly at least to thermal and photo-structural events

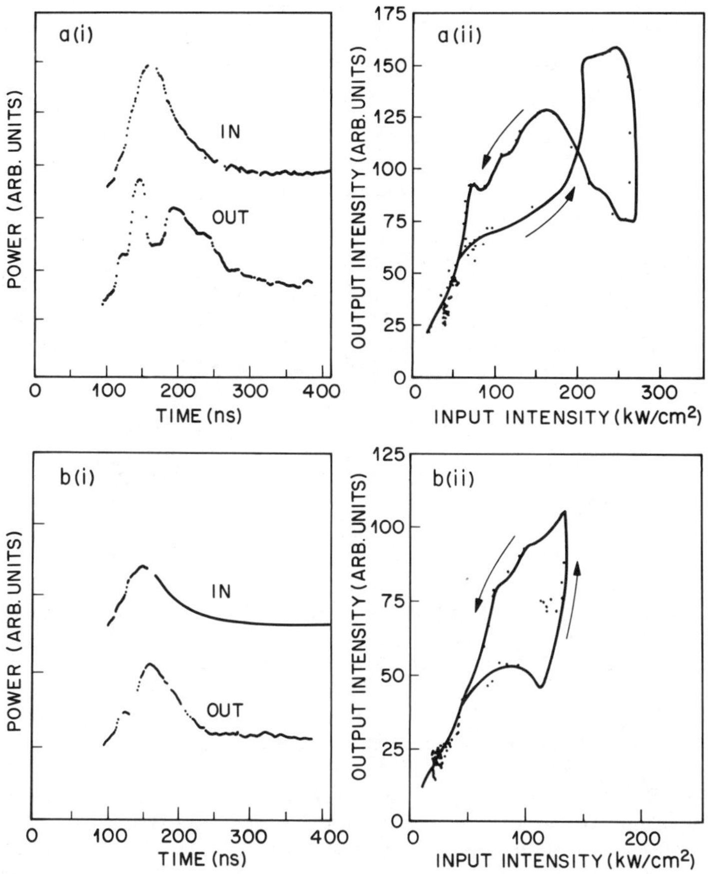

Fig. 9. Dynamic observations of refractive nonlinearity and optical bistability in InSb at 10.6 μm. System time constants ~1 ns.

and they also appear to be rather slow. It is likely, however, that similar non-linearities to those described in this paper will be partly responsible and several regimes of operation and device possibilities appear to exist.

CONCLUSIONS

Electronic non-linearities larger than 1 e.s.u. have been demonstrated for optical circuit element devices. These are usually characterized by natural time constants of the order of nanoseconds. The possibility of manipulating carrier lifetime suggests that picosecond time constants could be obtained with non-linearities of the order of $\chi^{(3)} \sim 10^{-5}$ e.s.u. The nature of band tails and consequent electron excitation processes is likely to be a fertile area for future research. For some recent reviews see Miller, Miller and Smith[23] and Abraham and Smith.[24]

REFERENCES

1. H. M. Gibbs, S. L. McCall and T. N. C. Venkatesan, Phys. Rev. Lett. **36** 1135 (1976).
2. T. Bischofberger and Y. R. Shen, Optics Lett. **4**, 40 (1979).
3. T. N. C. Venkatesan and S. L. McCall, Appl. Phys. Lett. **30**, 282 (1977).
4. P. W. Smith and E. H. Turner, Appl. Phys. Lett. **30**, 282 (1977); P. W. Smith, E. H. Turner and P. J. Maloney, IEEE J. Quant. Electron. **QE-14**, 207 (1978); P.W. Smith, I.P. Kaminow, P. J. Maloney and L. W. Stulz, Appl. Phys. Lett. **33**, 24 (1978); Appl. Phys. Lett. **34**, 62 (1979).
5. D. A. B. Miller, M. H. Mozolowski, A. Miller and S. D. Smith, Optics Comm. **27**, 133 (1978).
6. H. M. Gibbs, A. C. Gossard, S. L. McCall, A. Passner, W. Wiegmann and T. N. C. Venkatesan, Solid State Commun. **30**, 217 (1979).
7. D. A. B. Miller, S. D. Smith and A. Johnston, Appl. Phys. Lett. **35**, 658 (1979).
8. H. M. Gibbs, S. L. McCall, T. N. C. Venkatesan, A. C. Gossard, A. Passner and W. Wiegmann, Appl. Phys. Lett. **35**, 451 (1979).
9. C. Flytzanis in *Quantum Electronics*, Vol. 1A, eds. H. Rabin and C. L. Tang (Academic Press, New York, 1975) p. 9.
10. D. A. B. Miller and S. D. Smith, Appl. Optics **17**, 3804 (1978).
11. D. A. B. Miller, Ph.D. Thesis (Heriot-Watt University, Edinburgh, 1979).
12. D. Weaire, B. S. Wherrett, D. A. B. Miller and S. D. Smith, Optics Lett. **4**, 331 (1979).
13. A. Javan and P. L. Kelley, IEEE J. Quantum Electron. **QE2**, 470 (1966).
14. R. K. Jain and M. B. Klein, Appl. Phys. Lett. **35**, 454 (1979).
15. D. A. B. Miller, S. D. Smith and C. T. Seaton in *First International Conference and Workshop on Optical Bistability*, Asheville, North Carolina, U.S.A., 3-5 June 1980, to be published.
16. D. A. B. Miller, S. D. Smith and B. S. Wherrett, Opt. Commun. **35**, 221 (1980).
17. S. D. Smith, C. T. Seaton, M. E. Prise and W. J. Firth, XIIth IQEC 1982, Munich, June 22-25, Digest: Appl. Phys. **B28**, 132 (1982).
18. J. R. Hill, A. Miller and G. Parry, Opt. Commun., (to be published).
19. D. A. B. Miller, D. Chemla and P. W. Smith, XIIth IQEC 1982, Munich, June 22-25, Digest: Appl. Phys. **B28**, 96 (1982).
20. H. M. Gibbs, XIIth IQEC 1982, Munich, June 22-25, Digest: Appl. Phys. B**28**, 98 (1982).
21. A. K. Kar, J. G. H. Mathew, S. D. Smith, B. Davies and W. Prettl, Reported at XIIth IQEC 1982, Munich, June 22-25.
22. J. Hajto and I. Janossy, Phil. Mag., (to be published).
23. A. Miller, D. A. B. Miller and S. D. Smith, Advances in Physics **30**, 697 (1981).
24. E. Abraham and S. D. Smith, Rep. Prog. Phys. **45**, 815, (1982).

DISCUSSION

Optical Logic Devices

The discussion centered around two major areas. The first was the nature and underlying mechanisms for the nonlinearity, described by χ_3, in bistable devices. Professor Smith had pointed out that in certain cases when operating very near the band gap of a semiconductor, for example, effective values of χ_3 may be resonantly enhanced to approach 1 esu. This compares to intrinsic off resonant, nonlinear refractive index values as many as 9 to 12 orders of magnitude smaller.

In those cases in which this large effective value for χ_3 has been reached through real absorption as one of several multistep processes, a severe penalty is paid in terms of the time scale of related switching phenomena. In particular, if a real absorption event first excites carriers, which then as a secondary process modify the refractive index to cause tuning or detuning of a bistable Fabry-Perot, the recombination time of such carriers may severely degrade the rapid shut-off response for such a device. It was noted that experiments on optical bistability which measure only efficiency of switching or threshold energy required to cause a switching event are only asking half of the relevant questions. To gain a complete picture it is necessary to measure the time-dependent response of any bistable material or device in order to assess its utility in a working device.

The second point addressed the pros and cons associated with optical bistability in amorphous or disordered semiconductors as distinct from the crystalline materials studied to date. Smith's work in indium antimonide on several different samples with different histories and different doping levels appeared to reveal identical thresholds and efficiencies; in other words identical effective χ_3's. This led to the speculation that perhaps states tailing into the band gap were responsible, and further, that those states might find their origin in some small disorder in the otherwise crystalline host.

It was questioned whether extending this reasoning to incorporate the large densities of states found in the gaps of amorphous semiconductors might prove fruitful from two points of view as far as optical bistable devices are concerned. First, there is generally a greater density of states in the gap in the amorphous semiconductor. Therefore one might expect greater efficiency in enhancing χ_3. In addition, since the states generally tail further into the gap, a device operated in such a material might be efficient over a considerably broader range of optical frequencies than in the crystal. This is particularly true, for example, relative to the excitonically-enhanced resonances in gallium arsenide. Furthermore, lifetimes of excited carriers in amorphous materials are expected to be considerably shorter than in crystalline materials, thereby offering the possibility of greatly reduced shutoff times. On the other hand, one could argue that the increased number of states in the gap would serve only to increase the absorption and thereby reduce the finesse of the Fabry-Perot. Furthermore the lifetime reduction is beneficial only so long as the carriers are not trapped at defect sites, in which case the material would not return to its unpumped state as rapidly as the carrier lifetime would imply.

While none of these points was definitively resolved in the discussion it became clear that a path for fruitful experimental work had been identified here and that only such experimental work on a variety of disordered and amorphous semiconductors as active elements in bistable devices would answer some of these questions. The potential for improved performance, particularly at room temperature, of optical multistable devices based on amorphous material was certainly deemed real enough to warrant considerable experimental effort.

H. Gibbs, Chairman
P. A. Fleury, Discussion Leader

ELECTRON TRANSPORT IN CHALCOGENIDE GLASSES

A. E. Owen

Department of Electrical Engineering
University of Edinburgh
King's Buildings
Edinburgh EH9 3JL
Scotland

A brief general account of the band structure and transport mechanisms of amorphous semiconductors is first presented. Special features of the electronic band structure of chalcogenide glasses are reviewed, with particular emphasis on chemical bonding considerations and the formation of defect states associated with abnormal configurations. Experimental data on d.c. conductivity, thermopower, Hall mobility and drift mobility are described, with measurements on Se, As_2Se_3 and related compositions as examples. The data are interpreted in terms of band models in which relatively discrete defect levels within the mobility gap control the transport processes.

I. BACKGROUND

A. INTRODUCTION

There are sound arguments for the proposition that the one-dimensional one-electron density-of-states distribution in an ideal amorphous semiconductor is as shown schematically in Figure 1(a).[1,2] An ideal amorphous structure is a fully connected three-dimensional disordered network of atoms of finite size, i.e. a degree of short-range order is imposed but there are no unsaturated bonds. The label ES in Figure 1(a) means Bloch-type extended states above an energy E_C in the conduction band and below an energy E_V in the valence band. The label T indicates localized tail states which are split-off from the Bloch-type states of the valence and conduction bands but which retain the characteristics of their parentage; G indicates gap states of indefinite origin which are assumed present in low but uniform density across the energy gap (e.g. $\sim 10^{14}$ cm^{-3}). The tail and gap states are localized in the sense that electrons in these states have wave functions ψ_{loc} which decay exponentially in space, i.e. in one-dimension (x), -

$$\psi_{loc} \sim \exp - (\alpha x)$$

where α is a decay factor and is typically $\sim$0.1 A^{-1}.

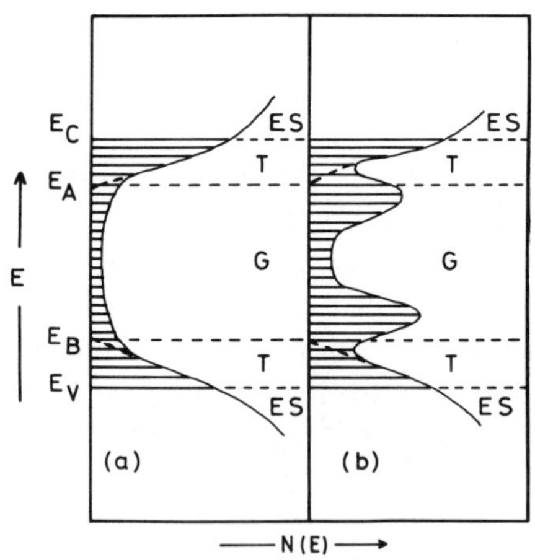

Fig. 1. Schematic one-electron one-dimensional density-of-states diagrams, (a) for an ideal amorphous semiconductor, and (b) for a real amorphous semiconductor.

A real amorphous semiconductor is thought to have a one-dimensional density-of-states more like that depicted schematically in Figure 1(b). There are still extended, tail and gap states but in addition there is evidence for energetically rather well-defined maxima in the gap state distribution and these features are attributed to specific structurally-related defects such as dangling (broken) bonds, deviations from stoichometry and/or wrong bonds in the case of compounds, and even perhaps to impurities.

B. THE MOBILITY GAP MODEL

Leaving aside the difficult question of the transition from extended Bloch states to localized tail states (is it gradual or is there a true discontinuity at some critical energy?), Figures 1(a) and 1(b) lead to the mobility-gap notion which has been the accepted basis for interpreting electronic transport in amorphous semiconductors since the early days of the subject. The mobility-gap idea is shown diagrammatically in Figure 2. In extended states just above E_C, or just below E_V, carrier transport is essentially a diffusive Brownian-type motion and hence the carrier mobility μ is approximately,[2]

$$\mu \approx \frac{1}{6}\frac{ea^2}{kT}\nu_{el} \tag{1}$$

where a is the average interatomic distance and ν_{el} is an electronic frequency of the order of $10^{15}\ s^{-1}$. The estimated drift mobility in extended states just above E_C or just below E_V is therefore about $1\ cm^2\ v^{-1}\ s^{-1}$ at room temperature. In localized states (T or G) transport can only occur by phonon-assisted hopping and the mobility is given by

$$\mu(E) \approx \frac{eR^2(E)}{kT}\nu_{ph}\exp(-2\alpha R)\exp(-W/kT) \tag{2}$$

where R is the average hopping distance which depends on the density-of-states distribution and is a function of energy (N.B. R may be greater than a), ν_{ph} is a phonon frequency ($\sim 10^{13}\ s^{-1}$) and W (the activation energy) is the energy difference between the initial and final localized states involved in the hopping motion. Close to E_C and E_V, $R \rightarrow a$ and $W \rightarrow 0$, hence $\mu \sim 10^{-2}\ cm^2\ V^{-1}\ s^{-1}$ at room temperature. Thus, near E_C and E_V the carrier mobility changes by three orders of magnitude or more and this defines the mobility gap, illustrated schematically in Figure 2(b) where μ is plotted as a function of energy.

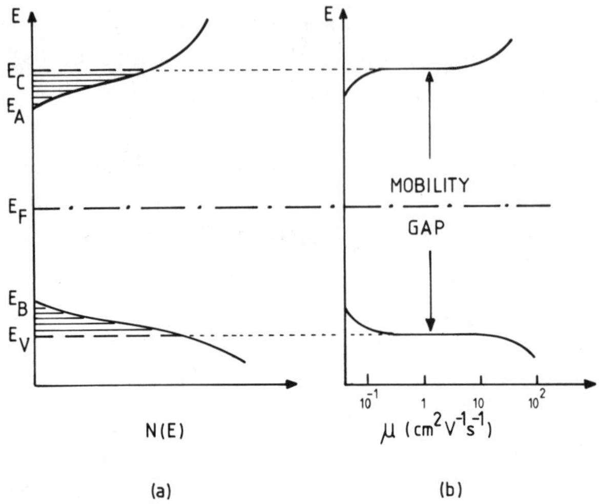

Fig. 2. Illustrating the mobility gap model for an amorphous semiconductor. (a) The one-electron density-of-states (the hatched regions indicate the tails of localized states in which the carrier mobility is low). (b) The mobility as a function of energy, corresponding to (a).

C. GENERAL CONDUCTION MECHANISMS - TEMPERATURE DEPENDENCE

Accepting the mobility gap idea, four mechanisms of conduction may be expected and each will dominate the d.c. conductivity in an appropriate range of temperature.[3] Starting at high temperatures, the four processes are as follows:

(a) Conduction by carriers excited into extended states just above E_C or just below E_V. In the case of hole transport, for instance, the conductivity will be given by

$$\sigma = C_o \exp[-(E_F - E_V)/kT] \tag{3}$$

Optical measurements usually show that the band gap of amorphous semiconductors decreases approximately linearly with temperature, i.e.

$$(E_F - E_V) = E(0) - \gamma T \tag{4}$$

where E(0) is the value of $(E_F - E_V)$ at $T = 0$ K and γ is its temperature coefficient. Thus,

$$C_o = \sigma_o \exp(\gamma/k) \tag{5}$$

In chalcogenide glasses, the temperature coefficient of the fundamental optical absorption edge, which is *approximately* equal (see Section 3) to the mobility gap, is usually found experimentally to be in the region of 4×10^{-4} to 8×10^{-4} eV deg^{-1}. Moreover, the Fermi level E_F is invariably situated near the middle of the gap and hence values of γ roughly half that magnitude are to be expected; the magnitude of $\exp(\gamma/k)$ is therefore likely to be in the range 10-100. The constant σ_o is generally equated with σ_{min} the so-called minimum metallic conductivity. Mott defined σ_{min} as the smallest non-zero value the conductivity can have at absolute zero, i.e. the lowest value of the conductivity contributed by carriers just at E_C (in the case of electrons) before the start of activated processes. Mott derived the simple relationship

$$\sigma_{min} = \text{Constant } (e^2/ha) \tag{6}$$

where a is the interatomic distance. The constant depends somewhat on structure and is in the region 0.03 to 0.1 so that σ_{min} is 200-800 ohm^{-1} cm^{-1}, if a = 3 Å. For conduction in extended states therefore the pre-exponential constant C_o (Eqs. (3) and (5)) should be roughly 10^3-10^4 ohm^{-1} cm^{-1}.

(b) Transport by carriers excited into the tail of localized states at energies close to E_A or E_B (see Figure 2) and migrating by a hopping mechanism. Assuming conduction by electrons again,

$$\sigma = C_1 \exp[-(E_F - E_B + W_1)/kT)] \tag{7}$$

where W_1 is the activation energy for hopping. It is not easy to make an estimate of C_1 but the lower mobility and the lower density-of-states near E_A compared with E_C, will make it several decades smaller than C_o. The energy difference $(E_F - E_B)$ is also expected to depend upon temperature but that is again difficult to determine.

(c) At low temperatures a significant number of carriers is not excited but if the density-of-states at the Fermi level is finite there will be a contribution from carriers with energies near E_F hopping between localized states. In this case,

$$\sigma = C_2 \exp(-W_2/kT) \tag{8}$$

where W_2 is the appropriate hopping energy and $C_2 < C_1$.

(d) At still lower temperatures it is probable that carriers will tend to hop beyond their spatially nearest neighbour states to states which are closer energetically. This is the so-called variable range hopping mechanism and Mott showed that if the density-of-states at E_F is $N(E_F)$, -

$$\sigma = C_3 \exp[-(T_o/T)^{1/4}] \tag{9}$$

with $T_o \simeq [18\,\alpha^3/kN(E_F)]$

D. POLARON TRANSPORT

If a charge carrier remains in the vicinity of a particular site long enough its field will tend to displace or polarize the surrounding atoms and in its bound state the carrier cannot move unless the polarization cloud also moves with it.[4,5] The trapped carrier and the surrounding polarized region can be treated as an entity known as a polaron; if the polarization cloud extends over only a few interatomic distances the particle is called a small polaron. The polaron has a lower energy than a free electron but a larger effective mass since it must carry the induced deformation when it moves from site to site; the decrease in energy relative to that of the electron in an undistorted lattice is called the polaron binding energy W_p.

In a crystal, the small polaron states may overlap sufficiently to form a polaron band in an analogous way to electron energy band formation in the undistorted lattice.[6] The small polaron band is usually narrow and its width decreases exponentially with temperature. In general therefore a small polaron can move by two different mechanisms. At low temperatures band conduction without phonon interaction is possible; at higher temperatures the small polaron can migrate only by hopping between equivalent sites. The deformation of the lattice to form equivalent adjacent sites requires energy from phonons and the hopping motion can therefore be regarded as phonon assisted tunnelling between sites.

Thus, polaron hopping depends on the occurrence of occasional structural fluctuations causing adjacent occupied and unoccupied sites to have momentarily coincident deformations. At each coincident event the carrier will have a certain jump probability, P, which can be written as the product of two terms:[6]

$$P = P_1P_2 \tag{10}$$

with P_1 the probability of occurrence of a coincidence event, and P_2 the probability of charge transfer during that event.

The probability of a coincident configuration can be written as

$$P_1 = (\omega_o/2\pi)\exp(-W_H/kT) \tag{11}$$

where $(\omega_o/2\pi)$ is a phonon frequency and W_H, the polaron hopping energy, is the minimum energy necessary to bring two adjacent sites into equivalence. The hopping energy is related to the polaron binding energy by $W_H = (W_p/2)$. Two particular cases are distinguished:

1. The adiabatic regime in which the carrier jumps between the coincident sites several times during the period that the two sites are equivalent in energy.
2. The non-adiabatic regime in which the carrier cannot follow the lattice vibrations and the time required for the carrier to hop is long compared with the duration of a coincidence event.

In the adiabatic regime the hopping probability is high, i.e. $P_2 \simeq 1$. In the non-adiabatic regime $P_2 \ll 1$.

According to Holstein, in the non-adiabatic regime,

$$P_2 = \frac{2\pi}{h\omega_o}\left(\frac{\pi}{W_H kT}\right)^{1/2} J^2 \tag{12}$$

where J is the electronic overlap integral between sites. Hence, writing for the conductivity mobility μ,

$$\mu = (ea^2/kT)P \tag{13}$$

using Eqs. (10), (11) and (12) gives

$$\mu = \frac{ea^2}{kT}\,\frac{1}{h}\left(\frac{\pi}{WkT}\right)^{1/2} J^2 \exp\left(-\frac{W_H}{kT}\right) \tag{14*}$$

* Note: Eq. (2) could also be used as an approximate expression for the polaron mobility, with $W = W_H$ and $R = a$. Eq. (14) is a more explicit relationship for the polaron case.

The polaron hopping mobility is therefore thermally activated at high temperatures but when $kT < W_H$, the pre-exponential term varying as $T^{-3/2}$ is predominant. The polaron hopping mobility is usually much less than $1\ cm^2\ V^{-1}\ s^{-1}$ at or near room temperature.

The possibility of small polaron formation in amorphous semiconductors and insulators has been propounded particularly by Emin. The theory developed for crystals still applies except that the site-to-site disorder energy W_D, characteristic of an amorphous solid, also contributes and the total hopping energy W' is given by

$$W' = W_H + \tfrac{1}{2} W_D \tag{15}$$

It must be recognized that the localized tail states and other gap states, which are the characteristic features of the band model of amorphous semiconductors described in the preceding sections, would be redundant features insofar as transport is concerned if polaronic mechanisms are applicable and it follows that the mobility gap model would have no basis.

Small-polaron energy levels typically lie within the optical energy gap of insulators or semiconductors, and the likely relationship between the optical gap and small-polaron levels for electrons and holes (E_C^{sp} and E_V^{sp} respectively) is illustrated schematically in Figure 3.[7] In crystals the polaron levels would be at discrete energies but, as noted above, in an amorphous solid they are distributed over a range of energy equal to the disorder energy W_D. The Franck-Condon principle implies that optical absorption associated with interband transitions is determined by the density-of-states in Figure 3(b) and the levels E_C and E_V. In Figure 3 small polaron states are shown for both electrons and holes but situations may occur in which only one species of carrier is self-trapped while the other moves in quasi-free Bloch states.

An important parameter in polaron mechanisms is the Hall mobility, μ_H. The calculation of the Hall mobility is difficult and it depends on the local geometry of sites. For a triangular lattice, Friedman and Holstein have derived for the non-adiabatic case,[8]

$$\mu_H = \frac{ea^2}{\pi}\left(\frac{\pi}{12kTW}\right)^{1/2} J.\exp\left(-\frac{W}{3kT}\right) \tag{16}$$

The theory also predicts a sign anomaly. If conduction is due to the hopping of small polaron *holes,* for example, the sign of the Hall effect is *negative* while the sign of the thermopower is positive (as expected from the polarity of the carrier).

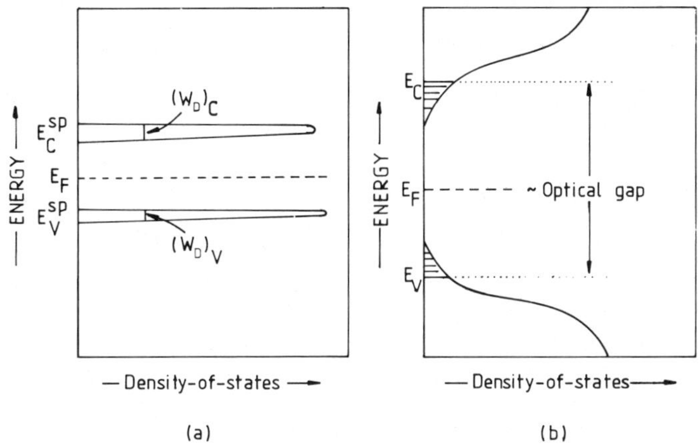

Fig. 3. (a) Proposed density-of-states diagram for small polarons in an amorphous semiconductor. The energy levels E_C^{sp} and E_V^{sp} are for electron and hole polarons, respectively, broadened by the small disorder energies $(W_D)_C$ and $(W_D)_V$. (b) Typical electron density-of-states diagram illustrating the likely relationship of the optical gap, to the polaron levels in (a) (See Ref. 7.)

II. THE ELECTRONIC BAND STRUCTURE OF CHALCOGENIDE GLASSES

A. INTRODUCTION

In all chalcogenide compounds the chalcogen atoms (S, Se or Te) are normally in two-fold coordination. The compounds of most interest are those formed with pnictide elements, such as arsenic, which normally bond in three-fold co-ordination but chalcogens and pnictides can vary their valency and atoms in abnormal coordination configurations are important, electronically, in the formation of defects.

The band structure of chalcogenide glasses is best approached in terms of normal chemical bonding however and this will be considered in the next Section, taking Se and As_2Se_3 as examples. Defect states are discussed in Section 2.3 and Section 2.4 is concerned with the effects of impurities.

B. BONDS AND BANDS

The outer electronic configuration of atomic As is $(4s)^2$ $(4p)^3$ and of Se $(4s)^2$ $(4p)^4$. The essential features of the band structure of solid Se and

As_2Se_3 in crystalline or glassy forms can be seen from simple molecular orbital considerations with slightly different results depending on whether or not hybridization of the atomic s- and p-states is assumed.

(a) Se

Figures 4(a) and 4(b) illustrate, in simplest terms, two possible molecular orbital configurations for Se. In Figure 4(a) it is assumed that the s- and p-states are first hybridized to form an sp^3-state. In bonding the atomic sp^3-state is split into a lower σ bonding state (b), an upper σ^* anti-bonding state (a) and the two doubly occupied sp^3-states form an intermediate lone-pair (LP) non-bonding (n) level in the molecular orbital scheme.

In Figure 4(b), there is no hybridization and the molecular orbitals are formed from pure s- and p-states. In bonding, the p-states again split into a σ lower bonding state (b), an upper σ^* anti-bonding state (a) and an intermediate non-bonding (n) lone-pair (LP) level. The atomic s-state forms a molecular level well below the bonding p-states. Note that in this case

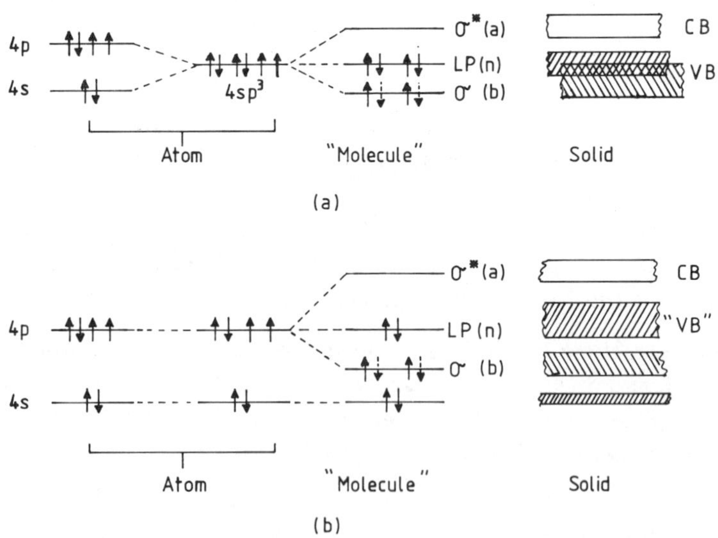

Fig. 4. Molecular orbital schemes for electronic conffgurations in Se. (a) Assuming sp^3 hybridization of atomic states. (b) Without hybridization. (see Refs. 9-12.)

however the LP level has half as many states (and electrons) as in the hybridized scheme.

Tutihasi and Chen,[9] and Chen[10] have proposed detailed molecular orbital models for the band structure of trigonal and monoclinic Se, and for amorphous Se containing rings (monoclinic) and chains (trigonal). An important feature of their calculations is that in solid Se (crystalline or amorphous) there is considerable overlap and mixing of the σ (b) and LP (n) molecular levels and hence both bonding and non-bonding states contribute to the uppermost *filled* band, i.e. to the valence band. This is also illustrated schematically in Figure 4(a).

Kastner,[11] following Mooser and Pearson,[12] proposed a band model for chalcogenide semiconductors based on the scheme of Figure 4(b), i.e. no hybridization. The important feature of this picture is that in the solid state, the lone pair levels remain well separated from the σ bonding states, and hence the uppermost filled band, conventionally called the valence band in semiconductor terminology, is made up entirely of non-bonding electrons. Strictly speaking the term "valence band" is a misnomer.

(b) As_2Se_3

The corresponding schemes for As_2Se_3, with and without sp^3 hybridization, are illustrated in Figure 5(a) and 5(b) respectively.[13] The numbers in parentheses correspond to the number of electrons in a "molecule" of As_2Se_3. Once again, calculations by Chen indicate strong mixing of bonding and nonbonding orbits,[10] and hence also in the uppermost filled (valence) band of solid As_2Se_3. By contrast, the picture based on pure s- and p-states leaves the lone pair (non-bonding) band separated from and above the σ-bonding band.

It is not all clear which of the two molecular orbital schemes - sp^3-hybridization (Figure 4(a) and 5(a)), or pure s- and p-state orbitals (Figures 4(b) and 5(b)) - is correct. Evidence from optical spectra favours pure p-state bonding (Figures 4(b) and 5(b)),[13] and this has certainly become the accepted basis for models of defect states associated with *chalcogen* elements in amorphous chalcogenide semiconductors, although sp^3-hybridization is invoked to explain corresponding defect states associated with pnictide atoms (Section 2.3). As Chen[10] points out however there is also evidence that light of band-gap energy can cause bond breaking in chalcogenide glasses, resulting for example in photo-dissociation and photo-induced crystallization. This seems more consistent with the mixing of bonding and non-bonding orbitals, from both the chalcogen and pnictide element, to form the valence band.

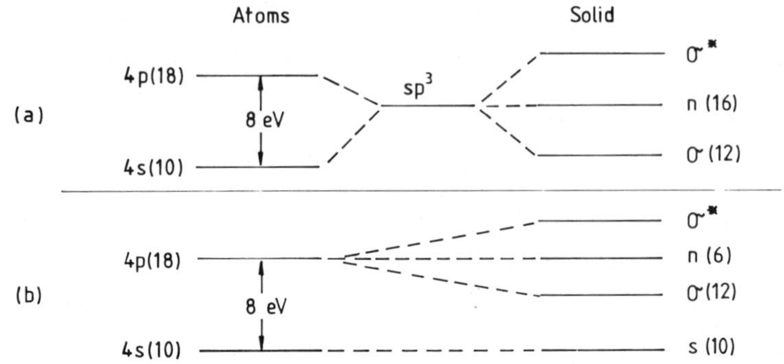

Fig. 5. Molecular orbital schemes for electronic configurations in As_2Se_3. (a) Assuming sp^3 hybridization of atomic states. (b) Without hybridization. (see Ref. 13.)

C. DEFECT STATES IN THE BAND GAP OF CHALCOGENIDE ELEMENTS AND COMPOUNDS

Chalcogenide glasses are diamagnetic and show no e.s.r. response, i.e. there appear to be no dangling bonds in the gap. On the other hand the Fermi-level appears to be pinned, suggesting a finite density of gap states. It was largely to resolve this contradiction that models of defect states were first proposed.

The earliest model was suggested by Mott, Davis, and Street (the so-called MDS model)[14,15,16] and it has proved successful in explaining a wide variety of phenomena in chalcogenides, such as the constant activation energy of the DC conductivity (i.e. the pinning of the Fermi-level), magnetism, luminescence, drift mobility and a.c. conductivity. In this model, bonding defects can have three charged states denoted D^+, D^- and D^o, associated with different local atomic configurations. The neutral dangling bond (D^o), say a singly-coordinated chain-end Se atom, would normally possess spin but it is unstable. Consider two such chain-end defects as in Figure 6. Provided the temperature is high enough so that some atomic movement is possible, one of the end atoms moves closer to and cross-links with the LP p-electrons of a Se atom in a neighbouring chain. The three coordinated Se atom now gives up an electron to the remaining dangling bond. The former becomes D^+, the latter D^-, and both are now diamagnetic. The reaction

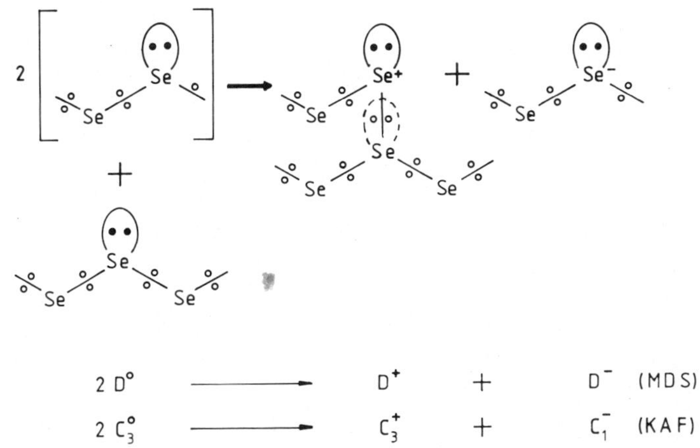

Fig. 6. An illustration of the formation of characteristic chalcogenide electronic defect states in elemental amorphous Se, corresponding to D^o, D^+ and D^- in the Mott, Davis and Street (MDS) scheme (Refs. 14-16) or C_3^o, C_3^+ and C_1^- in the Kastner, Adler and Fritzsche (KAF) model (Refs. 18-22).

$$2D^o \rightarrow D^+ + D^- \tag{17}$$

is exothermic because of a negative correlation (or Hubbard) energy, the potential energy decrease resulting from spin-sharing more than compensating the Coulomb-repulsive energy increase. The decrease in energy can also be regarded as the result of converting a non-bonding lone pair of electrons into a lower lying σ bonding state by dative bonding. The D^+ and D^- centres lie in the gap and determine the position of the Fermi-level, rather like donors and acceptors in a compensated semiconductor.

In the model of Kastner, Adler and Fritzsche (KAF) the chalcogen atom is denoted by C and its coordination by a subscript.[17–22] The dangling bond C_1^o is considered to cross-link with the nearest neighbour on an adjoining chain, so that its bonds are satisfied. It therefore becomes a normal C_2^o, and the neighbour must use one of its LP electrons to become the three-fold coordinated neutral C_3^o.

If atomic movement is possible while the glass is formed, this rearrangement is spontaneous. The reason is that C_1^o and its neighbour C_2^o are involved in only three bonds between them, but C_2^o and C_3^o form five bonds, a gain of two bond energies. In contrast with the MDS model, therefore, the common neutral defect in the glass is C_3^o. The unsatisfied fourth p-electron of C_3^o renders it paramagnetic however and still relatively unstable. For the same reasons as in the MDS model the exothermic reaction

$$2C_3^o \rightarrow C_3^+ + C_1^- \tag{18}$$

then leads to a "valence alternation pair" (VAP). Its members are identical to the D^+ and D^- centres of Eq. (17) and have the same properties.

Group V elements (pnictides) can also undergo valence alternation. The situation is more complicated than that of the chalcogens, because non-bonding s electrons become available for over-coordination only after hybridization. The bonding configuration of normally and defectively coordinated chalcogen and pnictide elements are compared in Figure 7. In a pnictide element, the lowest energy neutral defect, corresponding to C_3^o, is considered to be P_4^o formed by hybridizing the s and p states to create three equivalent sp^3 orbitals containing the 5 valence electrons, two of which remain in a non-bonding lone pair. In close analogy with the chalcogens, the lowest energy defects are charged VAPs, not neutral P_4^o. The negative

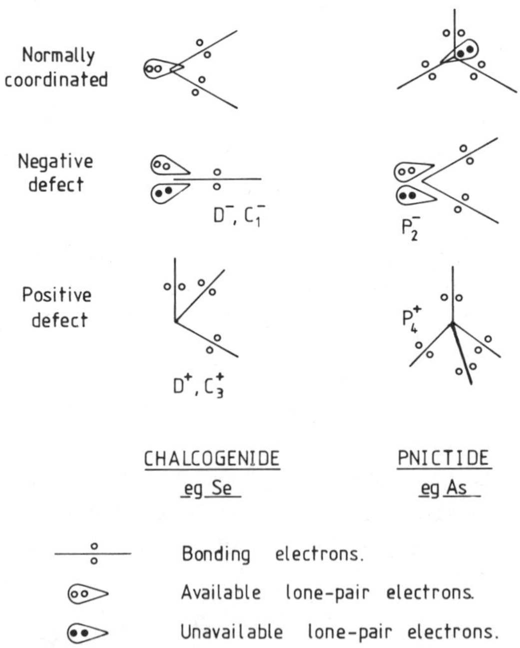

Fig. 7. Schematic representations of the normally coordinated atoms, negative defects and positive defects in chalcogenide and pnictide elements (Ref. 18).

correlation energy gained by transferring charge is larger in pnictides than in chalcogens, so that although the initial creation energy for P_4^o is greater than for C_3^o, the net energy gained in forming VAPs is comparable for the two group of elements. As a first step, P_4^+ and P_4^- defect states are created by transferring an electron from one P_4^o to another and this involves a positive correlation energy, but P_4^- is unstable and the following exothermic reaction takes place:

$$P_3^o + P_4^- \rightarrow P_2^- + P_3^o \tag{19}$$

This involves breaking a bond to one of the four neighbouring P_3^os, placing the two anti-bonding electrons into lone-pair (p-like) orbitals, and dehybridizing the P_4^o to make it a normal P_3^o. The complete exothermic reaction becomes

$$2P_4^o \rightarrow P_4^+ + P_2^- \tag{20}$$

the VAP P_4^+, P_2^- being the equivalent of D^+ and D^- in this case.

When both chalcogen and pnictide elements are combined in a glass such as As_2Se_3, all the stable defects, viz. P_4^+, C_3^+, P_2^-, C_1^-, will be present though in different concentrations. The following reactions will be simultaneously in equilibrium at a high temperature ($>T_g$ the glass transition temperature), assuming only chalcogen and pnictide bonds are present:

$$2C_2^o \rightarrow C_3^+ + C_1^-$$

$$2P_3^o \rightarrow P_4^+ + P_2^-$$

$$P_3^o + C_2^o \rightarrow P_4^+ + C_1^-$$

$$C_2^o + P_3^o \rightarrow C_3^+ + P_2^-$$

In alloys with non-stoichiometric composition the situation is even more complicated because three types of bond may exist (P–C and P–P or C–C) each with a different energy. To determine which reaction dominates, and which defects have the highest density, requires a knowledge of the relative magnitudes of the individual reaction energies. Kastner and Fritzsche[18] believe that C_1^- is always the dominant negative defect, but that either C_3^+ or P_4^+ could be the dominant positive partner depending on a number of considerations. They also argue that under preparation conditions where the defects are allowed to come into thermal equilibrium, P_4^+ and C_1^- will be the predominant VAPs although C_3^+ and P_2^- centres may also be present in lesser concentrations. Street and Lucovsky[23] have claimed that the predominance of P_4^+ and C_1^- is not proved.

Mott and Street have suggested the possible clustering of D^+, D^- centres due to their Coulomb attraction.[16] Such an overlapping D^+D^- pair has been called an "intimate" valence alternation pair (IVAP) and an IVAP is a neutral (actually a dipole) centre. IVAPs can annihilate each other even at temperatures well below T_g. Since separated VAPs represent charged centres, transport properties are more sensitive to VAPs than IVAPs. Even in a glass in which the ratio of VAP to IVAP density is small, the density of VAPs may be sufficiently high to pin the Fermi-level. Unlike VAPs, IVAPs are not likely to act as traps and therefore are not expected to play a large part in trap controlled transport or conductivity. Approximate calculations by Adler and Yoffa[24] suggest that the density of VAPs is of the order of 10^{18} cm^{-3} and of IVAPs, 10^{19} cm^{-3}.

Joannopoulos has calculated the energy levels of a variety of defect configurations in As_2Se_3 with the results shown schematically in Figure 8.[25]

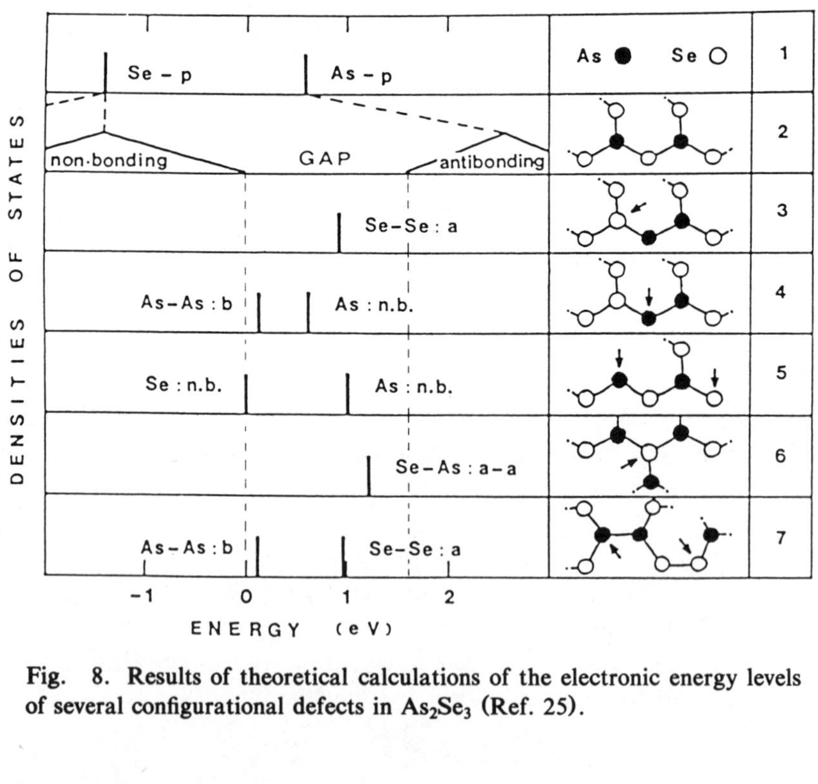

Fig. 8. Results of theoretical calculations of the electronic energy levels of several configurational defects in As_2Se_3 (Ref. 25).

The arrows in rows 3 to 7 indicate the atoms whose local density-of-states is shown in the left-hand part of the diagram. Rows 1 and 2 show the transition from atomic states to bonding states and bands in As_2Se_3. Rows 3 and 4 relate to a defect caused by an exchange of neighbouring As and Se atoms. The states in row 5 result from an As dangling bond (a p-orbital) which interacts strongly with a Se non-bonding orbital. Row 6 describes a three-fold Se defect which is very similar to the three-fold Se defect in elemental Se (Figure 6). In row 7 the defect configurations arise from like-atom bonds where the normal bonding coordination of each atom is reatined. Note that even in the last case the defect states occur at energies within the band gap.

D. THE INFLUENCE OF IMPURITIES

Mott has suggested that certain impurities either destroy bond centres of one sign, or form charged centres (D-centres) which are compensated by defect centres of the opposite sign.[26] In the case of metallic additives such as Mn, there will be very few defects of the same sign as the impurity atoms, by the law of mass action ($[D^+][D^-] = \text{const}$). The Fermi-level will become unpinned and the conductivity activation energy should decrease to 2/3 of its original value.

Kastner[27] and Fritzsche and Kastner[28] also consider the addition of foreign elements to the melt or annealing the adulterated chacogenide glass at T_g. Most of the foreign atoms would seek their lowest energy bonding configuration and remain neutral but some may be incorporated in the host material in a charged state. Thus both neutral and ionized impurities may be present simultaneously. Ionized impurities affect the density of VAP centres. In general, additives with electronegativity near that of the host atoms are less likely to ionize, whereas those which form ionic bonds, i.e. additives from the extreme ends of the periodic table (halogens, alkali metals), are expected to influence the VAP centre concentration strongly.

Kastner and Fritzsche[28] consider two separate cases which arise from different preparation conditions, viz. (a) when the charged impurity is allowed to equilibrate with VAPs in the melt or during annealing at T_g, and (b) when the charged impurity is not allowed to come to equilibrium with the VAPs, as for example in co-evaporation from separate sources onto a substrate below T_g.

In case (a) the charged impurity alters the total VAP density $\left(\left[C_3^+\right] + \left[C_1^-\right]\right)$ as a result of charge compensation. It is predicted that the conductivity activation energy remains essentially unchanged, although the magnitude of the conductivity may increase. For example, as the concentration of a positively charged impurity is increased, the concentration of negative VAPs rises, reducing the influence of impurities on conductivity in two ways. Firstly, they tend to prevent the occurrence of charged impurities. Therefore, the VAPs remain the source of the charge carriers. Unlike Mott,[26] Kastner and Fritzsche[28] do not predict a decrease of conductivity activation energy, except possibly at high temperatures and under exceptionally strong doping. Instead, under normal conditions and in equilibrium, the conductivity activation energy should remain constant.

In case (b), the charged impurities do not interact chemically with VAPs and therefore the VAP concentration remains unaltered. As long as the concentration of charged impurities is less than that of VAPs, the situation is not much different from case (a) but when it exceeds the VAP concentration, the conductivity activation energy decreases, and a much greater increase in conductivity is expected than in case (a). The Fermi-level becomes unpinned and determined by the impurities, moving towards one of the band edges. At sufficiently high doping level the chalcogenide behaves like an ordinary partially compensated and nearly degenerate semiconductor, with the charged impurities acting either as shallow acceptors or donors.

The traps which control the hole mobility in amorphous chalcogenides are thought to be the negatively charged centres C_1^-. In the case (a) of the KAF model, the addition of positively charged impurities, say A^+, would result in an increase in the C_1^- by the same amount, so that the mobility should vary as $(A^+)^{-1}$. The mobility activation energy is expected to remain constant however, except at very high concentrations when broadening of the impurity level may occur and provide a new transport level. Kastner and Fritzsche do not explicitly consider the effect on the hole mobility of A^+ impurities not equilibriated with the VAPs. Under these conditions, one would expect no change in C_1^- concentrations, nor therefore in mobility or its activation energy.

III. D.C. CONDUCTIVITY, THERMOPOWER AND HALL EFFECT

A.

Over a wide temperature range, the d.c. conductivity of most chalcogenide glasses obeys an equation of the form,

$$\sigma = C \exp(-E_\sigma/kT) \tag{21}$$

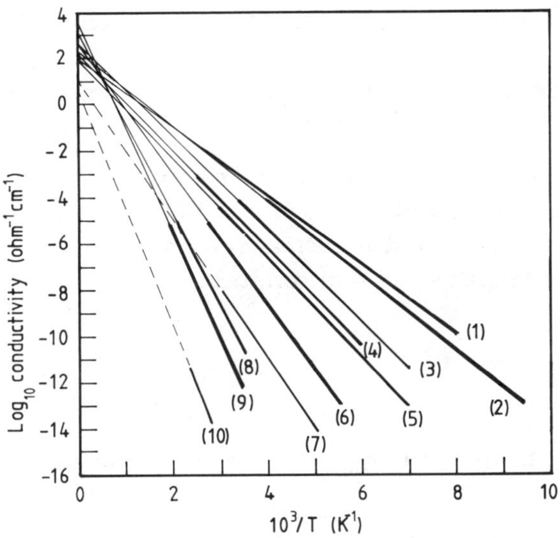

Fig. 9. Typical plots of the logarithm of conductivity vs. reciprocal temperature for a number of chalcogenide glasses, illustrating the characteristic straight line behaviour over a substantial range of temperature. (1) GeTe; (2) $As_2Te_3 \cdot Tl_2Se$; (3) As_2Te_3; (4) $4As_2Te_3 \cdot As_2Se_3$; (5) $Ge_{15}Te_{81}S_2Sb_3$; (6) $As_2Se_3 \cdot Tl_2Se$; (7) $As_{30}Te_{48}Si_{12}Ge_{10}$ (the so-called STAG glass); (8) $3As_2Se_3 \cdot 2Sb_2Se_3$; (9) As_2Se_3; (10) As_2S_3.

in which E_σ denotes the activation energy for conduction.[29,30,31] Several examples of typical results are shown in Figure 9, plotted in the usual log σ vs. (1/T) form and extrapolated to (1/T) = 0 (i.e. to give the pre-exponential constant C). Depending on composition, the activation energy E_σ is in the range of a few tenths of an eV to 1 eV or more, and almost invariably values of $2E_\sigma$ are close to the photon energy E_{opt} corresponding to the onset of strong optical absorption.[29–32*] With few exceptions C is 10^2 ohm^{-1} cm^{-1} or greater, and for many typical chalcogenide glasses such as Se and As_2Se_3 it is in the range $10^3 - 10^4$ ohm^{-1} cm^{-1}. Notable exceptions are As_2S_3 and the so-called STAG glass.

Also with few exceptions (some examples will be mentioned later), reasonably linear plots are normally obtained over the whole experimental temperature range. In particular variable-range hopping conduction, behaving even approximately according to Eq. (9), is not generally observed.

* Optical phenomena are considered in the article by P. C. Taylor, in this volume (ref. 32).

Nor, usually, is there any evidence of a lower activation energy at low temperatures although it must be recognized that the relatively low conductivities and large activation energies of the majority of chalcogenide glasses makes it difficult to extend measurements to very low temperatures.

The thermoelectric power has been measured for many chalcogenide glasses and it is always found to be *positive*.[29,30,31] It has a magnitude typical of semiconductors (mV K^{-1}), and it decreases with temperature according to the simple equation established for semiconductors, i.e.

$$S = -\frac{k}{e}\left(\frac{E_S}{kT} + A\right) \tag{22}$$

where E_S is the activation energy for thermopower and A is a constant (often ~1).

It was noted in Section I.C that values of the pre-exponential constant in the range $10^3 - 10^4$ ohm^{-1} cm^{-1} are consistent with conduction in extended states at the mobility edge. Thus, C and E_σ in Eq. (21) are to be equated with C_o and $(E_F - E_V)$ in Eq. (3); the evidence is therefore that in most chalcogenide glasses, the main transport mechanism involves holes at or close to the mobility edge and that the Fermi-level is close to the centre of the mobility gap. The smaller value of C for As_2S_3 and the STAG glass (Figure 9) presumably indicates conduction by hopping in localized states well away from E_V (or E_C).

A critical question is whether the activation energy for conduction, E_σ, is the same as that for thermopower, E_S. For As_2Se_3 the answer is inconclusive. The results of thermopower measurements on As_2Se_3 by Hurst and Davis,[33] Seager and Quinn,[34] and Chiu[35] are shown in Figure 10, and although there is some difference in the magnitude of S, Hurst and Davis and Chiu agree that $E_S = 0.90$ eV, which is essentially the same as E_σ, while Seager and Quinn find a substantially lower value of $E_S = 0.60$ eV. On the other hand there is well documented evidence for differences $(E_\sigma - E_S)$ of about 0.15 eV for glassy "alloys" in the $As_2Te_3Si_x$ and $As_2Te_{(2-x)}Se_x$ systems.[31,36,37]

The low conductivities and mobilities of chalcogenide glasses make Hall effect measurements extremely difficult but there are nevertheless several reports of the Hall-effect mobility μ_H. The Hall coefficient of these p-type materials is normally *negative* and this anomaly can be understood either in terms of the random-phase diffusive type transport of carriers in a 3-site motion through extended states at a mobility edge, or in terms of polaron hopping. Despite general agreement on the sign anomaly however, quantitatively the experimental situation for As_2Se_3 is again not clear.

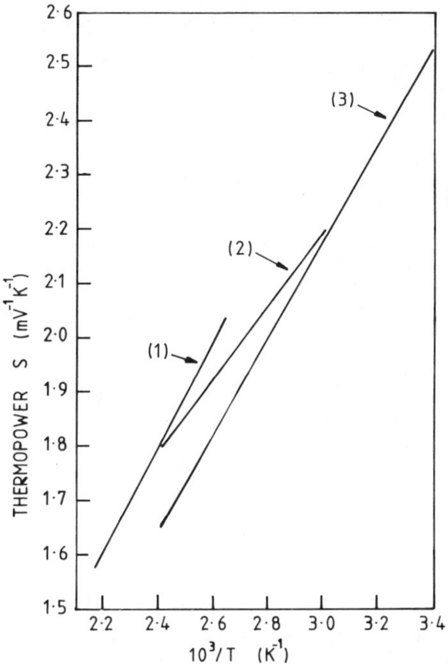

Fig. 10. Thermopower (S) vs. reciprocal temperature for vitreous As_2Se_3. Data from - (1) Hurst and Davis (Ref. 33), (2) Seager and Quinn (Ref. 34) and (3) Chui (Ref. 35).

According to Mytilineou and Roilos,[36] and Nagels et al,[31,37] μ_H in vitreous As_2Se_3 is small ($\sim 10^{-1}$ cm^2 v^{-1} S^{-1}) and unactivated, while Klaffke and Wood[38] report a small activation energy. For the $As_2Te_3Si_x$ and $As_2Te_{(3-x)}Se_x$ systems however there is general agreement that the Hall mobility is activated, i.e.

$$\mu_H = \mu_{Ho} \exp(-E_H/kT)$$

with E_H in the range of 0.03 - 0.05 eV. It is notable that these are the same materials for which consistent values of $(E_V - E_S) \sim 0.15$ eV are also reported.

At the present time therefore the weight of the evidence seems to be that in the case of As_2Se_3 Eq. (3) is applicable, that conduction occurs by hole transport in extended states close to the valence band mobility edge E_V, and that

$$E_\sigma = (E_F - E_V) = E_S = 2E_{opt} \,,$$

implying that the Fermi-level is close to the centre of the mobility gap.

However, in at least a number of chalcogenide glasses, of which the alloys $As_2Te_3Si_x$ and $As_2Te_{(2-x)}Si_x$ are typical, there is clear evidence that:

[i] The thermopower has a smaller temperature dependence than the d.c. conductivity with $(E_\sigma - E_S) \sim 0.15$ eV.

[ii] The Hall mobility is activated with an activation energy in the range 0.03 - 0.05 eV.

Two possible interpretations of these observations are currently considered to be likely explanations.

Nagels et al have proposed a two-path conduction process in which transport may occur almost simultaneously by holes in extended states *just* below E_V and by holes hopping in localized states *just* above E_V.[31,37] Using the subscript 1 to indicate extended state conduction and 2 for hopping conduction then

$$\sigma_1 = C_{01} \exp [-(E_F - E_V)/kT] \tag{23}$$

and

$$\sigma_2 = C_{02} \exp [-(E_F - E_B + W)/kT] \tag{24}$$

with the total conductivity σ,

$$\sigma = \sigma_1 + \sigma_2 \,. \tag{25}$$

Provided the rate constants of σ_1 and σ_2 do not differ greatly, a log σ vs. (1/T) plot will change its slope only gradually, and over at least a limited range of temperature it will not differ appreciably from a straight line. The thermopower is the weighted sum,

$$S = \frac{(S_1\sigma_1 + S_2\sigma_2)}{\sigma} \tag{26}$$

with

$$S_1 = \frac{k}{e}\left[\frac{(E_F - E_V)}{kT} + A_1\right] \tag{27}$$

and

$$S_2 = \frac{k}{e}\left[\frac{(E_F - E_B)}{kT} + A_2\right] \tag{28}$$

The expected form of log σ vs. (1/T) and S vs. (1/T) is shown schematically in Figure 11. Because the two thermopowers are similar for (1/T) = 0, the form of S in the transition region depends sensitively on the sharpness of the transition in σ and the slope of S vs. (1/T) has little significance in the transition region. The Hall mobility is also given by the weighted sum,

$$\mu_H = \frac{\mu_1\sigma_1 + \mu_2\sigma_2}{\sigma} . \tag{29}$$

Nagels et al assume that $\mu_2 = 0$, so that

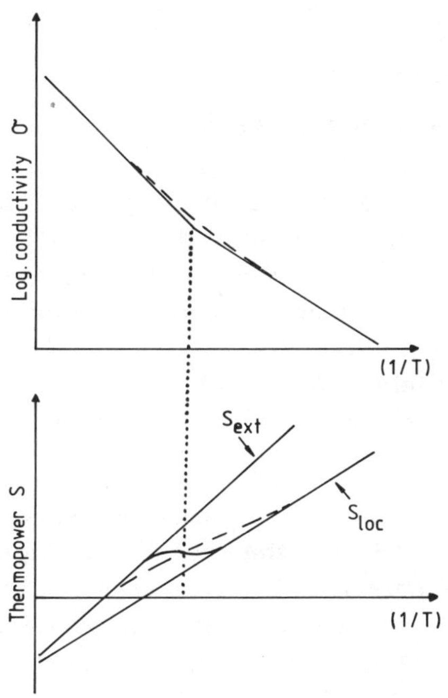

Fig. 11. Schematic plots of the logarithm of conductivity and thermopower vs. reciprocal temperature in the range of two-path conduction where transport changes from conduction in extended states above the mobility edge to hopping in localized states below the edge. The line S_{ext} is the thermopower for transport in extended states and S_{loc} for transport in localized states. (See Refs. 31 and 37.)

$$\mu_H = \frac{\mu_1 \sigma_1}{\sigma} . \tag{30}$$

When conduction occurs *mainly* in extended states μ_H is equal to μ_1, but when hopping also contributes the Hall voltage is generated by the few carriers remaining at the mobility edge E_V and from Eqs. (23), (24) and (30)

$$\mu_H = \mu_1 \left[1 + \frac{C_{02}}{C_{01}} \exp\left(E_B - E_V - W\right)/kT \right]^{-1} \tag{31}$$

The alternative interpretation, first proposed by Emin and coworkers,[5,39] and favoured by Seager and Quinn,[34] is that transport is by small polarons (see Section I.C). In that case the difference $(E_\sigma - E_S)$ gives the hopping energy W directly, and the activation energy for the Hall mobility is (W/3) (see Eq. (16)). From the available experimental data it seems as though the Hall mobility activation energy is approximately $[(E_\sigma - E_S)/3]$ and Emin[7] points out that C_o values of $\sim 10^3$ ohm^{-1} cm^{-1} are also compatible with small-polaron hopping transport.

Thus trap-modulated hopping of holes at or near the mobility edge E_V, OR polaron hopping, are equally valid interpretations of transport measurements on a number of the more complex chalcogenide glasses and a definitive experiment is lacking at the present time. Mott[40] suggests that indirect evidence *against* the polaron model is provided by the ON-state of threshold switching devices based on complex chalcogenide glasses. The high ON-state conductance implies mobilities of ~ 10 cm^2 V^{-1} S^{-1} and although it is not known which species (electrons or holes) is the major current carrier, it is clear that either electrons or holes, at least, are moving in quasi-free Bloch states.

The conclusions reached from measurements of d.c. conductivity, thermopower, Hall mobility and the optical absorption edge are summarized diagrammatically in Figure 12.* These experiments reveal nothing, directly, about states well within the mobility gap, such as the defect centres discussed in Section II.C, and the influence of gap states is considered in the next section.

* If carrier transport involves polaronic mechanisms however a band model like that of Figure 3(a) would be appropriate.

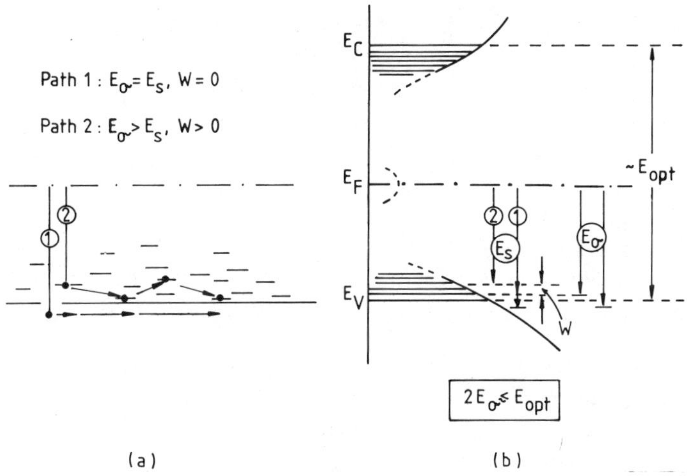

Fig. 12. A diagrammatic summary of the information derived on transport (a) and band structure (b) from measurements of d.c. conductivity, thermopower, Hall mobility and optical absorption (see text, Section III). E_σ is the activation energy for d.c. conduction (Eq. (21)), E_S is the activation energy for thermopower (Eq. (22)), W is a hopping energy (Eq. (7)), and the optical gap E_{opt} is obtained from the fundamental optical absorption edge.

IV. THE INFLUENCE OF STATES IN THE GAP

A. INTRODUCTION

Localized states in the mobility gap influence the electronic transport properties of chalcogenide glasses in a variety of ways, e.g.:[41]

1. Pinning of the Fermi level E_F at or close to the middle of the mobility gap. Fritzsche[42] has pointed out that the characteristic straight line plots of log σ vs. (1/T) over a wide range of temperature (see Figure 9) is evidence that E_F is fixed (pinned) by a large density of localized states, rather than of intrinsic conductions. Marshall and Owen[43,44] have proposed specific models for As_2Se_3, As_2Te_3 and the STAG glass (as in Figure 9), in which E_F is pinned midway between donor- and acceptor-like states.

2. The creation of a high concentration of space-charge density at interfaces and metal contacts. As a result the field effect conductance in

chalcogenide glasses is very small, and again this can be interpreted in terms of E_F positioned midway between large densities of localized states at relatively discrete energies.[44,45] A concomitant effect associated with a large density of gap states is a very short screening length (e.g. ~100 Å) and this explains the apparent ohmic behaviour of most metal-chalcogenide contacts.[46]

3. Localized gap states provide several possibilities for enhanced carrier hopping between neighbouring sites, i.e. for dipolar activity which contributes to the a.c. conductivity and dielectric behaviour.[47,48,49†]

4. The carrier drift mobility in chalcogenide glasses is determined by interactions with localized states and in some circumstances mobility measurements can give information on the energies and distributions of gap states which can be correlated with the models of defect states described in Section 2.3. Mobility data for amorphous Se and As_2Se_3 are discussed in Section 4.3 but first a brief account of the experimental technique is given.

B. TIME-OF-FLIGHT MEASUREMENTS OF DRIFT MOBILITY[50]

The chalcogenide glass sample is prepared in the form of a thin film (e.g. 0.1 - 10 μm thick) with metallic contacts on the opposite surfaces. The top contact is usually semitransparent and it is illuminated with an electron beam pulse or a pulse of strongly absorbed light to generate a thin sheet of excess carriers (electrons and holes) close to the contact. A bias is applied and, depending on its polarity, a unipolar current I(t) of electrons OR holes flows through the sample to be collected at the other electrode. In an ideal case the injected carriers drift through the sample as a coherent unbroadened sheet of charge which reaches the back electrode after a definite time transit time t_τ and a drift mobility μ can be derived from

$$\mu = \frac{\ell}{t_\tau E_{app}} \tag{32}$$

where ℓ is the sample thickness and E_{app} is the applied field. The I(t) vs. time plot for this ideal situation is illustrated in Figure 13(a). Normally the drift velocities of the individual carriers have a Gaussian distribution and the charge sheet, which propagates with a constant *mean*

† The a.c. conductivity of amorphous semiconductors is considered in detail in the article by E. A. Davis, in this volume (Ref. 49).

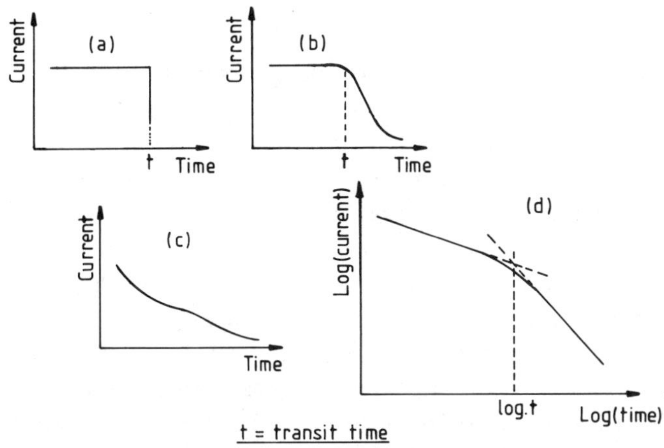

Fig. 13. Illustrations of typical transit current pulses observed in carrier mobility experiments by time-of-fiight techniques. (a) An ideal pulse obtained if the sheet of carriers drift without broadening. (b) Broadening due to a Gaussian distribution of mobilities. (d) Strong dispersive broadening. Note: In (a), (b) and (c) current (I) and time (t) are plotted linearly. (d) The current pulse (c) plotted in the form log (I) vs. log (t).

drift velocity, is broadened slightly. As a result the I(t) vs. t curves is rounded-off slightly, as shown in Figure 13(b), but a well-defined transit time can still be distinguished.

In some experiments on amorphous materials a *linear* I(t) vs. t plot has the rather featureless form shown in Figure 13(c). There is, at the most, a slight "knee" in the curve and there is not a well-defined transit time. Very often, however, a *logarithmic* plot [log I(t) vs. log (t)] yields two straight lines of different slopes - see Figure 13(d) - and the intersection of these lines is used to define a transit time. The mobility derived from this transit time often depends on the field strength and the sample thickness.

The fundamental cause of the logarithmic I(t) vs. t curves is that as the sheet of charge moves through the sample the mean velocity of the carriers decreases with time and the charge sheet becomes dispersed rather than drifting as a coherent packet. Various models have been proposed for this "dispersive" transport and a common feature is the presence of some stochastic variable which is responsible for the dispersion in the transit times. Possible mechanisms are as follows:

(1) HOPPING BETWEEN LOCALIZED STATES[51-54]

The temperature-independent hopping distance R, in Eq. (2), is assumed to vary stochastically. As R appears as an exponent in the tunnelling factor [exp $(-2\alpha R)$] a small change in R may cause an amplified change of the hopping probability from site to site, i.e. of the release time after a carrier is trapped. An injected carrier moves towards the collecting electrode with a constant hopping energy W (Eq. (21)) but occasionally it will be trapped at a site in which it remains for a long time. Therefore, only a small fraction of the injected carriers, i.e. the fastest ones, will reach the collecting electrode without passing a "difficult" hopping site, while most of the carriers are delayed by a different number of such hops. Scher and Montroll[52] showed that the probability for a carrier to jump to its next site at time t after having arrived at $t = 0$ is a slowly decaying function which can be approximated by

$$f(t) \propto t^{-(1-\alpha)} \tag{33}$$

where $0 < \alpha < 1$, and hence that the transient current decays algebraically as

$$I(t) \propto \begin{cases} t^{-(1-\alpha)} & \text{for } t < t_\tau \\ t^{-(1+\alpha)} & \text{for } t > t_\tau \end{cases} \tag{34}$$

On logarithmic scales, therefore, the current trace should appear as two straight lines intersecting at $t \simeq t_\tau$ with initial slope $-(1-\alpha)$ and final slope $-(1+\alpha)$, as in Figure 13 (d). A consequence of the algebraic distribution in Eq. (33) is that the mean displacement of the carrier sheet depends on time as t^α and hence that

$$t_\tau \propto \ell^{(1/\alpha)} \tag{35}$$

The drift of the carriers is spatially biased by the applied field and assuming that this asymmetry increases linearly with field strength, Scher and Montroll also derive

$$t_\tau \propto E_{app}^{-(1/\alpha)} \tag{36}$$

Thus, t_τ scales with sample thickness and applied field as

$$t_\tau \propto \left(\frac{\ell}{E}\right)^{1/\alpha} \tag{37}$$

and to derive a mobility from Eq. (32) would lead to a field and thickness dependent mobility such that

$$\mu \propto \left(\frac{E}{\ell}\right)^{(1-\alpha)/\alpha} \tag{38}$$

It is to be noted that as R is independent of temperature, hopping between localized states implies a temperature independent dispersion. If the hopping energy W (Eq. (2)) is included as a stochastic variable however, then the dispersion will be temperature dependent.

(2) TRAP-LIMITED BAND TRANSPORT[55-60]

In this case, carriers move in extended states (electrons just above E_C or holes just below E_V) but are occasionally trapped in localized states and a distribution of the energies of the trapping levels accounts for the dispersive nature of the transient current pulses. The dispersion is temperature dependent, i.e. the shape of the I(t) vs. t curve changes with temperature, otherwise the features described above apply. In particular, it has been demonstrated several times that the integration of individual trap release probabilities of the form

$$g(t) \propto \exp(-t/\tau)$$

over a distribution of traps decreasing exponentially with depth leads to results closely matching those of the Scher and Montroll theory.[58–60]

(3) TRAP-CONTROLLED HOPPING[54,61]

Charge transport occurs by hopping through localized states, as in (1), but with occasional trapping in deeper localized states. Generally, the deeper trapping states will be present in much lower density than the shallower transport states and hence one may expect temperature activated transport processes but a temperature independent dispersion.

C. AMORPHOUS Se AND As_2Se_3: A "DEFECT-STATE" APPROACH

Of the many chalcogenide glasses, transit-time drift mobility measurements have been applied most thoroughly to amorphous Se and As_2Se_3. Comprehensive reviews of the experimental data and their interpretation have been published (see Enck and Pfister,[62] and Owen and Spear[63]) and it is not necessary to repeat the details here but a brief account will be presented of a particular view which interprets the influence of gap states on transport phenomena in terms of specific electronic defect states of the kind discussed in Section II.C.

Amorphous Se is unusual amongst the chalcogenide glasses in that although holes are the more mobile carrier (i.e. Se is p-type, like most of the chalcogenides), the electron drift mobility is also easily measurable over a wide range of temperature. At least nine different groups have reported data on the electron or hole mobility and in general the agreement is surprisingly good.[62,63] Above about 200 K a well-defined transit time is observed (i.e. a transient current pulse as shown schematically in Figure 13(b)), but below 200 K the pulse shape becomes progressively more dispersive (as in Figure 13(c) and 13(d)). At about 300 K or a little higher, the hole mobility in amorphous Se tends to a constant (temperature independent) value, μ_o, in the region of 0.3 to 0.4 $cm^2\ V^{-1}\ S^{-1}$ while for electrons μ_o is about 0.05 $cm^2\ V^{-1}\ S^{-1}$. Below about 270 K the mobility for both electrons and holes is activated, i.e.

$$\mu = \mu_o \exp(-E/kT) \tag{39}$$

with E in the region of 0.28 - 0.30 eV for holes and approximately 0.33 eV for electrons.[62,63] It is important to note that plots of log μ vs. (1/T), or log (inverse transit time) vs. (1/T) are continuous, with the same values of activation energy, from relatively high temperatures (>270 K) where the transit pulses are Gaussian, to low temperatures (<270 K) where dispersive pulses are observed.

The μ_o value for holes is consistent with diffusive transport in extended states (see Section I.B), and while the value for electrons is lower it is still above the lower limit for diffusive mobility estimated by Cohen.[2] Thus, the evidence is that electron and hole motion in amorphous Se occurs by a process of trap-limited transport in extended states just beyond or very close to their respective mobility edges. The traps responsible for limiting the mobility could occur over relatively discrete range of energies (as in Figure 1(b)) or they could be part of a broad distribution of traps, decreasing in density with increasing depth. Marshall and Owen,[64] and Owen and Spear[63] have argued for the former possibility and this implies that at low temperatures (<270 K) a distribution of trap depth over a small range of energies is responsible for the dispersion of the transient current pulses. Combining the drift mobility experiments with d.c. conductivity, optical absorption and space-charge-limited current data, Owen and Spear have proposed the electronic density-of-states distribution shown in Figure 14(a). The mobility gap ($E_C - E_V$) = 2.1 eV is equated with the fundamental optical

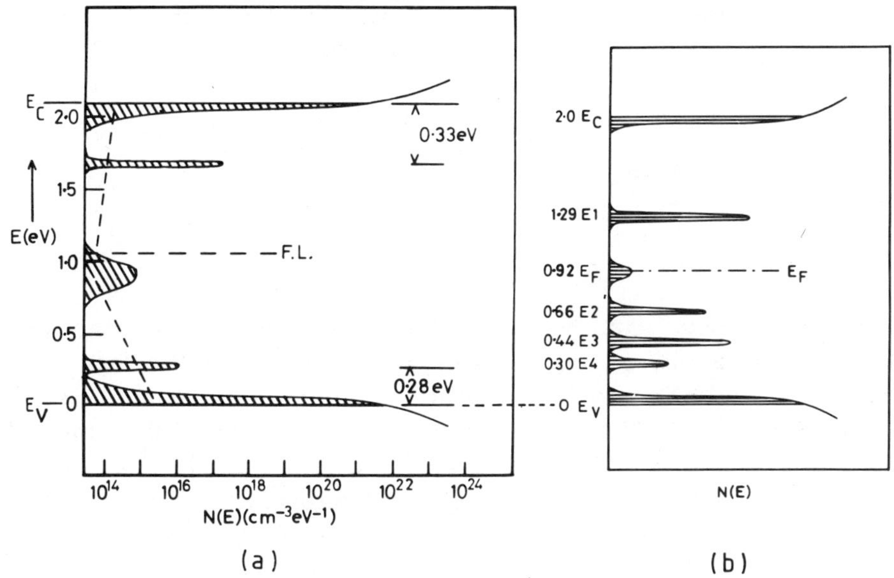

Fig. 14. Electronic density-of-states diagrams constructed from a variety of transport and optical measurements interpreted in the spirit of defect states. (a) Amorphous Se. (b) Amorphous As_2Se_3. (See Ref. 63.)

absorption edge E_{opt}, and the position of the Fermi level is determined by the activation energy for d.c. conductivity.

It is interesting to note that by the following argument Mott has tentatively correlated the electron and hole trapping states in Figure 14(a) with the defect states discussed in Section II.C (using the Kastner, Adler and Fritzsche notation): If valence alternation pairs (VAPs) act as traps, the capture of an electron or a hole produces the *same* centre, C_o. The trap depth for electrons, ϵ_1, is determined by the reaction

$$C_3^o \rightarrow C_3^+ + \text{electron}$$

and the trap depth for holes, ϵ_2, by

$$C_3^o \rightarrow C_1^- + \text{hole}$$

Thus an electron falling into a trap yields an energy ϵ_1 and a hole yields an energy ϵ_2, with the formation of two C_o centres. To return the system to normal, an energy E is gained through the reaction

$$2C_3^o \rightarrow C_1^- + C_2^+ \tag{40}$$

and

$$E + \epsilon_1 + \epsilon_2 = E_g \tag{41}$$

where E_g is the band gap. Note that the reaction in Eq. (40) is the same as that depicted in Figure 6. Mott equates the energy E with the chain scission energy of Se, less the energy to form the VAP. From NMR data, the Se chain scission energy is 1.4 eV at room temperature, or ~2 eV at 0 K; the atomic fraction of VAPs at 300 K is about 10^{-5} and hence $\exp(-W/2\,kT) \simeq 10^{-5}$, giving W = 0.5 eV for the energy of VAP formation. Thus E is 1.5 eV. Adding the trapping energies ϵ_1 and ϵ_2 (see Figure 14(a)) gives

$$E_{opt} = 1.5 + 0.33 + 0.28 = 2.1 \text{ eV}$$

in good agreement with the experimental value (see also Figure 14(a)).

Drift mobility data for As_2Se_3 are more difficult to interpret, on two counts. First, only hole transits are observable, and electron mobilities are not therefore determinable, and secondly even at the highest temperature at which measurements have been made (~370 K) the hole transit times are two to four orders of magnitude longer than in Se. Nevertheless, hole transport in amorphous As_2Se_3 is clearly an activated process and Marshall et al[55,65] have observed that the activation energy depends on the time scale of the measurement. For transit times <100 μs they report an activation energy of 0.43 eV, while for transit times >200 μs it is 0.63 eV. These trapping energies are in good agreement with the more detailed studies of steady-state and transient photoconductivity which it is possible to make on amorphous As_2Se_3, which also provide evidence for electron traps at about 0.71 eV from the conduction band mobility edge, E_C.[66]

The conclusion from Section III was that amorphous As_2Se_3 is a p-type semiconductor and that hole transport occurs in states at or very close to the valence band mobility edge E_V. Thus, the hole transport phenomena observed in drift mobility and photoconductivity experiments are trap-limited processes. Owen and Spear[63] have collected together the data from measurements of d.c. conductivity, optical absorption, drift mobility, photoconductivity and space-charge-limited current, and interpreted it in terms of the electronic density-of-states model shown in Figure 14(b). It is relevant to note here that transport experiments on amorphous Se with progressive additions of As indicate that some "Se-like" features are retained in the As—Se compositions, while very effective hole traps are also introduced by the addition of As.[63,67] There is, in addition, evidence that the electron traps at 0.33 eV in amorphous Se move farther away from the conduction band mobility edge E_C as As is added in concentrations of up to 8 atomic %.[68]

It is clear from Section II.C that there is a variety of likely electronic defect states in a chalcogenide compound such as As_2Se_3 and, as Joannopoulos's calculations suggest[25] (Figure 8), several of them are likely to be located within the band gap. At the present however it is not possible to attempt even the tentative correlation between the proposed band model and specific defect states as was described above for amorphous Se.

As mentioned at the beginning of this section, the interpretation presented here is a particular view derived from the notion that amorphous semiconductors may have structurally-related electronic defect states located at relatively discrete energies within the mobility gap. Density-of-states diagrams like those shown in Figure 14 were first proposed purely on the basis of a consistent analysis of experimental results on the assumption that such states exist.[43,66] The subsequent development of specific models for structurally-related defect states in amorphous semiconductors (Section 2.3) provided *post-facto* theoretical justification.[14–25] It should be noted that other data, such as that obtained from field-effect measurements, have been interpreted in a similar way.[44,45] Drift mobility experiments on pure and metal-doped amorphous As_2S_3 are also consistent with trap-limited transport involving defects of the same kind.[69–71]

It must be recognized however that other interpretations are possible. For example, Tiedje and Rose[60] have shown that the dispersive current transient pulses which seem to be so characteristic of transit-time mobility measurements on many, but *not all*, amorphous semiconductors can be explained by trap-limited transport in an exponential distribution of traps. This mechanism is obviously not applicable to amorphous Se however as the transient pulses in that material are only Gaussian-broadened at high temperatures (>200K). Moreover, the low temperature region, where dispersive pulses are observed in amorphous Se, seems to be continuous with the higher temperature range, i.e. there is no evidence of a change in the transport mechanism as the form of the current transient pulses change from Gaussian to dispersive.

Taylor and Ngai[72] have proposed that the dispersive hole transport observed in As_2Se_3 and Se (at low temperatures) can be explained by polaronic mechanisms. The balance of the evidence from d.c. conduction, optical absorption and thermopower however is that hole transport in amorphous As_2Se_3 occurs at or very close to the valence band mobility edge (see Section III and Figure 12).

V. CONCLUSIONS

Charge transport in the relatively simple amorphous chalcogenide semiconductors such as Se and As_2Se_3 most likely involves trap-limited processes. Carriers migrate via extended states or via states at or close to the mobility edges, E_C and E_V. The evidence from a variety of experiments is consistent with the view that the traps which limit carrier mobilities are located at relatively well-defined energies within the mobility gap, and that these traps originate from structurally-related electronic defects characteristic of various bonding abnormalities associated with the chemistry of chalcogenide elements and compounds.

The situation in more complex multi-component and/or non-stoichiometric chalcogenide glasses is more problematical. The additional compositional disorder associated with the chemical complexity will certainly affect transport mechanisms, perhaps encouraging multi-path conduction processes, polaron formation and probably also smoothing out the density-of-states distribution.

REFERENCES

1. N. F. Mott and E. A. Davis, "Electronic Processes in Non-Crystalline Materials" 2nd Edition, Chap. 1, pp. 39-52 and Chap. 6, pp 209-215 (Oxford University Press) (1979).
2. M. H. Cohen, J. Non-Cryst. Sol. **4**, 391 (1970).
3. Ref. 1. Chap. 6, pp. 219-222 (1979).
4. D. Emin, Adv. in Phys. **22**, 57 (1973).
5. D. Emin, in "Electrical and Structural Properties of Amorphous Semiconductors" Ed: P.G. Le Comber and J. Mort, pp. 261-328 (Academic Press) (1973).
6. T. Holstein, Ann. Phys. (NY) **8**, 343 (1959).
7. D. Emin, to be published in Comments in Solid-State Physics.
8. L. Friedman and T. Holstein, Ann. Phys. (NY) **21**, 494 (1963).
9. S. Tutihasi and I. Chen, Phys. Rev. **158**, 623 (1967).
10. I. Chen. Phys. Rev. **B8**, 1440 (1973).
11. M. Kastner. Phys. Rev. Lett. **28**, 355 (1972).
12. E. Mooser and W. B. Pearson, in "Progress in Semiconductors" Vol. 5, p. 104. (Heywood and Co., London) (1960).
13. G. Weiser, in "The Physics of Selenium and Tellurium" Ed: E. Gerlach and P. Grosse, p. 230, No. 13 of Springer Series in Solid-State Science (Springer-Verlag) (1979).
14. R. A. Street and N. F. Mott. Phys. Rev. Lett. **35**, 1293 (1975).
15. N. F. Mott, E. A. Davis and R. A. Street, Phil. Mag. **32**, 961 (1975).
16. N. F. Mott and R. A. Street, Phil. Mag. **36**, 33 (1977).
17. M. Kastner, D. Adler and H. Fritzsche, Phys. Rev. Lett. **37**, 1504 (1976).

18. M. Kastner and H. Fritzsche, Phil. Mag. **B37**, 199 (1978).
19. H. Fritzsche, "Proc. 7th Int. Conf. Amorphous and Liquid Semiconductors", p. 3. Ed: W. E. Spear (CICL, Edinburgh, U.K.) (1977).
20. H. Fritzsche, J. Phys. Soc. Japan **49**, Suppl. A, 39 (1980).
21. M. Kastner, J. Non-Cryst. Sol. **31**, 223 (1978).
22. M. Kastner, J. Non-Cryst. Sol **35-36**, 807 (1980).
23. R. A. Street and G. Lucovsky, Sol. State. Commun. **31**, 285 (1979).
24. D. Adler and E. J. Yoffa, Can. J. Chem. **55**, 1920, (1977).
25. J. D. Joannopoulos, J. Non-Cryst. Sol. **35-36**, 781 (1980).
26. N. F. Mott, Phil. Mag. **34**, 1101 (1976).
27. M. Kastner, Phil. Mag. **37**, 127 (1978).
28. H. Fritzsche and M. Kastner, Phil. Mag. **37**, 285 (1978).
29. A. E. Owen, Contemp Phys. **11**, 227 and 257 (1970).
30. See for example: Ref. 1 Chap 9, pp. 452-460.
31. P. Nagels, in "Amorphous Semiconductors," Ed. M. H. Brodsky, pp. 113-159 Vol. 36 of "Topics in Applied Physics" (Springer-Verlag) (1979).
32. P. C. Taylor - see this Volume.
33. C. H. Hurst and E. A. Davis, J. Non-Cryst. Sol. **16**, 343 (1974).
34. C. H. Seager and R. K. Quinn, J. Non-Cryst. Sol. **17**, 386 (1975).
35. D. M. Chui. "Photo- and Thermal Effects in Arsenic Chalcogenides" M.Sc. thesis (University of Edinburgh) (1976).
36. E. Mytilineou and M. Roilos, Phil. Mag. **B37**, 387 (1978).
37. P. Nagels, R. Callaerts and M. Denayer, in "Proc. 11th Int. Conf. Phys. of Semicond." Ed: M. Miasek, p. 549 (Polish Scientific Publishers, Warsaw) (1972).
38. G. R. Klaffke and C. Wood in "Proc. 4th Int. Conf. on Physics of Non-Crystalline Solids" Ed: G. H. Frischat, p. 236 (Trans. Tech. Publ.) (1977).
39. D. Emin, C. H. Seager and R. K. Quinn, Phys. Rev. Lett. **28**, 813 (1972).
40. N. F. Mott, J. Phys. **C13**, 5433 (1980).
41. See for example: Ref. 1, pp. 460-490.
42. H. Fritzsche, pp. 55-125 of Ref. 5.
43. J. M. Marshall and A. E. Owen, Phil. Mag. **24**, 1281 (1971).
44. J. M. Marshall and A. E. Owen, Phil. Mag. **33**, 457 (1976).
45. R. C. Frye and D. Adler, Phys. Rev. Lett. **46**, 1027 (1981).
46. A. Wallace, A. E. Owen and J. M. Robertson, Phil. Mag. **38**, 57 (1978).
47. Ref. 1. Chap. 6, pp. 223-235.
48. A. E. Owen, J. Non-Cryst. Sol. **25**, 370 (1977).
49. E. A. Davis - see this Volume.
50. F. K. Dolezalek, in "Photoconductivity and Related Phenomena" Ed: J. Mort and D. M. Pai Chap. 2, pp. 27-70 (Elsevier) (1976).
51. H. Scher and M. Lax, Phys. Rev. **B7**, 4491 and 4502 (1973).
52. H. Scher and E. W. Montroll, Phys. Rev. **B12**, 2455 (1975)
53. H. Scher, ref. 50, Chap. 3, pp. 71-116 (1976).
54. G. Pfister and H. Scher, Adv. in Phys. **27**, 747 (1978).
55. J. M. Marshall and A. E. Owen, Phil. Mag. **24**, 1281 (1971).
56. J. M. Marshall and A. C. Sharp, J. Non-Cryst. Sol. **35 and 36**, 99 (1980).
57. A. C. Sharp, J. M. Marshall and H. S. Fortuna, J. de Physique, Colloque C4, Suppl. No. **10**, **42**, 159 (1981).
58. M. Silver and L. Cohen. Phys. Rev. **B15**, 3267 (1977).
59. F. Schmidlin, Phys. Rev. **B16**, 2362 (1977).
60. T. Tiedje and A. Rose, Sol. St. Comm. **37**, 49 (1980).
61. G. Pfister and H. Scher, Phys. Rev. **B15**, 2062 (1971).
62. R. C. Enck and G. Pfister, Ref. 50, Chap. 7, pp. 297-302 (1976).
63. A. E. Owen and W. E. Spear, Phys. Chem. Glasses **17**, 174 (1976).
64. J. M. Marshall and A. E. Owen, Phys. Stat. Sol. (a) **12**, 181 (1972).
65. F. D. Fisher, J. M. Marshall and A. E. Owen, Phil. Mag. **33**, 261 (1976).

66. C. Main and A. E. Owen. Ref. 5, pp. 527-545 (1973).
67. J. Schottmiller, M. Tabak, G. Lucovsky and A. Ward, J. Non-Cryst. Sol. **4**, 80 (1970).
68. J. M. Marshall, F. D. Fisher and A. E. Owen, Phys. Stat. Sol. (a) **25**, 419 (1974).
69. M. Burman and J. Hirsch, J. Non-Cryst. Sol. **35-36**, 987 (1980).
70. M. Burman, J. Hirsch and T. Ramdean, J. Phys. C. **14**, 117 (1981).
71. M. Burman "Hole Mobility in Doped and Undoped As_2Se_3 Layers" Ph.D. Thesis (University of London) (1982).
72. P. C. Taylor and K. L. Ngai, Sol. State Commun. **40**, 525 (1981).

DISCUSSION

Electronics of Chalcogenide Glasses

The following points dominated the discussion on transport in chalcogenide glasses.

The interpretation of Hall effect in amorphous semiconductors still appears to be somewhat controversial. It was argued that the three-site random phase hopping model could still be applied when conduction is in extended states, for short mean free path values (Rivier/Davis).

The validity of the small polaron model has been questioned. Professor Owen again pointed out that this model may apply to certain glasses only. For example in amorphous Se, there is strong evidence against the formation of small polarons from the results of mobility measurements. The same conclusion also applies to some switching materials. The situation can however be different in other types of chalcogenide glasses which have different transport properties.

A comment was made to Professor Owen's suggestion that photostructural results favor Chen's sp hybridized bond model for chalcogens. It was argued that, in melt-quenched glasses, the shift of the absorption edge under illumination known as photodarkening effect, at least at low power, is not accompanied by any detectable structure change (this point was again emphasized by P. C. Taylor in his talk), and that it is thought to be due, not to bond breaking, but to changes involving nonbonding (lone pair) electrons. Only in films with "molecular"-type structures, do photostructural effects imply a large network reorganization (polymerization).

M. L. Theye, Chairwoman
J. Orenstein, Discussion Leader

OPTICAL PROPERTIES OF THE CHALCOGENIDE GLASSES

P. C. Taylor

Department of Physics
University of Utah
Salt Lake City, UT 84112

Glasses containing group VI elements, other than oxygen, are essentially covalent and exhibit thermally-activated electrical conductivities which are analogous to those observed in intrinsic crystalline semiconductors. In these so-called chalcogenide glasses the optical properties are dominated by strong electron-lattice interactions. Experimental examples of this coupling include a strongly Stokes-shifted photoluminescence (PL) peak and an optically-induced absorption which is nearly independent of energy and extends well below the band edge. Other absorption mechanisms common to all of the chalcogenide glasses include an Urbach tail to the band edge absorption and an optically-induced transient absorption which extends well into the gap. In addition, some of the more highly conducting chalcogenide glasses exhibit a free carrier absorption below the band edge which is thermally activated. Other PL processes observed in the chalcogenide glasses include several transient PL peaks which appear in all glasses studied to date.

Those optical effects, which appear to be "universal" features of all the chalcogenides, have been variously attributed to (1) a broad energy distribution of electronic states localized by the underlying disorder (the so-called negative U states), (2) defects, such as under- and over-coordinated chalcogen atoms, (3) polarons and localized or self-trapped excitons, and (4) impurities. The evidence for and against these explanations is critically examined.

I. INTRODUCTION

The chalcogenide glasses, which contain among other elements either S, Se or Te, have long been considered the archetypal covalent amorphous solids. Yet even for the simplest and most often studied compositions, such as Se or As_2Se_3, considerable controversy and uncertainty remains concerning the interpretation of the optical properties. For example, there is still some question as to which optical effects are due to "intrinsic" processes and which result from the presence of pervasive and inadvertent impurities. Also the importance of disorder in establishing the presence of certain optical effects is still very much a matter of debate. The present chapter will undoubtedly not resolve these controversies, but hopefully it will help to clarify the issues and the facts.

Although various theoretical models[1–6] have long suggested the existence of "intrinsic" localized electronic states which lie in energy between the valence and conduction band edges, unambiguous experimental verification of such states has been somewhat elusive. There is, of course, strong experimental evidence for the existence of localized levels within the band gap which are due to specific defects and impurities just as there is in the case of crystalline solids. The theoretical concepts proposed to explain localized electronic states in amorphous semiconductors will be outlined in Section II. In addition to electronic processes, vibrational processes also provide an important absorption mechanism at energies below the band gap. Although the present discussion is not concerned with vibrational absorption processes per se, some multiphonon and impurity absorption peaks occur in spectral regions which overlap those of the electronic processes.

Figure 1 presents a compilation of absorption data from several sources[7–22] for one of the most well characterized semiconducting glasses, As_2S_3. A casual inspection of Fig. 1 indicates two regions of relatively low absorption ($\bar{\nu} \leqslant 10\ \mathrm{cm}^{-1}$ and $10^3\ \mathrm{cm}^{-1} \leqslant \bar{\nu} \leqslant 10^4\ \mathrm{cm}^{-1}$ where $\bar{\nu}$ is the reduced frequency given by the inverse of the wavelength) separated by a region of strong vibrational absorption which peaks near $\bar{\nu} \sim 300\ \mathrm{cm}^{-1}$. Near $100\ \mathrm{cm}^{-1}$ there is a series of absorption peaks superimposed on a rapidly falling absorption tail. These peaks are due to multiphonon absorption processes from the major constituent As and S atoms in the amorphous network. Vibrational absorption peaks due to light impurities, such as oxygen or hydrogen, can

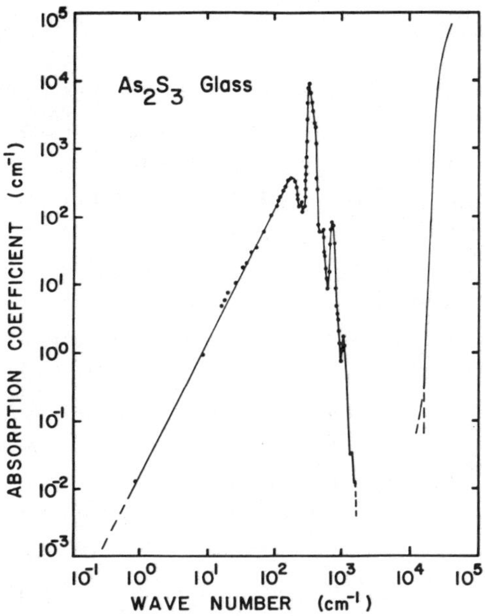

Fig. 1. Absorption coefficient in a representative chalcogenide glass as compiled from several sources.[7–22]

also exist in this region. The band edge ($\sim 10^4$ cm^{-1} in Fig. 1) is dominated by electronic absorption processes which yield the characteristic exponential, or Urbach, shape. Section V presents a discussion of the Urbach tail and of optically induced changes in absorption in this spectral region.

At energies greater than the band edge, electronic interband absorption processes dominate. These processes are very similar to those which occur in crystalline counterparts except that the usual critical-point structure disappears in the amorphous phase. These processes will be briefly summarized in Section III.

As already discussed, electronic transitions due to states deep in the gap, which are either "intrinsic" states, defects or impurities, also provide absorption mechanisms at energies well below the band edge. These topics, which constitute the main emphasis of the chapter, are described in Section IV.

II. THEORETICAL CONCEPTS AND MODELS

Since the pioneering work of Anderson,[23] which employed a simple model Hamiltonian to describe the effects of disorder in a periodic solid with a randomly fluctuating potential, both the existence and the nature of localized electronic states in amorphous solids have been much debated. Early models of semiconducting glasses (the so-called Mott-Cohen-Fritzsche-Ovshinsky or Mott-CFO model[1,24]) predicted tails of localized single-electron states extending so far into the gap from both the valence and conduction bands that they actually overlapped at mid-gap where they pinned the Fermi energy. But these paramagnetic states were never observed experimentally.[25] Recently, Anderson[2] suggested that localized electronic states which are intrinsic to amorphous solids do in fact exist within the gap, but that these states are always paired (doubly occupied or empty) in the ground state configuration due to an effective electronic correlation energy which is negative. In Anderson's original formulation these states were considered to be an intrinsic property of amorphous solids (both semiconductors and insulators) and not due to specific defects or impurities. However, since this proposal several specific models have been suggested which, although loosely based upon Anderson's original hypothesis, nonetheless postulate the existence of specific, well-defined, localized, diamagnetic defects whose ground state energies lie within the gap.[3–5] Although these defects invoke the concept of negative electronic correlation energies, they form only a small subset of those states first considered by Anderson, and thus they are not really within the spirit of the original model. In this section we first outline the more general approach to the problem and then describe the specific defect models which have also been proposed.

We follow the treatment of Anderson and consider an ensemble of localized sites which constitutes all of the electronic states which are of interest.[26] In the simplest tight binding approach the one electron energies are E_i and the Hamiltonian contains only the term

$$\sum_{i\sigma} E_i n_{i\sigma}$$

where i and σ are the site and spin indices, respectively.

As Anderson and others have pointed out, there is a tendency, even within the one electron framework,[2,4] for a solid to form a covalent network in which (doubly) occupied sites are lowered in energy by atomic displacements. This phenomenon is generally termed "bonding" although the process may involve non-bonding or "lone pair" orbitals. The result of such bonding is, of course, to produce a dip in the (1 electron) density of electronic states n(E) as a function of E.

More extended states are included in the one electron picture through the addition of a second term in the Hamiltonian which contains the hopping integrals V_{ij}:

$$H_o = \sum_{i\sigma} E_i n_{i\sigma} + \sum_{ij\sigma} V_{ij}\, c^+_{i\sigma} c_{j\sigma} \tag{1}$$

The energy spectrum of Eq. (1) essentially describes the Mott-CFO model[1,24] in which the band is narrowed by V_{ij} and the spectrum consists of extended states near the center with "localized band tails" at the edges. If the band tails are large enough, then the conduction and valence band tails overlap pinning the Fermi level near the center of the gap as in the original Mott-CFO model.

As mentioned earlier the one electron model fails to agree with several key experiments--most notably with the absence of paramagnetism in these solids. For these reasons Anderson added to the above model the coulomb term (in the Hubbard formulation) and postulated the existence of a strong electron lattice interaction.

$$H_c = \sum_i U_i n_{i\uparrow} n_{i\downarrow}\ . \tag{2}$$

$$H_{el\text{-}lat} = \sum_{i\sigma} C_i x_i n_{i\sigma} \tag{3}$$

In Eq. (3) the existence of a configuration coordinate x_i which is strongly coupled to each electronic state E_i has been assumed. This particular form for the electron-lattice interaction, in which the energy is assumed to be a short-range, linear function of the configuration coordinates, was first employed by Holstein.[27]

Although Eqs. (1) through (3) constitute the basic model, two additional simplifying assumptions[26] are useful both in reducing the complexity of the calculations and in illustrating the essential features of the model. First, one assumes a phonon Hamiltonian of the form

$$H_{ph} = \sum_i P_i^2/2m + \tfrac{1}{2}\, m\omega_{ph}^2 x_i^2 \tag{4}$$

which is basically a "polaron" approximation in that the phonons are described by the Einstein model. In view of the fact that the lattice dynamics of amorphous solids are so poorly understood this approximation at least avoids these complexities. Second, one assumes that the localized states near E_F come from a single E_i (i.e., the second term of Eq. (1) is unimportant).

This assumption means that one can identify localized electronic eigenstates near E_F with individual bonds.

To calculate the energies of the system given the above two approximations, consider the lattice as stationary and find the potential minimum in the configuration coordinates (i.e., $\partial E/\partial x_i = 0$) for the Hamiltonian given by Eqs. (1) through (4). This procedure yields for $E(n_i)$

$$\begin{aligned} E(0) &= 0 \\ E(1) &= E_i^{eff} \\ E(2) &= 2E_i^{eff} + U_i^{eff} \end{aligned} \tag{5}$$

where

$$\begin{aligned} E_i^{eff} &= E_i - C_i^2/2m\omega_{ph}^2 \\ U_i^{eff} &= U_i - C_i^2/m\omega_{ph}^2 \end{aligned} \tag{6}$$

The simplest convention is to measure all energies from the Fermi energy so that for the case of interest which is $U_i^{eff} < 0$ (the so-called "negative U" case), one has $E(0) + E(2) < 2E(1)$. We emphasize that these are the energies appropriate at "long times" over which the lattice can readjust.

For this case the two electron excitation spectrum (energy per pair of electrons) has the form shown schematically in Fig. 2a. States are filled with two electrons up to E_F above which states are empty and there is no energy

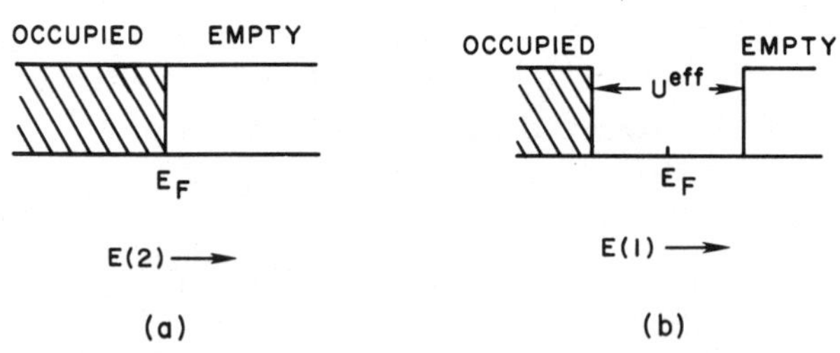

Fig. 2. (a) Density of states as a function of energy per pair of electrons (energy required per pair to change occupation between 2 and 0). (b) Density of states as a function of energy per electron (energy required to excite or de-excite an electron between an occupied and an unoccupied level).

gap in the two-electron spectrum. On the other hand, the one electron spectrum (excitation energy per electron) which is what is measured optically, does exhibit a gap as shown in Fig. 2b. In Fig. 2b U^{eff} is some average over all sites of U_i^{eff}.

The negative U concept is broader than the lattice relaxation envisioned at specific defect sites. All bonds are considered within the negative U framework, and distributions in U_i^{eff} and E_i^{eff} are an implicit, but vital, component of the model.

One of the most attractive features of the above model is that although there are broad distributions in both individual site energies and electron-lattice couplings, there are no deep states and no extensive band tails in the one electron spectrum. There are no deep states because of the assumption[2] that $|U^{eff}| > E_g$ so that when one electron leaves a site the second immediately follows. There are no extensive band tails if one assumes that the "hopping" term in the Hamiltonian is more important than the "polaronic" term (i.e., $V_{ij} > C_i^2/2m\omega_{ph}^2$). Of course the model also implies that both optical and electrical excitations involve chemical bonding changes and must, because of the broad distributions in E_i^{eff} and U_i^{eff}, also involve a wide range of relaxation times.

Emin[6,28,29] has examined the limit in which the polaronic term dominates and concluded that most of the observed optical properties can be explained by polaronic effects. It should be emphasized that this picture is qualitatively different from both the Anderson approach where one-electron polaronic effects are generally not important, and the defect models where lattice distortions are only important at specific defect sites. In the polaronic model it is assumed that the width of the band of small polaron levels is greatly increased in disordered materials over its typical values in crystalline solids. In fact the width is estimated[29] to be on the order of the site-to-site disorder energy Δ which may be on the order of the energy gap ($\Delta \sim E_g$). These hole and electron polaron bands effectively pin the Fermi energy near midgap.

We now examine the consequences of the negative U framework when specific bonding configurations are postulated. The first attempt to model "intrinsic" defects in semiconducting glasses was presented by Mott, Davis and Street[3,4] (MDS model). In this model, which in principle should be applicable for any group VI atom including oxygen, the prototype amorphous solid is the one containing only a chalcogen atom (i.e., glassy Se or S). In Fig. 3a we show schematically[30] the bonding in Se or S where each chalcogen atom is bonded to two nearest neighbors via p wave functions with the remaining p orbital doubly occupied but non-bonded. There is no s-p

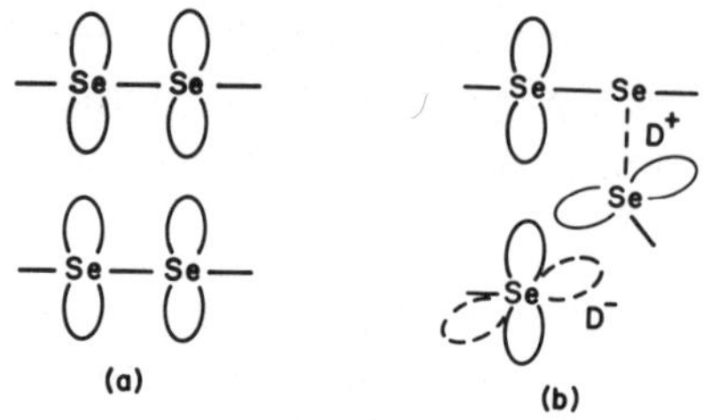

Fig. 3. (a) Schematic bonding diagram of two chalcogen chains; (b) Schematic bonding diagram of a pair of charged defects in a chalcogen.

hybridization in this simple scheme. The defects are assumed to form in pairs as depicted in Fig. 3b. Imagine that a bond is broken and that both electrons formerly in the bond are trapped by one of the chalcogen atoms which becomes a singly-coordinated, negatively-charged (paired-electron) defect (denoted D^-). The remaining chalcogen atom is considered to form a somewhat distorted or abnormal bond with another atom on a nearby chain by utilizing two electrons which were formerly non-bonding. The second (paired-electron) defect (denoted D^+ is triply coordinated and positively charged.

The essence of the MDS argument is that the formation of a D^+, D^- pair from two "dangling bonds" (D^o) is controlled by an exothermic reaction because of the effects of strong lattice distortions.

$$2D^o \rightarrow D^+ + D^- \tag{7}$$

A modification of this scheme has been presented by Kastner, Adler and Fritzsche[5] (KAF) who pointed out explicitly the coordination of the resulting defects (D^- was renamed C_1^- and D^+, C_3^+) and presented energetic arguments which suggested that the defects would be stable even without the effects of strong lattice distortions. These authors also distinguished between D^+ or D^- defects which are isolated and those which occur as close pairs. They suggested that the neutral, paramagnetic defect was three-fold and not one-fold coordinated.

We note that the defects as postulated are diamagnetic in the ground state (D^+ and D^-) in keeping with the experimental evidence.[25,31] However, it is envisioned that these states may be rendered paramagnetic by optical excitation. Thus the excitation of an electron from D^- (or its capture at D^+) yields a neutral paramagnetic species (D^o).

There are several significant difficulties with these simple defect models. In the prototype material, glassy Se, the metastable paramagnetic defect is found by optically-induced ESR to be singly coordinated in keeping with the suggested behavior according to MDS but in contrast to the description of KAF. More detailed tight binding calculations of Vanderbilt and Joannopoulos[32] also suggest one-fold coordination for D^o. In what is perhaps a more significant contradiction, recent calculations of Vanderbilt and Joannopoulos[33] suggest that the D^+, D^- defects in glassy Se exhibit positive U^{eff}.

Both the MDS and KAF models consider only those bonding configurations which can be obtained from broken and rearranged bonds within what we have termed the "normal bonding" configuration where the relative energies of the bonding, non-bonding and anti-bonding p orbitals are fixed in a given solid. This description is certainly accurate for the vast majority of the atoms in an amorphous semiconductor, but one may question the legitimacy of applying normal bonding energetics to those few (10^{16}-10^{17} cm^{-3}) highly distorted sites giving states that lie deep within the energy gap. Instead one may consider those, admittedly improbable, situations where an atom or group of atoms cannot relax to the normal bonding configuration because of conformal or topological constraints placed on these atoms by the surrounding glass network. At these distorted sites, abnormal bonding configurations are obtained in which the relative energies of the local bonding, non-bonding and anti-bonding orbitals can overlap or actually invert.

Ngai and Taylor[30] have considered all the possible bonding configurations, in the absence of s-p hybridization, for a single chalcogen atom which has at least one "normal" bond. This procedure amounts to considering all those abnormal bonding sites and oppositely-charged pairs of sites that can be obtained by breaking at most one normal bond in the ideal glass coordination and/or making at most one distorted bond between atoms that were previously not bonded to each other. If configurations involving the occupancy of more than one anti-bonding orbital on the chalcogen atom are ignored because they are less probable, then nine diamagnetic bonding configurations are obtained.

From these abnormal bonding configurations one may easily construct diagrams of the most probable deep gap states in elemental amorphous chalcogenides. One expects[5] positively charged and negatively charged sites to occur most often in pairs because of the Coulomb attraction, although one might envisage under certain circumstances these pairs "drifting" apart by alternate breaking and remaking of the appropriate bonds. Pairs of the form

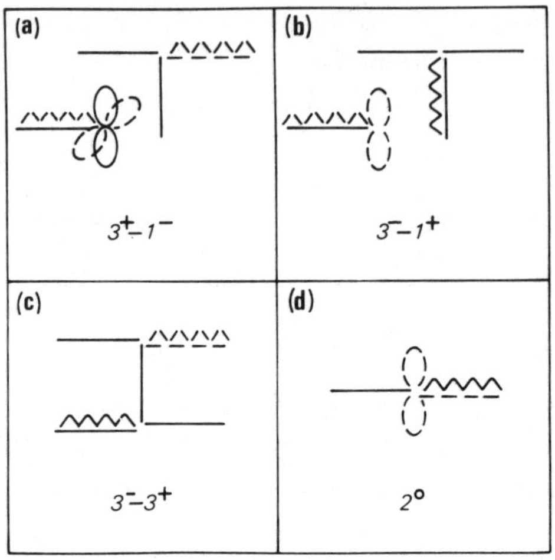

Fig. 4. Schematic diagrams of most probable deep gap states at "abnormal" bonding sites in elemental amorphous chalcogenides (after Ref. 30).

$1^- - 1^+$ (the number indicates the coordination and the superscript the charge) are unlikely for conformal reasons (see the detailed arguments in MDS[4] and KAF[5]) and will tend to rebond to form a 2° site. However, $3^+ - 1^-$, $3^- - 1^+$ and $3^- - 3^+$ pairs as drawn in Fig. 4 can be expected at abnormal sites in glassy Se or S. States derived from bonding, non-bonding and anti-bonding orbitals are represented by straight lines, "figure eights," and jagged lines, respectively. The vertices of the diagrams correspond to the chalcogen atoms at what we have termed an abnormal site while the termini denote chalcogens that are otherwise normally bonded. Two dashed orbitals culminating at a single vertex in Fig. 4 denote the two distinct bonding configurations obtained using either dashed orbital separately. Thus there are four possible ways of forming $3^+ - 1^-$ chalcogen pairs given the present set of assumptions (Fig. 4). In addition one expects the presence of the abnormal 2° sites also drawn in this figure.

Anderson[26] has suggested that doubly charged pairs of sites of the form $3^+ - 3^+$ or $1^- - 1^-$, which are obtained from stretched and compressed bonds, are likely candidates for localized electronic states in Se. The presence of these states may explain why the chalcogenides cannot be doped with H as amorphous Si can.

We caution that the diagrams of Fig. 4 are schematic and intended primarily as a procedure for cataloguing the possible configurations. For example, the specific $3^+ - 1^-$ pair of Fig. 4a, in which the 1^- derives from one bonding and two non-bonding orbitals, can exist as 3^+ and 1^- sites sharing a common normal bond instead of being unbonded as indicated in the figure. A similar possibility exists for the 3^- - 1^+ pair if the 1^+ derives from a bonding and a non-bonding orbital.

Perhaps the most striking feature of this abnormal bonding picture is that, even under our limiting set of assumptions, these exists an astonishingly large number of distinctly different deep gap states possible in a simple binary chalcogenide glass. For example in Se one can construct ten abnormal sites or pairs of sites involving only Se atoms. The situation is even more complicated for binary systems such as As_2Se_3. Certainly some of these sites will be less energetically favorable (and hence less probable) than others, but it may not be possible to anticipate which ones these will be by invoking arguments involving normal bonding energies. These considerations may help to explain why energetic arguments involving only dangling bonds created from "normal bonds" lead to apparent contradictions with experiment.

III. INTERBAND ABSORPTION

The interband absorption spectra have been calculated from visible and ultraviolet reflectivity and electro-reflectivity data for many chalcogenide glasses. For the purpose of illustration we restrict our attention in this section to a summary of the results for three of the most important prototypes, Se, As_2Se_3 and As_2S_3.

In all three materials[7,22,34–38] the interband absorption exhibits two peaks, one near 6 eV and a second near 10 eV. Representative spectra for As_2S_3 and As_2Se_3 are shown in Fig. 5. The two peaks are separated by a broad minimum which occurs near 7 or 8 eV. These features are remarkably similar to those which occur in the corresponding crystalline counterparts. There are two main differences between the crystalline and glassy spectra. The first difference is the disappearance in the amorphous phases of structure in the 2 to 7 eV range which results from singularities associated with symmetry points in the Brillouin zones of the crystals. The second difference is an overall broadening of the two-peak structure in the amorphous phases.

The close similarity between the interband optical absorption data of the glasses and crystals points out the relatively weak influence which the disorder has on the electronic density of states. This similarity results from the fact that these solids are molecular so that the gross features of the interband absorption spectra are determined by the molecules and their excited states. The influences of extra symmetries imposed by the long range crystalline order are minor.

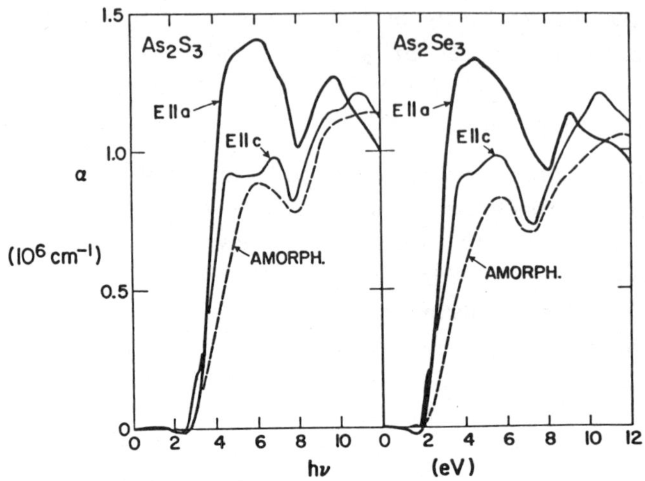

Fig. 5. Interband spectrum for glassy and crystalline As_2S_3 and As_2Se_3 at 300 K (after Ref. 7).

Using the sum rule and a Kramers-Kronig analysis of the reflectivity data, Drews et al.[7] have estimated the number of valence electrons contributing to each peak in the absorption spectrum. The first peak asymptotes near 3-3.5 electrons and the second <3 electrons. If we assume that there is essentially no s-p hybridization in either the As or S bonding, then there are ~3.2 non-bonding electrons/atom and ~2.4 bonding electrons/atom for As_2S_3 or As_2Se_3. It is thus reasonable to attribute the lower- and higher-energy onsets of absorption in Fig. 5 to thresholds for transitions originating from non-bonding and bonding valuence band states, respectively. This interpretation is consistent with the molecular nature of these solids and accounts nicely for the similarities between the crystalline and amorphous phases since the nearest neighbor coordination is identical in both forms.

IV. LOCALIZED ELECTRONIC STATES WITHIN THE ENERGY GAP

The identification of optical absorption and other related optical processes which occur at energies below the energy gap E_g with the presence of "intrinsic," disorder-induced localized states is still a matter of some debate. Although it is by no means foolproof, the first test for this classification is the simple requirement that the optical process be unique to the amorphous phase. In this section we review the various optical processes which have been observed below E_g and comment on their possible origins such as dangling bonds and other defects, departures from normal coordination,

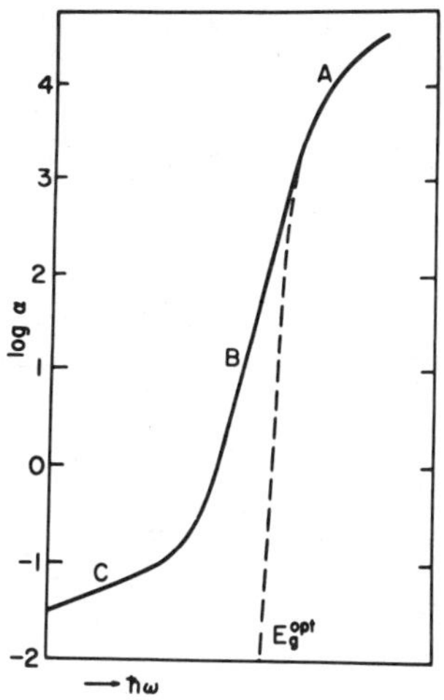

Fig. 6. Schematic diagram of a typical absorption edge in a chalcogenide glass showing the onset of interband transitions (A), the Urbach tail (B) and the residual below gap absorption (C) (after Ref. 39).

polaronic processes and impurity atoms. Where appropriate, comparisons with crystalline forms will be made.

The chalcogenide glasses exhibit highly reproducible optical absorption edges (see Fig. 6) which are relatively insensitive to preparation conditions, and under equilibrium conditions, the only observable absorption[39] within the gap is the low level tail of region C in Fig. 6 which has been shown to be ascribable to the presence of impurities. Regions A and B will be discussed in the next section. Furthermore, these glasses are highly diamagnetic and no ESR attributable to disorder induced localized states has been observed under equilibrium conditions.[25] Anderson's explanation[2] of the high transparency and diamagnetism of these materials in terms of paired electronic gap states has been discussed in Section II. It is possible to perturb optically these paired electronic states and to produce at low temperatures ($T \leqslant 100$ K) singly occupied localized states which are metastable and manifest themselves as an optically-induced ESR signal[31] and an optically induced absorption[31,40,41] (see Fig. 7). This absorption has an onset near mid-gap and is relatively flat up to the band edge. After optical excitation some of the electronic states in the glass are not in an equilibrium configuration, but because these states are metastable they do provide a sensitive probe of the local structural environment. These metastable states, as manifested in both the ESR and the below gap optical absorption, can be bleached with the

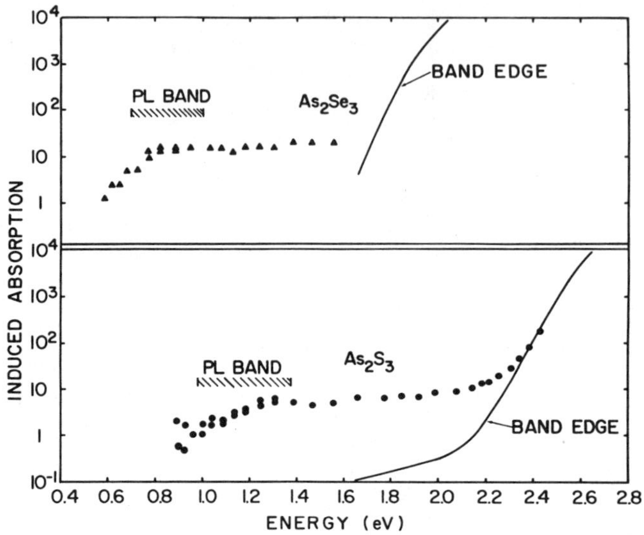

Fig. 7. Optically induced absorption spectra for glassy As_2S_3 and As_2Se_3 at ~6 K (after Ref. 31). PL bands are indicated for reference.

application of light below the band gap. Any energy between mid-gap and the energy at which the absorption coefficient α is less than $\sim 10\ cm^{-1}$ is effective in bleaching. This process eliminates most of the metastable electronic states and the system approaches its diamagnetic ground state configuration. Bleaching can also be accomplished thermally by cycling the sample to temperatures above ~150-200 K. Both the bleaching and inducing processes can be initiated with very low light intensities ($\sim 1\ mW\ cm^{-2}$).

Because the same conditions of optical excitation generate at the same rate (and bleach at the same rate) both an ESR response and the absorption shown in Fig. 7 for As_2Se_3 and As_2S_3, it is tempting to conclude that both effects result from the same electronic states. Furthermore, neither effect is observed[31,42] in *pure* crystals of As_2Se_3.

It has been concluded on the basis of the observed ESR spectra that two distinct paramagnetic centers are optically induced in the arsenic chalcogenide glasses: a hole center localized on a p orbital of a chalcogen atom, and an electron center localized predominantly on a p orbital of an arsenic atom.[31] Hence the optically induced paramagnetic states are not directly associated with impurity atoms; they are highly localized, and unique to the amorphous phase.

We pause at this point for an important caveat. Early optical and photoluminescence (PL) studies of the chalcogenide glasses conjectured that many additional effects to be described below also reflect "intrinsic" processes. However, without the direct tie to detailed local structural information such as that which the ESR provides, it is difficult to infer the nature of the electronic states strictly from the optical processes. As we shall see several optical effects which were originally thought to be intrinsic may very well be due to inadvertent and pervasive impurities in the host glasses.

Historically, the observation of PL in chalcogenide glasses[43,44] and related crystals[44] provided the first evidence for the existence of localized states within the gap in these materials. As shown by the PL spectrum and PL excitation spectrum in Fig. 8 for glassy As_2Se_3, irradiation of these glasses by light corresponding primarily to the Urbach tail in the absorption edge excites a broad PL band centered near mid-gap.[40] This PL band is commonly interpreted as due to at trapped hole whose energy is lowered to near mid-gap by a lattice distortion caused by a strong electron-phonon interaction.

In terms of the MDS and KAF models, the PL, mid-gap absorption and ESR centers are ascribed to various charge states of the one- and three-coordinated chalcogen defects. In the MDS approach, the shape of the PL excitation spectrum and the temperature dependence of the PL efficiency have prompted Street et al.[4,5] to postulate the existence of charged radiative recombination centers. The PL proceeds via radiative recombination of photoexcited electrons (or holes) from an isolated D^- (or D^+) and the metastable ESR is produced when the excited electrons (or holes) drift away before recombination leaving the center in an uncharged D^o configuration. In a recent iteration of the KAF model[46] the PL and ESR result only from uncharged, dipolar nearest neighbor pairs of defects ($C_1^- - C_3^+$ pairs). In both approaches the mid-gap optical absorption is due to D^o states which lie near

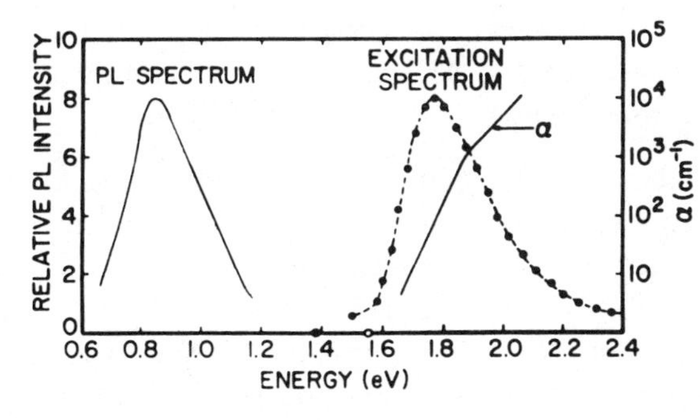

Fig. 8. PL and PL excitation spectra for glassy As_2Se_3 at ~6 K. The excitation spectrum is normalized to the number of incident photons (after Ref. 40).

mid-gap. This picture is complete for selenium which contains only chalcogen atoms, but for materials such as As_2Se_3 the models must also invoke defects associated with As atoms. The specific defect most often invoked is a two-fold coordinated As atom, although four-fold coordinated As atoms have also been proposed.[5,47,48]

In the small polaron model,[29] optical excitation of As_2Se_3 (with $h\nu \sim E_g$) at low temperatures generates a hole-like small polaron which yields an unpaired spin on a chalcogen atom and an electron-like small polaron which yields an unpaired spin on an arsenic atom. This self-trapping is long lived at those sites which are farthest apart. At low temperatures, the optically induced absorption results from absorption by small polarons which generates small polaron hopping and the spontaneous emission of phonons. The hole-like small polarons dominate the absorption[29] which is typically very broad with respect to E_g. The bleaching process occurs because the absorption of a photon with energy below the gap (i.e., within the optically induced absorption band) produces small polaron hopping which stimulates a return to equilibrium (i.e., electron-like and hole-like polarons recombine).

In addition to the creation of optically induced absorption and ESR, the application of light at the band edge (near the peak of the PL excitation spectrum) also produces a metastable decrease in the PL efficiency with time.[44,49,50] This decrease in efficiency, or fatigue, is reversible (i.e., the efficiency can be restored) either through the application of light with energy below the gap ($E_g/2 \leqslant h\nu < E_g$) or by cycling up in temperature to temperatures above about 100-150 K. One consequence of these fatiguing and restoration processes is that during PL excitation one may enhance[40] the fatigued PL by applying a second light source with $h\nu < E_g$. That is, one may study the dynamic equilibrium between the simultaneous fatiguing and restoring processes. Both the fatiguing and the restoring processes for the PL parallel the growth and decay processes for the ESR and the optically-induced below gap absorption. The rates and the temperature dependences are similar for all three effects.

As we cautioned above these parallels do not necessarily imply that all three processes directly involve the same localized electronic states. In fact, there is good evidence to suggest that at least the PL fatiguing process must be separated from the other two. First of all, both PL and PL fatigue occur in the best crystalline samples which exhibit neither the optically induced below-gap absorption nor the optically induced ESR.[31,42] Second, since the temperature dependence of the PL in As_2Se_3 is very different from that of the fatiguing process, these two effects must be independent.[51] Third, several experiments[52–55] on samples of As_2Se_3, As_2S_3 or Se which contain controlled amounts of impurities suggest that the fatiguing process is due to the production of competing non-radiative centers and not to a change in the charge state of an existing PL-active center.[3–6] The evidence which implicates competing non-radiative processes in the PL fatigue comes from several sources. Specifically, two PL bands which are very different in shape, in temperature dependence and in sensitivity to impurities occur in Se at 0.8 and 0.57 eV, but these two peaks always fatigue at identical rates.[52,53] Also in As_2S_3 doped with Fe the PL efficiency is reduced but the fatigue rate is unchanged.[54] Finally in crystalline As_2S_3 all PL bands, narrow or broad, fatigue at the same rate.[55]

If PL fatigue is due to the development of a competing non-radiative process, then the question remains as to the origin of this process. Since both iron and copper are pervasive impurities in the chalcogenide glasses and crystals, they have both been suggested as possible centers which contribute to the PL fatigue.[54,55] The general idea is that one can optically alter the existing charge state of either of these transition metals through a process which results in intra-configurational d-d transitions.[56] The process cannot be a simple $Fe^{2+} \rightarrow Fe^{3+}$ conversion because one would observe this conversion by ESR,[54] but $Cu^{1+} \rightarrow Cu^{2+}$ is possible because the resulting ESR response would be buried under the usual optically induced ESR signal.[55]

Even before optical excitation at low temperatures, there is a contribution to the absorption below the gap in the chalcogenide glasses[21] (region labeled C in Fig. 6). Although these "absorption tails" were first thought to result from "intrinsic" electronic states,[16] both low temperature magnetic susceptibility[57,58] and optical absorption[21,59] experiments on doped samples have demonstrated that the absorption is due to the presence of iron. Figure 9 shows the results[59] for three samples of glassy As_2S_3, two of which have been doped with Fe and one of which is nominally undoped. The absorption in the "pure" As_2S_3 sample is in fact due to[59] $\sim$5 ppm Fe which occurs at about this level or higher in all chalcogenide glass samples prepared with existing technologies. From the measurements of Fig. 9 and low frequency ESR results[59] Tauc and coworkers concluded that the absorption is probably due to charge transfer transitions of the form $Fe^{2+} \rightarrow Fe^{3+} + e$.

Tails similar to those observed for the chalcogenide glasses (Figs. 6 and 7) have also been observed in crystalline chalcogenides, such as As_2S_3 and As_2Se_3, by PL excitation[60] and optical absorption[61] spectroscopy. Street and coworkers[60] suggested that the optical transitions which give rise to the PLE tails initiate from the recombination centers themselves. If this assertion is correct then one might further speculate that the PL itself involves an impurity species, although this is certainly not the most popular interpretation.

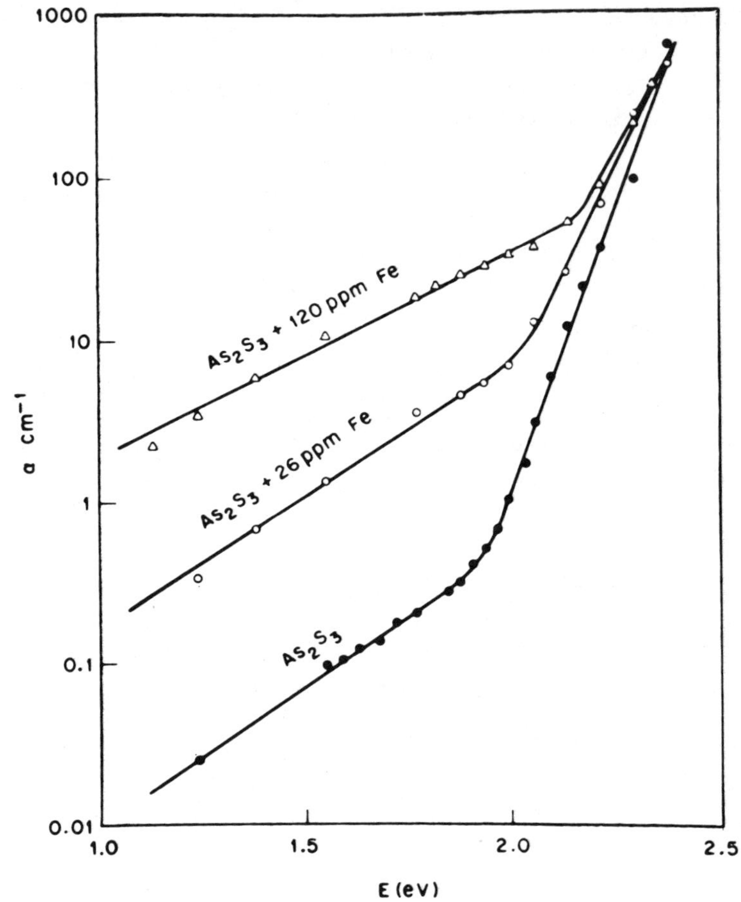

Fig. 9. Optical absorption coefficient as a function of energy for nominally pure and Fe-doped As_2Se_3 glass at 300 K (after Ref. 59).

Most of the recent information concerning the optical properties of the chalcogenide glasses below the band gap has come from time resolved measurements. The majority of these measurements has concerned the transient photoluminescent processes[62] in As_2S_3 on a time scale ranging from ~10 nsec to ~10 msec. At short times some authors[63,64] have observed two PL peaks at 1.5 and 1.7 eV in As_2S_3. In some cases the relative amplitudes of these two peaks depend on the sample preparation techniques.[63] Other authors[65,66] observe one broad feature extending from less than one to greater than 7 eV. Because the PL spectrum is so broad, the exact shape depends on the corrections made for the detector and the source and whether the result is presented per unit frequency (energy) or per unit wavelength.[67] Figure 10 shows representative spectra at various times after excitation with 2.5 eV light for ~10 nsec.[67] The spectra of Fig. 10 do not show any resolved features.

Whether or not two features are seen, the bulk of the PL intensity at times less than ~1 μsec lies well above the CW PL peak at 1.1 eV (see Fig. 10), and the peak intensity shifts to lower energy in time.[64,67–70] At the longest times (~10 msec) some authors[68,69] report that the peak energy of this PL feature crosses below the CW PL peak while others[64,70] suggest that the peak energy asymptotically approaches the CW PL peak.

When two peaks are seen, the higher energy peak at 1.7 eV appears to decay rapidly with delay time[64] and temperature.[69] There is a shift in the relative amplitudes of the two peaks with excitation energy.[63,71] At lower excitation energies the 1.5 eV spectrum is relatively stronger. When only one broad feature is seen in the PL spectrum, the center of gravity of this spectrum is also observed to shift down in energy with excitation energy.[70] There also appears to be a more rapid change in both the width of the PL spectrum[70] and the shape of the decay[72] near an excitation energy of ~2.3 eV.

The sensitivity of the PL spectrum to sample history and sample-to-sample variations in the observed structure suggest again the possibility that impurities may play at least an indirect role in some of the time resolved PL

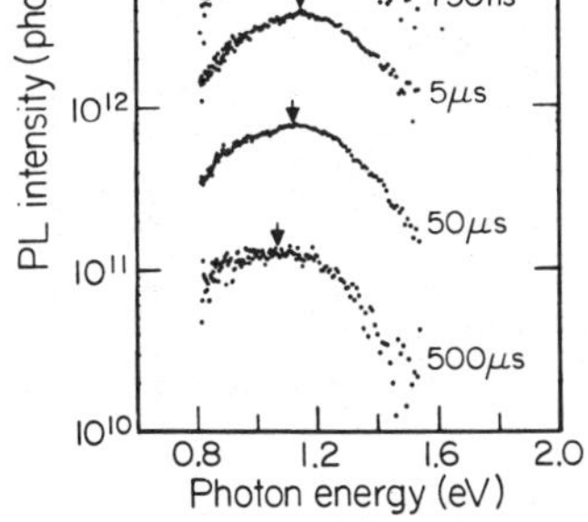

Fig. 10. PL spectra of glassy As_2S_3 at 2 K for various delay times after pulsed excitation at 2.5 eV (~0.2 Mw cm^{-2}) for ~10 nsec. Arrows are an indication of the mean PL energy at each delay (after Ref. 67).

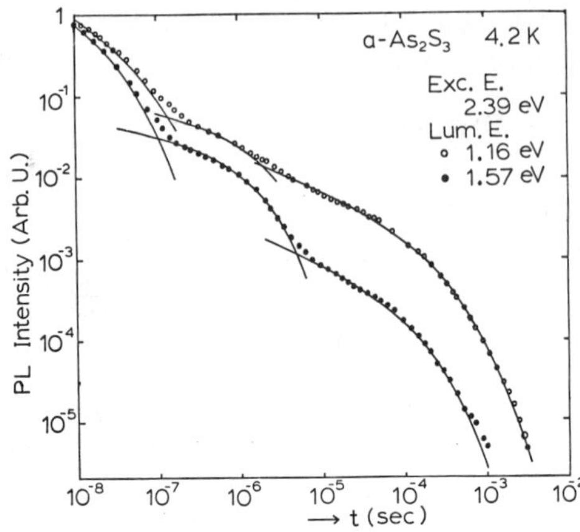

Fig. 11. PL Decay at selected energies in glassy As_2S_3 as a function of time after excitation at 2.39 eV for ~10 nsec. Data were taken at 4.2 K (after Ref. 73).

processes. However, this possibility has not been considered seriously in the current interpretations.

One can also look at the time decay of the PL either at selected energies[73] or integrated over most of the band.[67] At low energy excitation ($E \leqslant 2.4$ eV) either procedure yields three distinct features as shown[73] in Fig. 11. These three features correspond to times on the order of 2×10^{-8}, 2×10^{-6} and 2×10^{-4} sec. At higher excitation energies ($E > 2.4$ eV) the structure washes out when the integrated PL intensity is plotted, and the decay curve becomes nearly linear on a log-log plot (power law decay).[67]

Murayama and coauthors[71,74] have measured the polarization of the PL after exciting with polarized light and observed that at short times the PL remains polarized parallel to the exciting light. This "polarization memory" persists for times up to $\sim 10^{-4}$ sec (i.e., for the first two features in the decay curves of Fig. 11), but decays rapidly after this time. The PL which decays on a 2×10^{-4} sec time scale is probably what is observed in the CW measurements because it is the most intense.

Recently Higashi and Kastner[75] have shown that the PL near 2×10^{-6} sec time delay is strongly excited by below gap light. Although the interpretation of this result is not obvious,[75] the possible role of impurities in this excitation process should be considered.

In addition to the broad PL response covering the range indicated in Fig. 10, there have been reports[76–79] of fast PL processes which occur at higher energies ($E \gtrsim 2$ eV). Because it is most likely that these effects are the result of surface contaminants on the samples,[80] we shall not consider them further.

Several of the time resolved PL effects observed in glassy As_2S_3 have parallels in crystalline As_2S_3. Two PL peaks[55,81–84] at 1.2 and 1.6 eV are observed in crystalline As_2S_3. The lower energy peak is more strongly excited by higher energy light and vice versa.[81,55] The temperature dependences[81] and the time dependences[75] of these two processes parallel those observed in glassy As_2S_3 at similar PL energies.

Since the results of the time resolved PL in As_2S_3 are somewhat confused and conflicting, it is not surprising that the interpretations are particularly disparate. It is reasonably clear that the time decay points to at least three distinct PL processes and the PL spectrum to at least two, but beyond these simple facts the interpretations quickly diverge. Street[62] has suggested that the slow PL ($\tau \sim 10^{-4}$ sec) involves recombination at charged centers. Specifically a D^- defect is presumed to capture a hole after which the recombination proceeds via band tail electrons. Bosch and Shah[64] have speculated that donor-acceptor pairs (probably D^+-D^- pairs) are responsible for both the faster ($\tau \sim 2 \times 10^{-6}$ sec) and slower PL processes. Shah[69] has suggested that the slower process involves a D^- defect and band tail electrons in agreement with Street while the faster process involves a D^+ defect which captures an electron and recombines with band tail holes. Murayama[73] supposes that the fastest PL process ($\tau \sim 2 \times 10^{-8}$ sec) results from localized excitons (presumably localized at defects) which are strongly interacting with phonons. According to Murayama[73] the other PL processes with $\tau \sim 2 \times 10^{-6}$ and $\tau \sim 2 \times 10^{-4}$ sec are due to recombination of electron-hole pairs at defects and self-trapped excitons, respectively. The latest interpretation of Higashi and Kastner[75] suggests that the slowest PL process involves excitons trapped at defects. These authors further speculate[70,75] that the relatively abrupt changes which occur in several experimentally measured quantities near $E \sim 2.3$ eV indicate the division between PL processes where excitons are localized and those where they are mobile. Bosch, Epworth and Emin[85] attribute all three processes to recombination between electron and hole small polarons or self-trapped excitons without recourse to the influence of defects. Finally, Phillips[86] has suggested that internal surfaces play a major role in these PL processes.

It is fair to say that a coherent picture of all of these time resolved PL processes is yet to emerge, although support for the importance of excitonic mechanisms in some of the PL processes has recently come from optically detected electron spin resonance measurements on[71,87–88] glassy As_2S_3 and

As_2Se_3. The role of defects in these PL processes is still unclear, and no one has seriously addressed the possible influences of impurities.

A second set of transient experiments involves optical absorption processes in the semiconducting glasses. Most experiments have probed the absorption on a nanosecond time scale[89–92] although some results on a picosecond time scale have been reported.[93,94] Figure 12 shows the spectral dependence of the transient photo-induced absorption as a function of delay time after the excitation pulse (Fig. 12a) and as a function of temperature at fixed delay (Fig. 12b) for glassy As_2Se_3. At short times (10^{-5} sec) the induced absorption is at most $\sim 1\ cm^{-1}$ and this absorption decays with time. The spectra at longer delay times have been normalized by factors of 1.6, 2.7, and 9.3 for 10^{-4}, 10^{-3}, and 10^{-2} sec, respectively. It is apparent from Fig. 12a that the absorption decreases in magnitude and shifts to higher energies with the time delay.

At a constant delay time the optically induced absorption becomes more and more constant in energy as the temperature decreases (Fig. 12b). At the lowest temperature the absorption approaches the metastable absorption as shown in the top half of Fig. 7. This correspondence has led Orenstein and Kastner[89] to suggest that the same mechanism applies at all temperatures. These authors have interpreted this absorption process within the framework of a multiple trapping model where the electrons become trapped and are subsequently thermally released from an exponentially decreasing density of states at the band edge. The most significant difficulty with this explanation

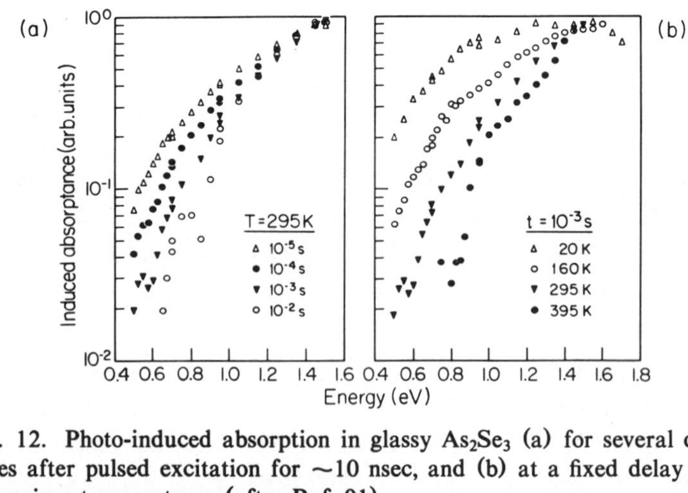

Fig. 12. Photo-induced absorption in glassy As_2Se_3 (a) for several delay times after pulsed excitation for ~ 10 nsec, and (b) at a fixed delay time for various temperatures (after Ref. 91).

is that some of the important predictions require an exponential density of traps whereas a Gaussian distribution is more physically plausible.

In the picosecond regime the experimental restrictions require that the excitation and probe energies be identical; therefore, no spectral dependences of induced absorption are possible. Perhaps the most significant conclusion drawn from these measurements is that the optically excited carriers thermalize on a time scale $\lesssim 1$ psec.[93,94] Two decay processes are observed depending upon the excitation energy. At high excitation energy[93] a long-lived (> 300 picoseconds) induced absorption, which is excited by two-photon processes, is observed, and at low excitation energy the absorption decays on a picosecond time scale, which can be explained by geminate recombination of the thermalized carriers.[94]

We close this section with a brief mention of an absorption mechanism which occurs below the gap in "highly conducting" chalcogenide glasses. This mechanism, which has been attributed to free carriers,[95] has been studied in glassy $T\ell_2SeAs_2Te_3$ for which the dc conductivity is $\sim 3 \times 10^{-3}\ \Omega\ cm^{-1}$ at 300 K. The absorption coefficient for $T\ell_2SeAs_2Te_3$ glass below the energy gap is essentially frequency independent but strongly temperature dependent. At temperatures above 340 K the absorption coefficient in this glass is thermally activated with an activation energy which is the same as that for dc conductivity.

The near equivalence of the thermal activation energies for the dc and optical conductivities is strong evidence that the thermally activated infrared absorption or conductivity mechanism involves thermally generated carriers in extended band states. As such, this phenomenon constitutes the nearest counterpart of a free carrier optical absorption yet observed in an amorphous semiconductor. However, the extremely low carrier mobilities which characterize amorphous semiconductors cause the frequency or wavelength dependence of this "free" carrier absorption to differ drastically from the classical plasma edge observed in crystalline semiconductors.

V. THE BAND EDGE

In the chalcogenide glasses and other amorphous semiconductors the absorption edge (region B of Fig. 6) depends exponentially on energy. Similar behavior in ionic solids was first described empirically by Urbach[96] in the following form:

$$\alpha(\omega) \sim \exp\left[\sigma(\hbar\omega - \hbar\omega_o)/kT\right] \tag{7}$$

where σ is a constant of order unity and ω_o is a constant which roughly corresponds to the lowest excitonic angular frequency. Although the chalcogenide glasses generally exhibit temperature dependences more complicated than that predicted by Eq. 8, the exponential form of the edge is observed in essentially all materials. This edge is broader than in crystalline solids.

Although there is no detailed understanding of the relationship between disorder and the Urbach tail in the chalcogenide glasses, the mechanism is generally thought to arise from electric-microfield-induced ionization of excitons as suggested by Dow and Redfield.[97,98] In this model the source of the ionizing electric field is phonons, but the shape of the absorption edge is insensitive to the details of the microfield distribution. Because the edge is so broad in the chalcogenide glasses, the Dow-Redfield model requires large internal electric fields[99] whose existence is not well established.

All chalcogenide glasses exhibit an optically induced shift of the exponential absorption edge to lower energies after excitation with band gap light.[100] This so-called photodarkening effect is generally larger at lower temperatures. Typical photodarkening shifts for glassy As_2Se_3 are shown in

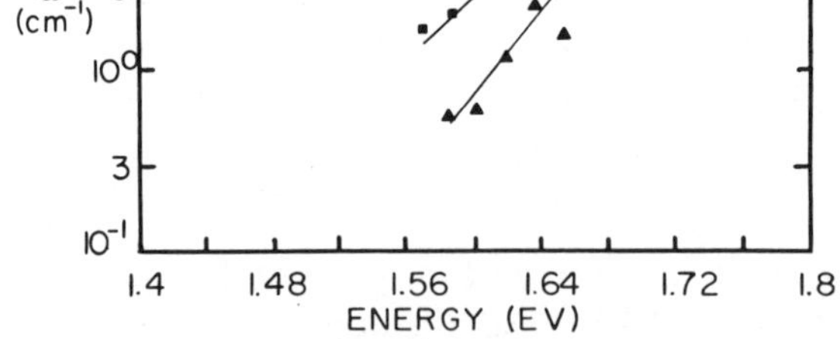

Fig. 13. The absorption coefficient of bulk glassy As_2Se_3 as a function of energy in the region of the electronic band edge. The open circles refer to measurements at room temperature and the solid symbols to measurements at 77 K. Triangles represent as-received samples; circles represent samples after irradiation at 1.8 eV; solid squares represent samples irradiated, warmed to 300 K and cooled to 77 K for remeasurement (after Ref. 101).

Fig. 13 at 77 K (solid points) and at 300 K (open points).[101] Larger effects are typically observed in As_2S_3 than in As_2Se_3. These shifts can be reversed by thermal annealing at or below the glass transition temperature, and some of the photodarkening can be reversed optically with the application of below-band-gap light.[102] The excitation spectrum for photodarkening peaks at higher energy ($\alpha \sim 10^3$ cm^{-1}) than the corresponding excitation spectrum for the below gap absorption ($\alpha \sim 10^2$ cm^{-1}) discussed in Section IV.

It is clear from the properties of the photodarkening process that the metastable optically induced absorption below the gap and the metastable photodarkening of the band edge are distinct and separable processes. In addition to different spectral dependences, excitation spectra, thermal stabilities, and inducing and bleaching rates, the total integrated changes in absorption in the photodarkening processes are orders of magnitude greater than those involved in the below-gap absorption. Also, the photodarkening process appears to require non-bonding (lone-pair) p-electrons[103] while the optically induced below gap absorption and ESR do not.[91] For example, both of the latter two effects occur in amorphous As which is entirely three coordinated with no lone-pair p electrons,[31] but photodarkening is not observed in this amorphous solid.[103]

The photodarkening effect has been attributed to gross structural changes such as the breaking of bonds,[100,102,104] but recent evidence[101] suggests that more subtle changes involving mainly non-bonding electrons are involved. (In fast evaporated films of the chalcogenide glasses, whose structure is different from the bulk, irreversible gross photostructural changes are observed[100,102,104] in addition to the thermally reversible photodarkening effects.)

If the photodarkening process does not involve gross bonding changes, then its explanation within the Dow-Redfield framework involves an increase in the characteristic exciton "binding energy" after photo-excitation, but the mechanism for this increase remains a mystery. The defect models interpret the photodarkening process as due to different subsets of the defects proposed to explain the metastable below gap absorption discussed in Section IV. For example, the KAF defect model[5] ascribes the below gap absorption to nearest-neighbor or second-nearest-neighbor pairs of oppositely charged defects (D^+-D^- pairs) while the photodarkening is envisioned to involve mainly physically separated defects. In the MDS model[4,105] the below gap absorption is attributed to isolated charged defects and the photodarkening to close pairs of defects. It has also been suggested, within the framework of the MDS model, that the photodarkening may correspond to the increased density of ESR spins which is observed after higher power laser excitation.[105]

VI. SUMMARY

Several models have been proposed to explain the optical properties of the chalcogenide glasses. Anderson has suggested that the localized electronic states in amorphous solids are either doubly occupied or empty in the ground state configuration as the result of strong electron-lattice interactions. In a Hubbard model formulation this situation results from broad distributions of individual site energies and effective electron-electron correlation energies which are negative (negative U). Defect models, which involve under- and over-coordinated chalcogen atoms have also been suggested to explain the optical properties. Although loosely based upon the negative U concept, these models only invoke strong electron-lattice effects to localize specific electronic states at specific defect sites. Polarons and localized or self-trapped excitons have been suggested as a third possible explanation for the optical properties of the chalcogenides. This approach assumes strong electron-lattice coupling but without regard for defects. Finally, the possibility that some of the commonly observed optical properties of the chalcogenide glasses are due to the presence of pervasive impurities must be considered.

Regardless of the explanation, the most important feature of the optical properties in the chalcogenide glasses is the dominance of strong electron-lattice interactions. Experimental examples of strong electron-lattice effects include a Stokes-shifted PL peak occuring near mid-gap and a meta-stable optically induced absorption which extends from the band edge to the mid-gap region. At least two additional PL processes are observed in the chalcogenide glasses under transient conditions (10 nsec - 10 msec). Other absorption mechanisms common to the chalcogenide glasses include an Urbach tail to the band edge absorption and an optically-induced transient absorption which extends well into the gap. An analog of free carrier absorption is also observed near 300 K in some more highly conducting chalcogenide glasses and at higher temperatures in the more insulating glasses.

REFERENCES

1. M. H. Cohen, H. Fritzsche and S. R. Ovshinsky, Phys. Rev. Lett. **22**, 1065 (1969).
2. P. W. Anderson, Phys. Rev. Lett. **34**, 953 (1975).
3. R. A. Street and N. F. Mott, Phys. Rev. Lett. 35, 1293 (1975).
4. N. F. Mott, E. A. Davis and R. A. Street, Phil. Mag. **32**, 961 (1975).
5. M. Kastner, D. Adler and H. Fritzsche, Phys. Rev. Lett. **37**, 1504 (1976).
6. D. Emin in *Amorphous and Liquid Semiconductors*, ed. W. E. Spear (University of Edinburgh, Edinburgh, Scotland, 1977), p. 261.
7. R. E. Drews, R. L. Emerald, M. L. Slade and R. Zallen, Solid State Commun. **10**, 293 (1972).

8. J. A. Savage and S. Nielsen, Infrared Phys. **5**, 195 (1965).
9. M. Tanaka and T. Minani, Japan J. Appl. Phys. **4**, 1023 (1965).
10. M. S. Maklad, R. K. Mohr, R. E. Howard, P. B. Macedo and C. T. Moynihan, Solid State Commun. **15**, 855 (1974).
11. M. Onomichi, T. Arai and K. Kudo, J. Non-Cryst. Solids **6**, 362 (1971).
12. P. A. Young, J. Phys. C**4**, 93 (1971).
13. J. T. Edmond and M. W. Redfearn, Proc. Phys. Soc. **81**, 380 (1963).
14. S. Tsuchihashi and Y. Kawamoto, J. Non-Cryst. Solids **5**, 286 (1971).
15. D. Treacy and P. C. Taylor in *Optical Properties of Highly Transparent Solids*, S. S. Mitra and B. Bendow, eds. (Plenum Press, New York, 1975), p. 261.
16. J. Tauc, A. Menth and D. L. Wood, Phys. Rev. Lett. **25**, 749 (1970).
17. G. Lucovsky, Phys. Rev. B**6**, 1480 (1972).
18. J. P. Mathieu and D. Poulet, Bull. Soc. for Mineral Crystallogr. **22**, 532 (1970).
19. P. C. Taylor, S. G. Bishop, D. L. Mitchell and D. Treacy, in *Amorphous and Liquid Semiconductors*, J. Stuke and W. Brenig, eds. (Taylor and Francis, London, 1973), p. 1267.
20. P. B. Klein, P. C. Taylor and D. J. Treacy, Phys. Rev. B**16**, 4511 (1977).
21. D. L. Wood and J. Tauc, Phys. Rev. B**5**, 3144 (1972).
22. R. Zallen, R. E. Drews, R. L. Emerald and M. L. Slade, Phys. Rev. Lett. **26**, 1564 (1971).
23. P. W. Anderson, Phys. Rev. **109**, 1492 (1958).
24. N. F. Mott, Phil. Mag. **24**, 935 (1971).
25. S. C. Agarwal, Phys. Rev. B**7**, 685 (1973).
26. P. W. Anderson in *La Matie're mal Condensée/Ill-Condensed Matter*, R. Balian et al., eds (North Holland, Amsterdam, 1979), p. 161.
27. T. Holstein, Ann. Phys. (N.Y.) **8**, 325 (1959); 343 (1959).
28. D. Emin, C. H. Seager and R. K. Quinn, Phys. Rev. Lett. **28**, 813 (1972).
29. D. Emin, Adv. Phys. **24**, 305 (1975).
30. K. L. Ngai and P. C. Taylor, Phil. Mag. B**37**, 175 (1978).
31. S. G. Bishop, U. Strom and P. C. Taylor, Phys. Rev. B**15**, 2278 (1977).
32. D. Vanderbilt and J. D. Joannopoulos, Phys. Rev. Lett. **42**, 1012 (1979); Phys. Rev. B**22**, 2927 (1980).
33. D. Vanderbilt and J. D. Joannopoulos, Phys. Rev. Lett., **49**, 823 (1982).
34. L. B. Zlatkin and E. K. Ivanov, J. Phys. Chem. Solids **32**, 1733 (1971).
35. M. L. Belle, B. T. Kolomiets and B. V. Pavlov, Fiz. Tech. Poluprovod. **2**, 1448 (1968).
36. A. G. Leiga, J. Opt. Soc. Am. **58**, 1441 (1968).
37. G. Weiser and J. Stuke, Phys. Stat. Solidi **35**, 747 (1969).
38. J. Stuke, J. Non-Cryst. Solids **4**, 1 (1970).
39. J. Tauc in *Amorphous and Liquid Semiconductors*, ed. J. Tauc (Plenum, New York, 1974), p. 159.
40. S. G. Bishop, U. Strom and C. S. Guenzer in *Amorphous and Liquid Semiconductors*, eds. J. Stuke and W. Brenig (Taylor and Francis, London, 1974), p. 963.
41. J. Cernogora, F. Mollot and C. Benoit á la Guillaume, *The Physics of Semiconductors*, ed. M. H. Pilkuhn (Teubner, Stuttgart, 1974), p. 1027.
42. P. C. Taylor, U. Strom and S. G. Bishop, Solar Energy Materials **8**, 23 (1982).
43. B. T. Kolomiets, T. M. Mamantova and A. A. Babeev, Phys. Stat. Sol. **27**, K15 (1968).
44. R. A. Street, Advances in Physics **25**, 397 (1976).
45. R. A. Street, T. M. Searle, and I. G. Austin, Phil. Mag. 29, 1157 (1974).
46. M. Kastner and H. Fritzsche, Phil. Mag. B**37**, 199 (1978).
47. W. B. Pollard and J. D. Joannopoulos, Phys. Rev. B**19**, 4217 (1979).
48. D. Adler, Phys. Rev. Lett. **41**, 1755 (1978).
49. J. Cernogora, F. Mollot, and C. Benoit á la Guillaume, Phys. Stat. Sol. A**15**, 401 (1973).
50. R. A. Street, T. M. Searle, and I. G. Austin, J. Phys. C**6**, 1830 (1973).
51. M. Kastner and S. J. Hudgens, Phil. Mag. **37**, 665 (1978).
52. S. G. Bishop, U. Strom and P. C. Taylor, in *The Physics of Selenium and Tellurlum*, eds. E. Gerlach and P. Grosse (Springer-Verlag, New York, 1979), p. 193.

53. S. G. Bishop and P. C. Taylor, J. Non-Cryst. Solids **35** & **36**, 909 (1980).
54. S. G. Bishop and P. C. Taylor, Phil Mag. **B40**, 483 (1979).
55. S. G. Bishop, B. V. Shanabrook, U. Strom and P. C. Taylor, J. de Physique **42**, C4-383 (1981).
56. D. J. Robbins and P. J. Dean, Adv. Phys. **27**, 499 (1978).
57. F. J. DiSalvo, A. Menth, J. V. Waszczak and J. Tauc, Phys. Rev. B6, 4574 (1972).
58. D. U. Gubser and P. C. Taylor, Phys. Lett. **40A**, 3 (1972).
59. J. Tauc, F. J. DiSalvo, G. E. Peterson and D. L. Wood, in *Amorphous Magnetism*, ed. H. O. Hooper and A. M. de Graaf (Plenum Press, New York, 1973), p. 119.
60. R. A. Street, T. M. Searle and I. G. Austin, Phil. Mag. **32**, 431 (1975).
61. R. A. Street, T. M. Searle, I. G. Austin and R. A. Sussman, J. Phys. C7, 1582 (1974).
62. R. A. Street, T. M. Searle and I. G. Austin in *Amorphous and Liquid Semiconductors* eds. J. Stuke and W. Brenig (Taylor and Francis, London, 1974), p. 953.
63. K. Murayama, T. Ninomiya, H. Suzuki and K. Morigaki, Solid State Commun. **24**, 197 (1977).
64. M. A. Bösch and J. Shah, Phys. Rev. Lett. **42**, 118 (1979).
65. G. S. Higashi and M. Kastner, J. Phys. **C12**, L821 (1979).
66. G. S. Higashi and M. Kastner, J. Non-Cryst. Solids **35-36**, 921 (1980).
67. G. S. Higashi and M. Kastner, Phys. Rev. **B24**, 2295 (1981).
68. R. A. Street, Solid State Commun. **34**, 157 (1980).
69. J. Shah, Phys. Rev. **B21**, 4751 (1980).
70. G. S. Higashi and M. Kastner, Phys. Rev. Lett. **47**, 124 (1981).
71. K. Murayama, H. Suzuki and T Ninomiya, J. Non-Cryst. Solids **35-36**, 915 (1980).
72. M. A. Bösch, Phys. Rev. Lett. **48**, 649 (1982). (Comment).
73. K. Murayama and T. Ninomiya, Jap. J. Appl. Phys. **21**, L512 (1982).
74. K. Murayama, K. Kimura and T. Ninomiya, Solid State Commun. **36**, 349 (1980).
75. G. S. Higashi and M. A. Kastner, Phil. Mag. (1982), in press.
76. J. Shah and M. A. Bösch, Phys. Rev. Lett. **42**, 1420 (1979).
77. J. Shah and M. A. Bösch, Solid State Commun. **31**, 769 (1979).
78. M. A. Bösch and J. Shah, Phys. Lett. **74A**, 446 (1979).
79. J. Shah and P. M. Bridenbaugh, Solid State Commun. **34**, 101 (1980).
80. B. A. Wilson in *Tetrahedrally Bonded Amorphous Semiconductors* eds. R. A. Street, D. K. Biegelsen and J. C. Knights (AIP, New York, 1981) p. 273.
81. K. Murayama and M. A. Bösch, Phys. Rev. **B23**, 6810 (1981).
82. S. G. Bishop, U. Strom, E. J. Frickle and P. C. Taylor, J. Non-Cryst. Solids **32**, 359 (1979).
83. B. T. Kolomiets, T. N. Mamantova and A. A. Babeev, J. Non-Cryst. Solids **35-36**, 915 (1980).
84. F. Mollot, J. Cernogora and C. Benoit á la Guillaume, Phys. Stat. Solidi A**21**, 281 (1974).
85. M. A. Bösch, R. W. Epworth and D. Emin, J. Non-Cryst. Solids **40**, 587 (1980).
86. J. C. Phillips, J. Non-Cryst. Solids **41**, 179 (1980).
87. S. P. Depinna and B. C. Cavenett, Phys. Rev. Lett. **48**, 556 (1982).
88. S. P. Depinna and B. C. Cavenett, Phil. Mag. B (1982), in press.
89. J. Orenstein and M. Kastner, Phys. Rev. **43**, 161 (1979).
90. M. Kastner and J. Orenstein in *Physics of Semiconductors-1978* ed. B. L. H. Wilson (Institute of Physics, London, 1979), p. 1301.
91. J. Orenstein and M. Kastner, Phys. Rev. Lett **46**, 1421 (1981).
92. J. Orenstein, M. A. Kastner and V. Vaninov, Phil. Mag. **B46**, 23 (1982).
93. R. L. Fork, C. V. Shank, A. M. Glass, A. Migus, M. A. Bösch and J. Shah, Phys. Rev. Lett. **43**, 394 (1979).
94. D. E. Ackley, J. Tauc and W. Paul, Phys. Rev. Lett. **43**, 715 (1979).
95. D. L. Mitchell, P. C. Taylor and S. G. Bishop, Solid State Commun. **9**, 1833 (1971).
96. F. Urbach, Phys Rev. **92**, 1324 (1953).
97. J. D. Dow and D. Redfield, Phys. Rev. **B1**, 3358 (1970).
98. J. D. Dow and D. Redfield, Phys. Rev. **B5**, 594 (1972).

99. I. Z. Kostadinov. J. Phys. C **10**, L263 (1977).
100. J. P. de Neufville in *Optical Properties of Solids--New Developments* ed. B. O. Seraphin (North Holland, Amsterdam, 1976), p. 437.
101. D. J. Treacy, P. C. Taylor, and P. B. Klein, Solid State Commun. **32**, 423 (1979).
102. K. Tanaka in *Structure and Excitations of Amorphous Solids*, ed. G. Lucovsky and F. L. Galeener (AIP Conf. Proc. #31, AIP, N.Y., 1976), p. 148.
103. E. E. Mytilineau, P. C. Taylor and E. A. Davis, Solid State Commun. **35**, 497 (1980).
104. K. Tanaka and M. Kikuchi, Solid State Commun. **11**, 1311 (1972).
105. D. K. Biegelsen and R. A. Street, Phys. Rev. Lett. **44**, 803 (1980).

DISCUSSION

Optics of Chalcogenide Glasses

The optical properties of the chalcogenides exhibit little variation from sample to sample for melt-quenched glasses, except for some impurity effects which may appear in the absorption edge tail. Vapor deposited films can present quite different properties. Annealing, although reducing the discrepancy, does not succeed in restoring the bulk properties. In extreme deposition conditions, films can even have a molecular-type structure. Taylor emphasized that there is no real discrepancy between the published two sets of data on photoinduced ESR (Bishop, Strom, and Taylor, Biegelsen and Street) because the experiments were not performed under the same illumination conditions (time and power). Illumination can indeed have two distinct effects: first, it may change the charge state of existing defects; second, at longer time and/or higher power, it may create new defects.

Professor Taylor pointed out that there was strong evidence for two, and possibly three, photoluminescence processes in chalcogenide glasses. It was suggested that the existence of several processes could be explained by analogy with the behavior in crystals. There the several processes correspond to exciton recombination, both bound and free, and to donor-acceptor pair recombination. The process observed in glasses can be fit into this picture, particularly with the aid of ODMR results (see Cavenett and Depinna). As for the model of the self-trapped exciton proposed to explain part of the time-resolved experimental data, it was emphasized that the model does not imply that the exciton is related to a local defect (D^-, for example). It was suggested that investigations on a-Se, which is expected to present an exciton behavior, should help to discriminate between the models.

M. L. Theye, Chairwoman
J. Orenstein, Discussion Leader

COMMENTS

Pnictides: A Bridge Between the Two Classes of Defects in Amorphous Semiconductors

Discussions are given above by Owen and Taylor of the concept of spin-pairing through the negative effective correlation energy (U_{eff}) mechanism proposed by Anderson (1975), and in particular the application of these ideas to the case of structural point defects in chalcogenide glasses. In this specific application, spin-pairing mediated by a negative U_{eff} mechanism and driven by the lattice relaxation is characterized by the *over-coordination* of certain chalcogen centers (C_3^+) facilitated by the availability of lone-pair chalcogen-derived p-states lying at the top of the valence band; in addition conjugate under-coordinated centers (C_1^-), are also formed. The properties expected of a negative U_{eff} system are the following: no ESR signals observable in the dark, but only after the application of near band-gap light; photoluminescence (PL) considerably Stokes-shifted from the excitation energy; no variable range hopping, and d.c. conductivity exemplified by a $T^{-1/4}$ temperature dependence.

On the other hand, the other class of point defects which occurs in amorphous semiconductors are those which suffer a positive correlation energy, i.e. for which spin-pairing is energetically unfavorable, and which are exemplified by the class of group IV amorphous semiconductors, such as Si or Ge. Even the addition of hydrogen does not appear to change the correlation energy characteristics of the defects significantly, as discussed in the paper by Knights. The properties expected of positive U_{eff} systems are: ESR signals observable even in the dark; little Stokes-shift of the PL, reflecting the relatively small electron-phonon coupling present; and variable-range hopping of the d.c. conductivity. Over-coordination at defect centers *cannot* occur in these materials (if d-hybridization is neglected). Therefore a negative U_{eff} mechanism similar to that proposed for chalcogenide materials cannot take place, and the stable defect is thus expected to be the paramagnetic center, T_3^0. It has been suggested[2] that spin-pairing nevertheless could occur in a-Si or a-Ge, by a somewhat different mechanism, namely driven by changes in hybridization energy with a T_3^- center relaxing towards a p-bonded configuration and a T_3^+ center relaxing towards an sp^2 configuration, the assumed close proximity of such centers ensuring the exothermic nature of the spin-pairing by means of the Coulomb interaction.

It is of interest to inquire what may be the nature of the structural defects in those elements bridging the chalcogens in group VI and the tetrahedral materials in group IV, i.e. the pnictides, P, As and Sb. We will concentrate first on a typical pnictide material, a-As. This has been found to exhibit an ESR signal in the dark (which moreover increases with increasing temperature) but also exhibits an additional signal upon irradiation by near band-gap light; Stokes-shifted PL is also observed. Further, variable-range hopping d.c. conductivity is observed in certain circumstances, e.g. after the application of high pressure in bulk material, or in thin films vapor deposited onto low-temperature substrates. (A comprehensive survey of the physical properties of amorphous arsenic is given in the review by Greaves, Elliott, and Davis[4].) It is clear, therefore, that on the basis of the experimental evidence, a-As appears to behave as a "mixed" positive/negative U_{eff} material. The question thus arises as to how this circumstance can be understood.

Spin-pairing resulting from the over-coordination mechanism proposed by Mott, Davis, and Street[7] or Kastner, Adler, and Fritzsche[6] for chalcogenides can only take place in pnictides if sp^3-hybridization occurs, forming a P_4^+ center. This is because the only available nonbonding electrons are the s-electrons lying deep in the valence band, all three p-electrons being utilized in forming σ-bonds in the trigonal pyramidal arrangement characteristic of these materials. Thus, to form a P_4^+ center requires the sp^3 hybridization energy, and this endothermic process (including the true correlation energy U needed to place a second electron at a site producing the conjugate P_2^- center) is ameliorated to only a small degree by the extra bonding energy gained by the formation of an extra bond to give P_4^+. Hence, whether or not the overall process $2P_2^0 \rightarrow P_4^+ + P_2^-$ is exothermic (negative U_{eff}) or endothermic (positive U_{eff}) is a finely balanced matter. We have proposed[3] that it is only at those highly distorted sites, where the bond-angle is near the tetrahedral angle of 109° and consequently where the hybridization energy has already been supplied effectively by the local lattice distortion, that a negative U_{eff} will obtain. All other sites will be potential positive U_{eff} centers, because now the hybridization energy must be provided which makes the process of spin-pairing endothermic. Thus, the fact that a-As appears to be a mixed positive/negative U_{eff} system becomes understandable. The increase in the density of paramagnetic centers with increasing temperature as monitored by ESR, is explicable by assuming that those P_4^+, P_2^- centers for which U_{eff} is negative but small, disproportionate in a back-reaction forming paramagnetic P_2^0 centers with increasing temperature.

Having seen that a-As does indeed nicely bridge the gap between chalcogenide materials on the one hand and tetrahedral materials on the other by exhibiting characteristics common to both classes, it is of interest to inquire what systematic trend, if any, might be expected as one descends group V, i.e. on going from P to As to Sb. If our arguments concerning the nature of those sites in pnictides most likely suffer a negative U_{eff} are correct, it would be expected that examination of the mean bond-angle of the series would reveal the trend to be expected. Diffraction data have shown that the bond-angle varies from 102° to 98° to 96° for P, As, and Sb, respectively. Recalling that bond-angle fluctuations of 5-10° are indicated by the experimental data and predicted by CRN models, it is immediately apparent that one would expect more sites in P than Sb to be near the tetrahedral configuration, and concomitantly expect that more sites in P to be potential negative U_{eff} centers than would be the case for Sb. Thus we expect P to be more "chalcogen-like" and Sb to be more "tetrahedral-like".

Experimental evidence appears to bear out this contention. PL and ODMR experiments in a-P (see e.g. Cavenett's paper below) appear to confirm the existence of charged defect centers, whereas no PL is observed in a-Sb, presumably due to both the low level of radiative charged centers and the high concentration of neutral, nonradiative centers. A.C. conductivity experiments have shown too that fewer P_2^0 centers exist in a-P than in a-As (see e.g. Davis, above and Greaves et al.[4] Furthermore variable-range d.c. hopping conductivity is readily observed in a-Sb and is not easily annealed away[6] unlike the situation in a-As, again a sign that paramagnetic P_2^0 centers are most stable in Sb. Thus, pnictides offer a most fascinating transition between the defect-controlled properties exhibited by chalcogenides on the one hand and tetrahedral materials on the other.

S. R. Elliott
Dept. of Physical Chemistry
University of Cambridge
Cambridge, England

REFERENCES

1. P. W. Anderson, Phys. Rev. Lett. *34*, 953 (1975).
2. S. R. Elliott, Phil. Mag., *B38*, 325 (1978).
3. S. R. Elliott and E. A. Davis, J. Phys. C. *12*, 2577 (1979).
4. G. N. Greaves, S. R. Elliott, and E. A. Davis, Adv. Phys. *28*, 49 (1979).
5. M. Kastner, D. Adler, and H. Fritzsche, Phys. Rev. Lett. *37*, 1504 (1976).
6. A. MacIntosh and A. D. Yoffe (1982). To be published.
7. N. F. Mott, E. A. Davis, and R. A. Street, Phil. Mag. *32*, 961 (1975).

Optical Modification of Tunneling Systems in As_2S_3 Glass

A metastable gap in the low-energy density of states of As_2S_3 is created by irradiation of the glass with band-edge photons at low temperatures. Electric resonance and echo studies indicate that $\sim 10^{16}$ cm^{-3} tunneling centers are annihilated, a number which corresponds to the density of paramagnetic electronic centers created by low-level optical excitation. These findings associate tunneling activity with defective or low-density regions in glass and lead to a microscopic model for a tunneling center.

A microscopic description of the low-energy tunneling systems which occur in glasses and structurally disordered solids has not yet been found. The ubiquity of the atomic tunneling phenomenon has argued for a picture which is relatively general and not tied to specific atomic structures or bonding configurations.[1,2] Statistical reasoning has suggested that the large numbers of configurational states frozen into a glass as the liquid is evolved through its glass transition are supportive of widespread low-energy tunneling. The relationship of tunneling to theories of the glass transition has been explored.[3]

Alternatively, there have been attempts to relate tunneling to defect sites in glasses. Such theories are necessarily structure-specific, but since defects occur in all solids, there is some hope that there may be some generality to this picture. Two criteria are available for comparison of theory and experiment. First, there should be a correspondence between the number density of defects with tunneling centers, the latter believed to be between 10^{19} and 10^{20} cm^{-3}. Second, there must be unusually strong defect-phonon coupling, as evinced by direct measurements of tunneling center deformation potentials $\sim$1-2 eV. This latter criterion eliminates most structural and impurity-related defects in glasses which have been probed by esr.

A particular class of defects which has often been invoked as candidates for tunneling centers are the electronically active states which inhabit amorphous chalcogenide semiconductors.[4–7] Specifically, there exist two-electron centers which are stabilized by an effective negative correlation energy originating in atomic relaxation in the vicinity of the center. The two electron center can be ionized by a photon of band-gap energy, a process which is accompanied by a strongly-Stokes shifted photoluminescence and the formation of induced optical absorption at half the band-gap. The photoluminescence and induced absorption are time- and energy-dependent, but at temperatures below $\approx$80K a metastable, paramagnetic center persists indefinitely.[8] The mid-gap one-electron center may be removed, and the original properties of the glass restored, by thermal annealing above $\sim$150K or by optical bleaching with half band-gap radiation. Paramagnetic resonance[8] has shown that the maximum density of optically induced metastable centers is $\sim 10^{17}$ cm^{-3}.

When a typical chalcogenide glass, As_2S_3, is optically irradiated at low temperatures the properties of its tunneling systems are modified.[9] In particular, band-gap radiation can annihilate very low energy tunneling centers, thereby opening a gap in the tunneling density of states $\overline{P}$ below 5 GHz (20 μeV). This corresponds to the removal of $\sim 10^{16}$ cm^{-3} tunneling systems, a density in good accord with the density of paramagnetic centers induced under comparable irradiations. The tunneling systems can be restored by thermal annealing above 150K, or partially restored by bleaching with half band-gap irradiation.

In this discussion, we review low temperature electric resonance experiments which provide information on $\overline{P}$ in three low energy spectral regions.

Electric Echoes

At temperatures below 0.1K, phase memory times are sufficiently long that electric echoes, the analog of spin echoes, are observable as microwave radiation from tunneling systems in a-As_2S_3. Echo amplitudes are proportional to $\overline{P}(\hbar\omega_0)\mu^2$, where μ is the induced electric dipole moment of the tunneling system and $\overline{P}(\hbar\omega_0)$ is the tunneling density of states at energy $\hbar\omega_0$. Fig. 1 shows the striking decrease in the echo amplitude at 5 mK as a

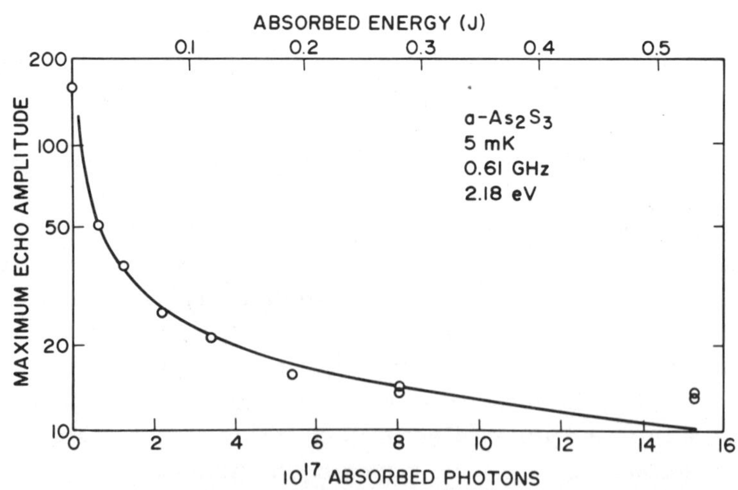

Fig. 1. Change of electric echo amplitude on step-wise irradiation of As_2S_3 glass with band-edge light. The line is proportional to $n^{-1/2}$.

function of absorbed band-edge photons, n, at 2.18 eV. The echo decreases by an order of magnitude after $\sim 10^{18}$ photons are absorbed. Bleaching with 1.16 eV light increases the echo amplitude by a factor of 3.

Dephasing Time T_2'

The echo amplitude modification does not unambiguously tell one whether it is $\bar{P}$ or μ which is changed by light. The echo decay with time is a measure of the dephasing time T_2' and is independent of μ. The rate $1/T_2'$ is governed by elastic dipole-dipole interactions between the resonantly excited tunneling species at $\hbar\omega_0$ with the thermally excited species at energy $\lesssim 2\ k_BT$. In the spectral diffusion model of Black and Halperin,[10] $1/T_2'$ depends on the concentration of centers as $\bar{P}^{1/2}(2k_BT)$ in the Gaussian decay regime. Fig. 2 shows results for optical modification of T_2' by excitation and bleaching at optical wavelengths. The effect of excitation is to *increase* T_2' by a factor of 2, which is consistent with the *decrease of* $\bar{P}$ by at least a factor of 4 at energies corresponding to a frequency of 0.2 GHz (5 mK). Bleaching restores the original T_2'. These results clearly indicate that $\bar{P}$ is substantially modified at very low energies.

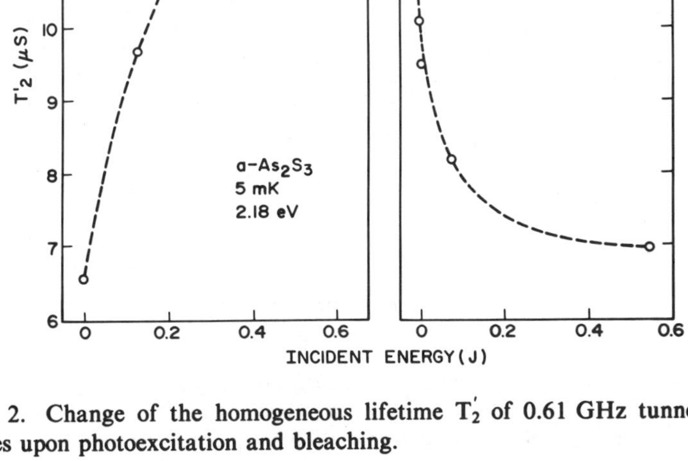

Fig. 2. Change of the homogeneous lifetime T_2' of 0.61 GHz tunneling states upon photoexcitation and bleaching.

Dielectric Constant $\epsilon(\omega,T)$

The resonant interaction of electric fields with the broad, flat distribution of tunneling systems produces a temperature-dependent contribution to the microwave dielectric constant of the form $\bar{P}(2k_BT)\mu^2 \ln T/T_0$, where T_0 is a reference temperature, and where $\hbar\omega_0 << 2k_BT$. Thus, the logarithmic slope of $\epsilon(\omega,T)$ is proportional to $\bar{P}(2k_BT)$. Fig. 3 shows $\Delta\epsilon/\epsilon$ at 0.61 GHz for temperatures between 60 and 700 mK and exhibits the expected ln T dependence. The surprising feature of these data is their insensitivity to optical irradiation. It is possible to reconcile this result with the echo data only if it is appreciated that the dielectric data at these temperatures probe a much higher energy region, $\gtrsim$5 GHz. A careful numerical calculation of $\epsilon(\omega,T)$ for various model densities of states has shown that a gap in $\bar{P}$ between 0 and 5 GHz will not produce observable effects in the temperature region shown in Fig. 3.

The conclusions of the three experiments are summarized in Fig. 4 which illustrates the optical modification of $\bar{P}$ below 5 GHz. We believe it is significant that the number of states removed, 2×10^{16} cm^{-3}, is close to the number of paramagnetic centers induced by similar irradiations. The opening of a gap and not simply a redistribution of the states in energy

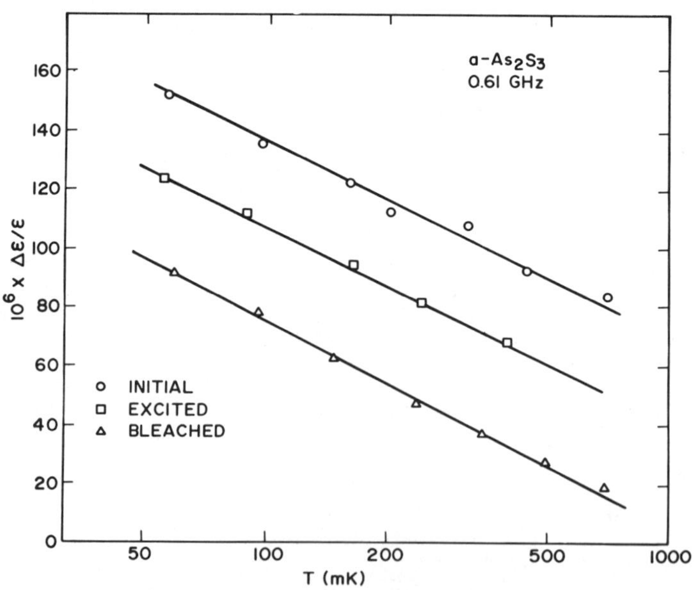

Fig. 3. Temperature dependence of the dielectric constant of As_2S_3 glass at 0.61 GHz. The curves have been offset vertically for clarity. No change in the slope is observed within experimental accuracy.

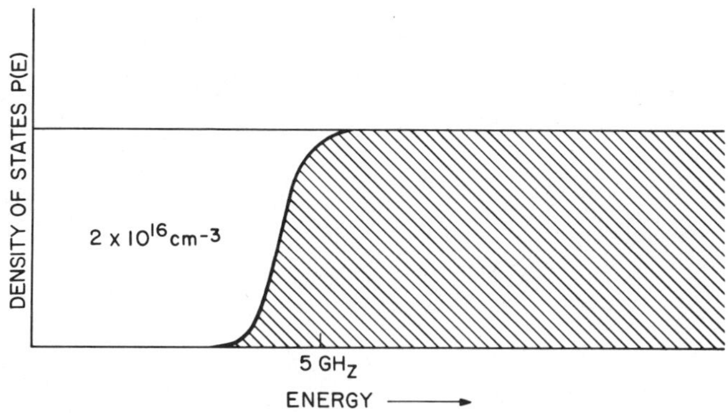

Fig. 4. Schematic of the hole burned into the tunneling system density of states by band-edge optical irradiation of As_2S_3 glass.

implies a spatial correlation of tunneling systems with electronic centers. A possible mechanism for modifying the gap is strain coupling. Photoexcitation of an electronic center would sufficiently change the environment of a tunneling system so as to shift it completely out of the tunneling spectrum.

A specific model for the tunneling center would associate it with a low mass, undercoordinated atom eg., a C_1^-. The tunneling coordinate would be essentially orthogonal to a bond axis. On photoexcitation the C_1^- is converted to a doubly coordinated C_2^0, for which a less favorable environment for tunneling would exist owing to a greater number of bonding constraints. In this picture, the tunneling site is clearly associated with a low-density, defective region in the glass network. The density of defects, or charged defect pairs (IVAP's),[11] is related to their energy of formation and the glass transition temperature, at which point the equilibrium defect density is frozen into a solid. Thus, the association of tunneling centers with defect sites provides a microscopic picture for recent theories on the relationship between $\overline{P}$ and T_g.[3]

In closing we note the observation of an as yet unexplained phenomenon associated with very slow recombination processes in As_2S_3. Fig. 5 shows the recovery of the echo amplitude at 5 mK after bleaching. The recovery follows a $\ln t$ behavior over several decades in time. The possibility that this effect is associated with a slow cooling of the sample not detected by our thermometer can be eliminated. If T_2' is monitored during this process it is observed to decrease with time. Since T_2' increases with decreasing T, this phenomenon reflects a time-dependent $\overline{P}$.

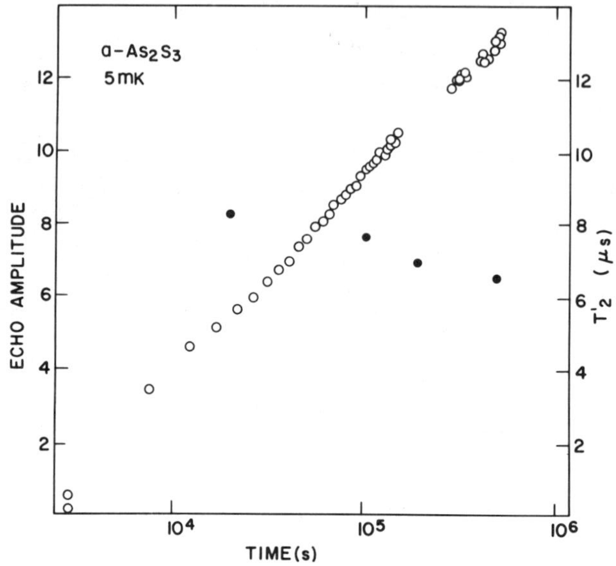

Fig. 5. Post-bleaching changes in As_2S_3 echo amplitude (open circles) and T_2' (solid circles). The temperature is constant at 5 mK. Note the increase in amplitude coincident with the decrease in T_2' indicative of a logarithmically time-dependent P(E).

We thank W. H. Haemmerle, P. D. Lazay, M. Stavola, and J. E. Graebner for their contributions to this work.

B. Golding
D. L. Fox
Bell Laboratories
Murray Hill, New Jersey 07974 USA

REFERENCES

1. P. W. Anderson, B. I. Halperin, and C. M. Varma, Philos. Mag. *25*, 1 (1972).
2. W. A. Phillips, J. Low Temp. Phys. *7*, 351 (1972).
3. M. H. Cohen and G. S. Grest, Phys. Rev. Lett. *45*, 1271 (1980).
4. P. W. Anderson, Phys. Rev. Lett. *34*, 953 (1975).
5. R. A. Street and N. F. Mott, Phys. Rev. Lett. *35*, 1293 (1975); N. F. Mott, E. A. Davis, and R. A. Street, Philos. Mag. *32*, 961 (1975).
6. M. Kastner, in *Proceedings of the Seventh International Conference on Liquid and Amorphous Semiconductors, Edinburgh, Scotland, 1977*, edited by W. E. Spear (G. G. Stevenson, Dundee, Scotland, 1977), p. 504.
7. E. N. Economou, K. L. Ngai, and T. L. Reineke, Phys. Rev. Lett. *39*, 157 (1977).
8. S. G. Bishop, U. Strom, and P. C. Taylor, Phys. Rev. B*15*, 2278 (1977).
9. D. L. Fox, B. Golding, and W. H. Haemmerle, Phys. Rev. Lett. *49*, 1356 (1982).
10. J. L. Black and B. I. Halperin, Phys. Rev. B*16*, 2879 (1977).
11. M. Kastner, D. Adler, and H. Fritzsche, Phys. Rev. Lett. *37*, 1504 (1976).

DEFECT STATES IN TETRAHEDRAL AMORPHOUS SEMICONDUCTORS

J. C. Knights

Xerox Palo Alto Research Center
Palo Alto, CA 94304, U. S. A.

The issue of defect states in tetrahedral amorphous semiconductors is addressed from both structural and electronic viewpoints. Extended structural inhomogeneities are identified as the principal structural defects while localized dangling bonds are the principal electronic defects. Correlations between the two in the specific case of a–Si:H are discussed in detail and the issue of the interrelationship between defects and impurities described.

I. INTRODUCTION

Tetrahedral amorphous semiconductors comprise a group of materials that is currently receiving considerable attention due in part to a number of important technological applications and in part to its utility as a 'test-bed' for the improvement of basic understanding of amorphous materials. The canonical material is hydrogenated amorphous silicon - papers on this and its derivatives have dominated recent amorphous semiconductor meetings. It should be recognized that the other members of the group - a–Ge, a–C, a–GaAs and other III-V materials - are being studied and do show a number of unique features, but the sheer volume of activity on a–Si:H has pushed understanding of this material well beyond the others. As such, discussion of work on a–Si:H will form the bulk of this paper, although many of the models for defects and their electronic properties have wider applicability.

It is also appropriate to mention at this point that the study of a—Si:H is entering a new era, comparable to that experienced by crystalline silicon 30-40 years ago in which issues relating to surfaces and interfaces are beginning to play a major role both in the basic science and in the technology. The implications of this for the study of defect states are considerable - spatial distributions of defects with respect to interfaces, interface defects themselves and band-bending effects due to interfaces and surfaces being among the more important.

This paper will attempt to cover the subject of defect states from a novel perspective, namely a structural one. The starting premise is that many of the issues regarding defects emanate from the fact that no tetrahedral amorphous semiconductors are glass formers and that all the materials currently being studied must be prepared in thin film form by some form of deposition process that does not permit thermal equilibrium to be achieved, i.e. there is a considerable kinetic component in the forces that shape the structure of the materials. Following a background review, the structural defects observed to date will be described and compositional inhomogeneity associated with these defects discussed. The electrical and optical results that relate to defect states will be described and the reasons for the assignment of the dominant defect to the silicon dangling bond discussed. Finally results pointing to the existence of interface dangling bonds and to impurity-defect complexes will be described.

II. BACKGROUND

As stated above, tetrahedral amorphous semiconductors cannot be prepared other than by some form of thin film deposition process. It is also a fact of life that, to date, all materials prepared in the absence of hydrogen or some other monovalent additive such as fluorine or chlorine have shown, in their as-deposited states, very high densities of defect states in the forbidden gap. The evidence for these states comes from electron spin resonance (ESR) measurements indicating high ($\geqslant 10^{20}$ cm^{-3}) densities of unpaired electrons,[(1)] the absence of appreciable photoconductivity,[(2)] optical absorption measurements that show strong absorption tails below the band-edges[(1)] and D.C. conductivity measurements that indicate $\ln\sigma\alpha T^{-1/4}$,[(3)] i.e. variable range hopping in a high density of states at E_F, the Fermi level.

The first indications that materials with much lower densities of states in the forbidden gap could be prepared came from the work of Spear and coworkers[(4)] and Lewis et al.[2] In the case of Spear and his collaborators, amorphous silicon was obtained, following work by Chittick,[(5)] through the use of glow discharge decomposition of silane. Lewis et al.[2] studied the addition of hydrogen to sputtered a–Ge. Both groups found that the materials behaved like intrinsic semiconductors with well defined conductivity activation energies, high room temperature resistivities and optical absorption edges shifted to considerably higher photon energies than those of conventionally evaporated or sputtered materials. Fig. 1 illustrates the optical results of Lewis et al.[2] together with an absorption spectrum that identified, for the first time, the presence of covalently bonded hydrogen via the Ge—H vibrational mode absorptions at 0.07 and 0.23 eV. Subsequent to these results, the presence of hydrogen in the silicon films deposited by glow-discharge decomposition of silane was clearly identified both by infrared vibrational spectroscopy[6] and by measuring the pressure of hydrogen evolved on heating the material to crystallization[7] (~650-750°C). It was assumed that the role of hydrogen was to act as a terminator of dangling bonds, removing the unpaired electron and transforming the localized state in the forbidden gap into states degenerate with the conduction and valence band extended states. What came as a surprise to

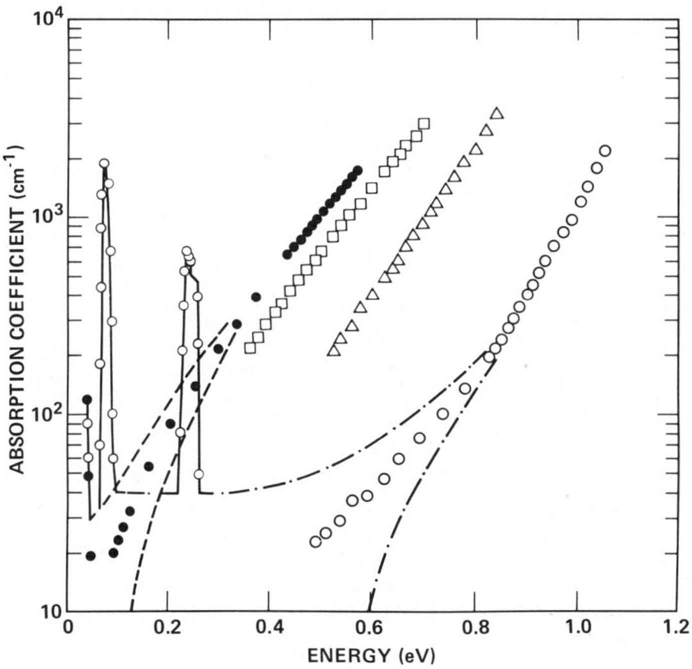

Fig. 1. Absorption coefficient, as a function of photon energy for sputtered a–Ge:H films as a function of hydrogen partial pressure (•, ×, Δ, O = P_{H2} of 0, 1.0, 2.2, 6.5 mtorr) from A. J. Lewis, Jr., G. A. N. Connell, W. Paul, J. R. Pawlik and R. J. Temkin in AIP Conf. Proc. **20**, 27 (1974) with permission.

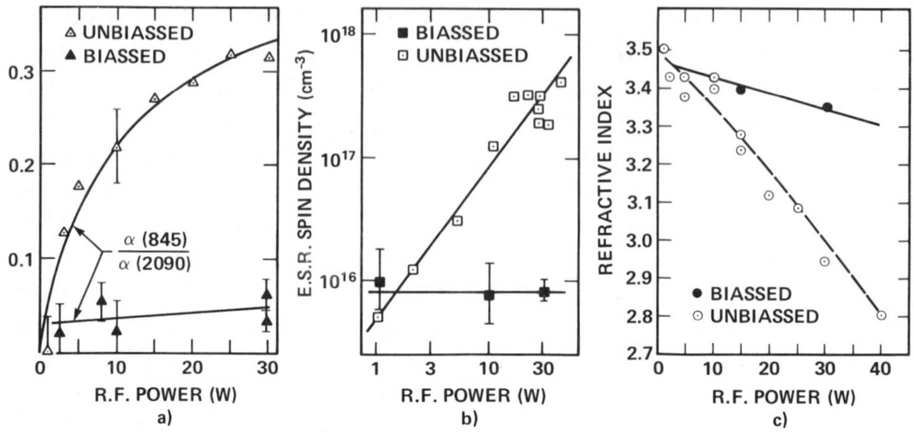

Fig. 2. Ratio of intensities in the 845 cm^{-1} to 2909 cm^{-1} SiH vibrational modes in a–Si:H films as a function of RF power and substrate bias (b) electron spin density in the same films. (c) refractive index of the same films.

early workers in the field was that the amount of hydrogen in the a–Si:H was ~5-35 atomic percent, far in excess of the amount calculated to be necessary to satisfy the broken bonds that occur as a result of cumulative bond angle distortion in the construction of continuous random network models. Following the discovery of the presence of hydrogen, considerable efforts were made to identify correlations between trends either in overall hydrogen content or in local bonding configurations (SiH, SiH_2, SiH_3, etc.) and trends in the residual defect density. Surprisingly, again, the correlation with hydrogen content was the inverse of that one might expect - the more hydrogen, in general, the higher the residual defect density.[(8)] This general correlation was soon reinforced by a strong correlation between the intensity of a specific I.R. vibrational mode, characteristic of an $(SiH_2)_n$ chain,[(9)] the defect density as measured by electron spin resonance and the density of the material as reflected in the refractive index. This is illustrated in Figure 2(a), (b) and (c). This association of high hydrogen content and defect densities with low physical density led in turn to a structural examination of the 'defective' material with transmission and scanning electron microscopy.[(10)] The results of these studies and subsequent studies of the spatial distribution of hydrogen form the core of the next section.

III. STRUCTURAL AND COMPOSITIONAL INHOMOGENEITY

Fig. 3 comprises scanning and transmission electron micrographs of, respectively, fracture surfaces parallel to the film growth direction and thin films perpendicular to the growth direction of hydrogenated amorphous

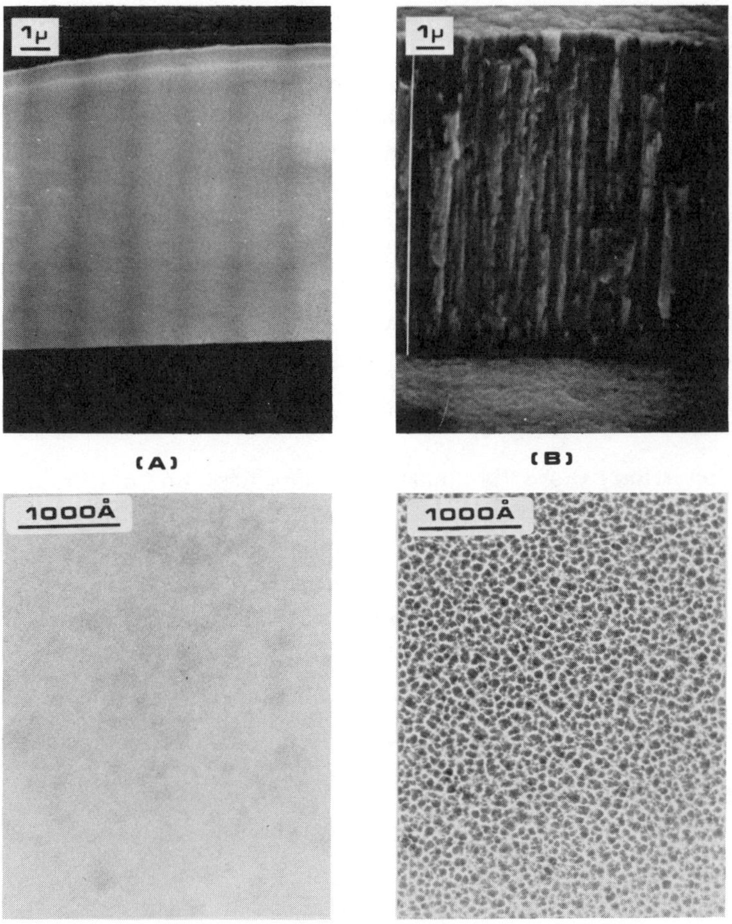

Fig. 3. Scanning and transmission electron micrographs of the fracture surfaces of thick a–Si:H films and of thin (~500 Å) films respectively illustrating high and low defect density material.

silicon produced by plasma deposition. Clearly there is inhomogeneous lateral morphology in one of the thin films that is associated with columnar growth morphology in the corresponding thick film. This anisotropic inhomogeneity is confirmed by neutron scattering measurements[(11)] that, combined with annealing studies, suggest that this material can be considered as a bundle of ~60-100 Å rods separated by voids. This type of film is the highly hydrogenated, high defect density material discussed above. The lowest defect density material in contrast contains no structural inhomogeneity observable by any technique including neutron scattering. The inhomogeneity has been associated by a number of authors[(10,12,13)] with nucleation of the films at discrete points followed by a failure to coalesce and subsequent columnar growth. The precise mechanisms for this process are beyond the scope of this article, but it is important to recognize that columnar growth morphology is widely observed in many thin film growth processes and is characteristically attributed to shadowing of the interstitial regions from a flux of arriving atoms by asperities that develop during growth. It is thus concluded that this type of structural defect, which is characteristic of the kinetics of the deposition process and not intrinsic to the material, is a primary class of defect in the tetrahedral amorphous semiconductors. It has been observed in a–Si:H prepared by both plasma deposition[(10)] and sputtering,[(12)] in a–Ge:H[(14)] and a–Ga:As[(15)] prepared by sputtering and its presence has been inferred in a–Ge:H prepared by plasma decomposition of germane.[(16)]

The fact that there is material that has no detectable structural inhomogeneity might be construed as indicating that the 'ideal' material is in fact a continuous random network. There is, however, strong evidence (a) that there is compositional inhomogeneity involving hydrogen and (b) that this inhomogeneity is universal, being easily detectable in the same material that shows no structural inhomogeneity.

Fig. 4 shows the nuclear magnetic resonance line generated by protons in a–Si:H.[(17)] It can be decomposed as shown into two components that, via

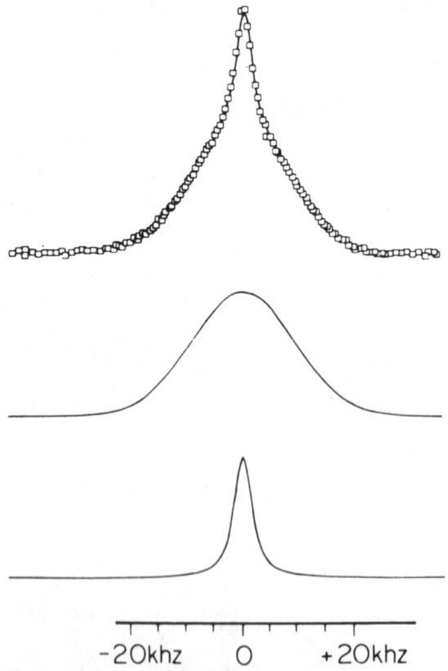

Fig. 4. Proton magnetic resonance line in a–Si:H film showing fit of sum of Gaussian and Lorentzian components.

either hole-burning[18] or dipolar echo experiments,[19] can be shown to be spatially isolated from each other in the sense that spin diffusion is very slow between the two components despite rapid spectral diffusion within the broad component. The lineshapes can be further analyzed to yield the following information: First, both lines owe their widths to static homonuclear dipole-dipole interactions, i.e. to random fields generated by neighboring protons. These linewidths can be compared to calculations of model spatial distributions to yield the information that the narrow line only originates from dispersed monohydride (SiH) groups while the broad line can originate from a number of groupings such as SiH_2, $(SiH_2)_n$, SiH_3 and heavily hydrogenated silicon surfaces. Second, the absence of fast spin diffusion indicates that the regions of material occupied by these two 'phases' must be separated by hydrogen-free regions $\geqslant 7$ Å deep.[17] This behavior is characteristic of all a–Si:H be it plasma-deposited,[17,18] sputtered,[20,21] low defect density or high defect density and it has also been observed, with some qualitative and quantitative differences in a–C:H and a–SiC:H.[22] Fig. 5 shows the interesting trend in hydrogen content distribution between the

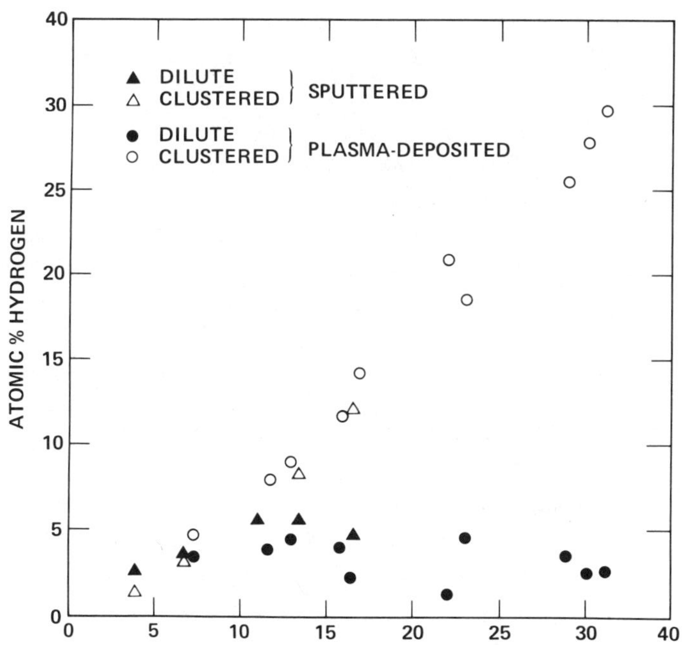

Fig. 5. Distribution of hydrogen between the dilute and clustered environments in plasma-deposited and sputtered a–Si:H after the work of Reimer et al. and Jeffrey et al.

two 'phases' in both sputtered and plasma-deposited a–Si:H. Up to a hydrogen content of ~5-10 atomic percent, the hydrogen is distributed with slightly more in the dilute phase as opposed to the clustered phase. Above this point the amount in the clustered phase increases monotonically while that in dilute saturates and even shows indications of declining. The fact that the linewidths do not change in the same range of hydrogen content indicates that the spatial extent of the clustered phase is increasing. It seems most probable that the clustered phase corresponds to hydrogenated internal surfaces - hydrogen evolution on heating occurs initially from this phase at temperatures so low that Si–Si bond reconstruction must occur to recover some of the energy of Si–H bond breaking.

Summarizing the structural and compositional data, it is clearly established that a major structural defect is a void or microcrack that may or may not be anisotropic and oriented parallel to the growth direction. It is equally well established that hydrogen clustering does exist and that a picture of a–Si:H as a compositionally inhomogeneous solid over the entire composition range is consistent with all observations. The issue of whether the two inhomogeneities are commensurate and the sequelae should they be, such as the presence of one requiring the presence of the other are not fully explored.

IV. ELECTRONIC DEFECTS

Before describing the various results concerning the electronic states of defects, it is helpful to define what is an electronic defect. Starting with a 'perfect' continuous random network, it is now accepted that Anderson localization due to potential fluctuations will cause localization of the electron states in the band-tails.[23] These localized states, depending on their distribution in energy, can act as traps and as such might be characterized as 'defect' states. In a–Si:H, however, there is good evidence that in the best material, despite dispersive transport of both electrons and holes associated with multiple trapping and release from these band-tail states, essentially no carriers are lost into these traps in time-resolved drift mobility measurements. These states are not, therefore, considered as defect states. The defect that has been clearly identified both by its trapping and recombination characteristics and by electron spin resonance is the single silicon dangling bond. The **g** value of the observed resonance is 2.0055, a value equal to that calculated from a combination of observations on known silicon dangling bonds at Si–SiO_2 interfaces.[24] The various charge states of this defect and its role as a recombination center have been detected by a combination of ESR light-induced ESR and luminescence measurements.[25,26,27] Fig. 6 shows the relationship between luminescence intensity at low temperatures and the

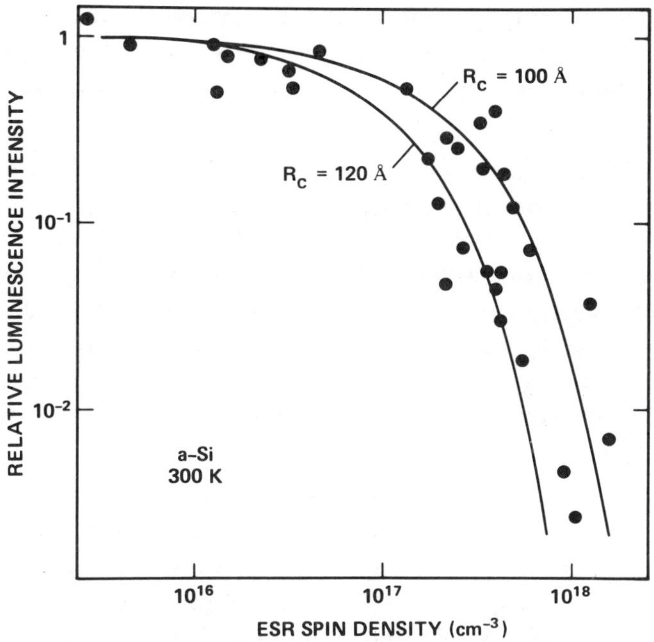

Fig. 6. Luminescence intensity versus electron spin density in undoped a–Si:H.

dangling bond density measured by ESR. The luminescent transition involves band tail to band-tail transitions while the competing non-radiative transition involves tunneling to the dangling bond with a rate p given by

$$p = \nu_0 \exp(-2R_D/R_0) \tag{1}$$

where R_D is the distance to the defect, $\nu_0 \sim 10^{12}$ sec^{-1} and R_0 is the effective Bohr radius of the electron hole pair determined from the radiative lifetime to be ~10 Å. This non-radiative transition will dominate when R_D is less than R_C where

$$R_C = 1/2\ R_0 \ln(\nu_0 \tau_R) \tag{2}$$

Assuming a random distribution of defects the luminescence efficiency is given by

$$Y_L = \exp(-4\Pi R_C^3 N_s/3) \tag{3}$$

Putting an experimentally determined mean value of $\tau_R \sim 10^{-3}$ sec and $R_0 = 10$ Å into (2) gives a value of $R_C \sim 100$ Å. The lines in Fig. 6 are the luminescence efficiency calculated using equation (3) with $R_C = 100$ Å and 120 Å.

The fact that a model based on a random distribution fits the data quite well raises the question of whether the electronic defects and the structural defects bear any relationship to each other. The answer is that these data do not permit a real comparison. Defect densities of 10^{15} and 10^{18} cm^{-3} correspond to average separations between defects of ~500 Å and 50 Å respectively. These separations are greater than or comparable to the scale of the structural inhomogeneity, hence it is unlikely at these densities that any inhomogeneity in defect distribution commensurate with the structural inhomogeneity would be detectable.

The charge states of the dangling bond can be detected by a combination of doping and optical excitation at low temperatures.[(27)] Doping low defect density a–Si:H with either boron or phosphorus produces the following results: the equilibrium spin density is reduced to below detection limits with either boron or phosphorus doping. The luminescence is sharply reduced at high doping levels while a resonance line identical to the equilibrium ESR line appears upon optical excitation in doped material. The explanation of the ESR behavior that best fits the observations has been proposed by Street and Biegelsen.[(27)] They propose that the neutral dangling bond has a positive correlation energy for the addition of a second electron, subsequent optical absorption[(28)] and resonance measurements[(29)] indicate that this energy is ~0.3-0.4 eV. Doping performs two functions: it shifts the equilibrium Fermi level, either emptying (boron) or doubly occupying (phosphorus) the dangling bond state and in addition it creates new dangling bonds. The result is that there are no equilibrium neutral dangling bonds in doped material, but a large non-equilibrium population can be generated under optical excitation. In addition at high doping levels and in high defect density undoped material a new luminescence peak is observed at 0.9 eV (the normal peak is at −1.3 eV) that has been associated with a radiative transition directly into a dangling bond state.

The association of doping with increased defect densities raises the question of whether the new defects are associated directly with the dopant atoms or are created as a result of the position of the Fermi level. Results on compensated material[(30)] suggest strongly that the position of the Fermi level is the dominant factor. Fig. 7 shows the 1.3 eV luminescence band intensity and light-induced ESR spin density as a function of increasing boron compensation of a sample with a fixed phosphorus concentration. Clearly the

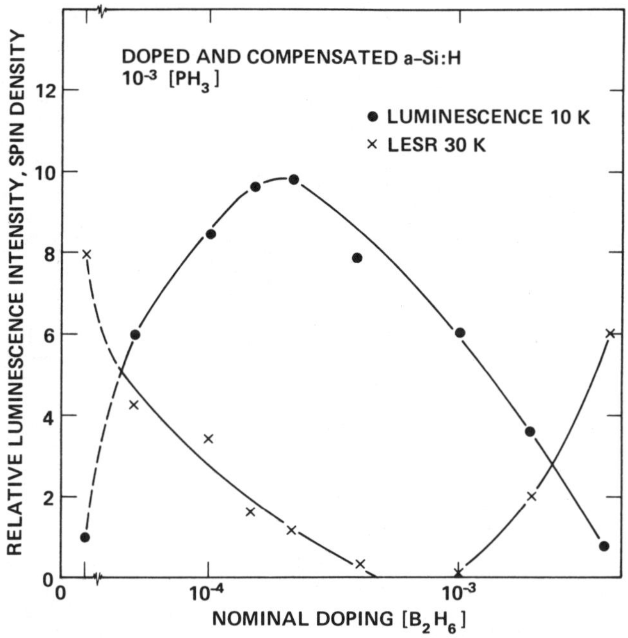

Fig. 7. Luminescence intensity and dangling bond light-induced ESR spin density in 10^{-3}[PH_3] doped a–Si:H as a function of boron compensation.

number of dangling bonds is reduced by compensation. The implication is that as the Fermi level is moved close to the band tails, an autocompensation mechanism takes over that increases the dangling bond density and prevents further movement of the Fermi level. It has been proposed that this mechanism can arise from an impurity-defect complex in which the dopant atom retains it normal coordination, i.e. 3, and a vacancy is created with a silicon dangling bond state. This is consistent with the observation that the majority of incorporated dopant atoms, at least in the case of arsenic at 1% concentration, are threefold coordinated.[31] Again the question can be asked, is there any evidence that structural defects are associated with the electronic defects. The answer is that there certainly is a correlation. High levels of doping, particularly with boron, are found to cause inhomogeneity that can be detected either by direct imaging microscopy or by enhanced low-temperature hydrogen effusion. Establishing a direct association in clearly difficult, but perhaps studies of microstructure and proton NMR in compensated material might be illuminating.

In summary, the dominant defect state is a silicon dangling bond in one of its charge states. The electrons in those states are strongly localized and

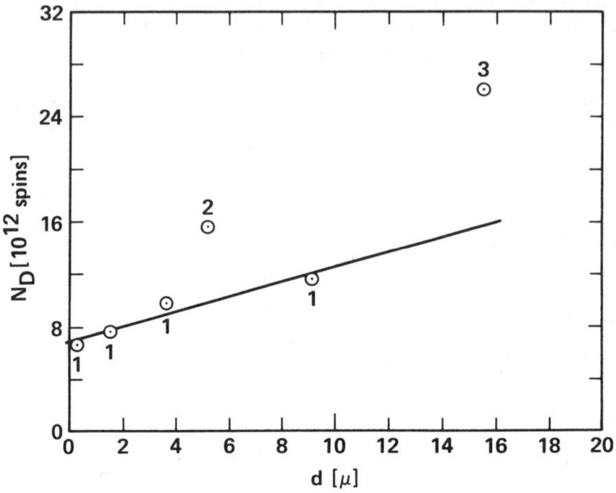

Fig. 8. Equilibrium dangling bond spin density in a–Si:H as a function of thickness, points marked 2 and 3 correspond to samples that have the corresponding number of layers of material obtained by additional depositions following exposure to atmosphere.

although there is a correlation between the presence of high densities of these states and large internal surfaces that are heavily hydrogenated, there is no direct evidence suggesting that the dangling bonds are at these surfaces. Doping introduces both structural and electronic defects, but again there is only a correlation, not a direct connection.

V. INTERFACE OR SURFACE DEFECTS

As the bulk defect densities have been reduced to levels $\sim 10^{15}$ cm^{-3}, it has become possible to address the issue of surface and interface defects directly. They have been detected by examining thickness scaling of equilibrium ESR in samples of low bulk defect density material,(32,33) and by studying oxidation of material with large areas of internal surfaces due to columnar growth.(34) Fig. 8 illustrates the results of thickness scaling of defect density measured by ESR. Two significant points are (a) there is a zero thickness intercept $\sim 10^{13}$ dangling bonds and (b) that multiple depositions preserve this signal in each layer. The results of photothermal deflection spectroscopy(33) on the defect absorption has extended this result to thinner films lower bulk density material where it is found that the surface or interface defect density is $\leqslant 10^{12}$ cm^2. Finally the study of internal surface oxidation(34) clearly identified dangling bond states at the SiO/Si interface in densities $\sim 3 \times 10^{11}$ cm^2.

In these studies the nature of the defect is found to be the same as the bulk, i.e. the silicon dangling bond. At present there is no direct evidence to suggest that there are any other defect types involved. There is, however, pervasive evidence that all the thin film deposition processes used introduce high concentrations (10^{18}-10^{21} cm^{-3}) of oxygen, carbon and nitrogen at the beginning of deposition and the possibility of impurity-defect complexes must be raised.

VI. CONCLUSION

Considering a–Si:H in isolation, it is clear firstly that the material is in general structurally and compositionally inhomogeneous although there are questions as to the scale and extent of both these inhomogeneities. It is also clear that the dominant electronic defect, both in the bulk and at the surface is the silicon dangling bond. The fact that impurities both enhance electronic defect levels and enhance the structural and compositional inhomogeneity is strongly suggestive of a link between the two types of defect, as is the overall trend to higher electronic defect densities in undoped material with greater inhomogeneity, but no direct link has been shown.

Considering now the whole range of tetrahedrally coordinated amorphous semiconductors, the point made in the introduction concerning the non-equilibrium deposition process suggests that the inhomogeneities are likely to be present in all these materials. In the presence of hydrogen it is also likely that single dangling bonds will be the major electronic defect in all the materials. Confirmation of some of these predictions is already available for a–Ge:H, a–C:H and a–GaAs:H. These are, however, possibilities for other defect states associated for instance with B–H–B and Ga–H–Ga groupings that have already been detected in the III-V materials.[(35)]

Finally, it should be recognized that there is strong circumstantial evidence from work with a–Si:H solar cells[(35)] that other impurities such as oxygen, nitrogen and chlorine have significant effects on transport properties suggesting that impurity-defect relationships, as in the case of crystalline semiconductors, will be a subject for future scrutiny.

ACKNOWLEDGEMENTS

Much of the work discussed here is part of a broad collaborative study of a–Si:H undertaken over the past seven years at Xerox PARC. I would particularly like to acknowledge the contributions of R. A. Street, D. K. Biegelsen, R. J. Nemanich, G. Lucovsky, J. A. Reimer, W. B. Jackson and R. A. Lujan.

REFERENCES

1. M. H. Brodsky and R. S. Title, Phys. Rev. Lett. **23**, 581 (1969).
2. A. J. Lewis, G. A. N. Connell, W. Paul, J. R. Pawlik and R. J. Temkin, AIP Conf. Proc. **20**, 27 (1974).
3. A. H. Clark, Phys. Rev. **154**, 750 (1967).
4. W. E. Spear, Proc. 5th Int. Conf. on Amorphous and Liquid Semiconductors, edited by J. Stuke and W. Brenig (Taylor & Francis, London 1974) p. 1.
5. R. C. Chittick, J. Non-Cryst. Sol. **3**, 255 (1970).
6. M. H. Brodsky, M. Cardona and J. J. Cuomo, Phys. Rev. B **16**, 3556 (1977).
7. A. Triska, D. Dennison, H. Fritzsche, Bull. Am. Phys. Soc. **20**, 392 (1975).
8. J. C. Knights and G. Lucovsky, Critical Rev. in Solid State and Mater. Sci. **9**, 211 (1980).
9. J. C. Knights, Jap. J. Appl. Phys. **18**, Supp. 18-1, 101 (1979).
10. J. C. Knights and R. A. Lujan, Appl. Phys. Lett. **35**, 244 (1979).
11. A. J. Leadbetter, A. A. M. Rashid, R. M. Richardson, A. F. Wright and J. C. Knights, Solid State Commun. **33**, 973 (1980).
12. R. C. Ross and R. Messier, J. Appl. Phys. **52**, 5329 (1981).
13. W. Paul, J. Physique **42**, C4-1165 (1981).
14. S. V. Krishnaswamy, R. Messier, Y. S. Ng, T. T. Tsong and S. B. McLane, J. Non-Cryst. Sol. **35/36**, 531 (1980).
15. D. K. Paul, J. Blake, S. Oguz and W. Paul, J. Non-Cryst. Sol. **34/35**, 501 (1980).
16. G. Lucovsky, R. J. Nemanich and J. C. Knights, Phys. Rev. B **19**, 2064 (1979).
17. J. A. Reimer, R. W. Vaughan and J. C. Knights, Phys. Rev. Lett. **44**, 193 (1980).
18. J. A. Reimer, R. W. Vaughan and J. C. Knights, Phys. Rev. B **24**, 3360 (1981).
19. W. E. Carlos and P. C. Taylor, Phys. Rev. Lett. **45**, 358 (1980).
20. W. E. Carlos, P. C. Taylor, S. Oguz and W. Paul, AIP Conf. Proc. **73**, 67 (1981).
21. F. R. Jeffrey, M. E. Lowry, M. L. S. Garcia, R. G. Barnes and D. R. Torgeson, AIP Conf. Proc. **73**, 83 (1981).
22. J. A. Reimer, R. W. Vaughan, J. C. Knights and R. A. Lujan, J. Vac. Sci. Technol. **19**, 53 (1981).
23. N. F. Mott and E. A. Davis, *Electronic Processes in Non-Crystalline Materials* (Clarendon Press, Oxford 1971).
24. D. K. Biegelsen, Solar Cells **2**, 421 (1980).
25. R. A. Street, J. C. Knights and D. K. Biegelsen, Phys. Rev. B **18**, 1880 (1978).
26. R. A. Street and D. K. Biegelsen, J. Non-Cryst. Sol. **35/36**, 651 (1980).
27. R. A. Street and D. K. Biegelsen, Solid State Commun. **33**, 1159 (1980).
28. W. Jackson, Solid State Commun. (in press).
29. J. D. Cohen, J. P. Harbison and K. W. Wecht, Phys. Rev. Lett. **48**, 109 (1982).
30. R. A. Street, D. K. Biegelsen and J. C. Knights, Phys. Rev. B **24**, 969 (1981).
31. J. C. Knights, T. M. Hayes and J. C. Mikkelsen, Jr., Phys. Rev. Lett. **39**, 712 (1977).
32. W. Jackson, D. Biegelsen, R. Nemanich and J. Knights, to be published.
33. R. A. Street and J. C. Knights, Phil. Mag. B **43**, 1091 (1981).
34. Z. P. Wang, L. Ley and M. Cardona, Phys. Rev. B **26**, 1987 (1982).
35. A. Delahoy and R. W. Griffith, J. Appl. Phys. **52**, 6337 (1981).

DISCUSSION

Defects in Amorphous Semiconductors

Knights' concluding remarks that the structure of the surface of a-Si:H would be the next important area of interest led to the initial discussion remark that it is difficult to begin to understand the structure of the surface in a system where bulk is so poorly understood. In particular the way in which hydrogen is incorporated in the dilute phase, revealed by NMR, is unknown. It was mentioned, in response to a question that a-Si:H is two-phase at all H concentrations that have been studied thus far.

In the discussion of doping it was noted that many unanswered questions remain. For example, P does not create as many defects as B. Also, compensation is complicated because the dopant molecules are known to interact in the glow discharge process.

The remainder of the discussion centered on amorphous III-V materials. For several reasons the structure is more complicated even than a-Si. In the III-Vs there is also evidence for columnar structure. On the microscopic scale there is additional chemical disorder. This arises from wrong bonds (e.g. Ga-Ga), and from the possibility of 3-fold, rather than tetrahedral, coordination. As a result it is not obvious that alloying with H will compensate a-III-Vs.

It was pointed out that structural inhomogeneities, which can be viewed as density fluctuations, can also be found in nonhydrogenated amorphous tetrahedrally coordinated semiconductors. This has been shown (for example, by electron microscope and small angle X-ray scattering experiments) on a variety of a-Ge and a-Si films prepared by different methods. The results are strongly dependent on the method of deposition, the substrate conditions, contamination, etc. These films, grown in strongly nonequilibrium conditions, exhibit important structural relaxation effects just after deposition, as well as during subsequent annealing. This relaxation, which, at least during its first stages, proceeds by collective motion of neighboring atoms over short distances, not only eliminates dangling bonds, but also leads to network reorganization on a larger scale, with probable redistribution of dihedral angles and strains. The presence of such structural inhomogeneities which cannot be entirely eliminated by suitable changes of deposition conditions on annealing, could for example play a role in the absolute values of the a.c. conductivity, which can widely vary from sample to sample while its temperature behavior remains identical.

Another class of tetrahedrally coordinated amorphous semiconductors, i.e. amorphous III-V compounds, was briefly discussed. Although the gross features of their structure, optical and transport properties have been determined, very little has been achieved up to now by way of characterizing the defects in these amorphous materials. The defect situation indeed appears to be more complex than in a-Ge or a-Si. Defects related to dangling bonds are also likely to be present but novel situations can be expected: i) both components can take their preferred, three-fold coordination; ii) defects equivalent to outside defects in the crystalline compounds (see c-GaP), where an atom at a given site is replaced by an atom of the other constituent (antisite defects), as well as different configurations of wrong bonds, i.e. bonds between two atoms of the same component, should occur. These defects in their different charge states are expected to introduce a variety of localized states near the band edges and throughout the pseudo-gap. Although the incorporation of hydrogen, in both sputtered and glow-discharge samples, has been shown to produce the same kind of blue shift of the absorption edge as in a-Ge or a-Si, it has yet to be proved that hydrogen can effectively compensate dangling bonds related to both components, so as to produce a significant decrease of the density of localized states in the pseudo-gap.

Another characteristic of the amorphous III-V compounds seems to be that at least in some cases (for example GaSb), a large excess of one component (here Sb) can be incorporated into the network in a self-compensated mode, leading to a continuous variation of the optical gap, transport properties, etc. (alloying effects).

J. Orenstein, Chairman
M. L. Theye, Discussion Leader

OPTICALLY DETECTED MAGNETIC RESONANCE IN AMORPHOUS SEMICONDUCTING GLASSES

B. C. Cavenett and S. P. Depinna

Department of Physics
The University, Hull
HU6 7RX, U.K.

Although it has been assumed for many years that photoluminescence in materials such as the chalcogenide glasses is due to a single recombination process, recent optically detected magnetic resonance (ODMR) studies have shown that both exciton and pair processes are involved. This paper will briefly outline the techniques and principles of ODMR as applied to both crystalline and amorphous materials, and illustrate the importance of the method by a consideration of recombination at triplet exciton states in glasses. Amorphous phosphorus has become the model material for these investigations, and the relationship between the ODMR, luminescence lifetime and time resolved results will be used as a basis for discussing parallel results in the chalcogenide glasses, where, in particular, new luminescence measurements of both the crystals and the related glasses can be compared with the ODMR results. The investigations establish the ODMR technique as an important method for analyzing the recombination processes in these and other amorphous semiconductors.

I. INTRODUCTION

Studies of recombination in amorphous semiconductors, such as the chalcogenide glasses, have been stimulated by the rapid growth of technological interest in these materials over the last two decades. The experimental techniques of luminescence spectroscopy, involving both cw and time resolved studies, excitation spectra and PL fatigue measurements, are direct and sensitive probes of recombination processes, and many aspects of this work have recently been reviewed by Street[1] and Mott and Davis.[2] Luminescence studies alone, however, can give relatively little information as the precise identity of the recombination centres, and are further limited to the observation of those recombination centres which are *radiative*, although the presence of competitive non-radiative processes can often be inferred indirectly from PL decay measurements.[3] In parallel with the luminescence studies, the technique of ESR forms a valuable probe of the microscopic material structure[4] and can in principle provide detailed information about the nature and bonding structure of intrinsic and extrinsic defects. ESR is also limited, however, in certain respects. Firstly, it is not as sensitive as luminescence (due to the use of microwave detection), and secondly, it is not limited to the observation of recombination centres (which are of particular technological interest), but detects signals due to any form of paramagnetic species within the material, and one must rely on additional information (e.g. correlation with PL or photoconductivity data) to determine which ESR centres, if any, are involved in recombination.

The results of photoluminescence (PL) and ESR studies have recently been complemented by the technique of optically detected magnetic resonance (ODMR), which essentially combines the advantages of both ESR and PL into a single experiment. In the ODMR technique, the magnetic resonance of both radiative and linked non-radiative centres is directly observed by monitoring the change in the luminescence intensity induced by magnetic resonance of a recombination centre. Such studies are now widely known in both crystalline and amorphous semiconductors, and the subject has been recently reviewed by Cavenett.[5] The detailed results will be discussed later. Meanwhile, we note that characteristic ODMR signals have been observed for triplet exciton recombination in many semiconductors (e.g. GaS, GaSe, GaP, SiC, a–P, a–As_2Se_3, c–As_2Se_3, a–As_2S_3), as have the characteristic ODMR signals related to distant pair recombination (in materials such as GaS, GaP, a–P, a–As_2Se_3, c–As_2Se_3, a– and c–As_2S_3, ZnS, ZnSe and CdS), and recombination at deep traps (such as O in GaP, Cr in GaAs and PP_4 in GaP). In many of these cases, it has been possible to provide a microscopic identification of the recombination centres involved from a consideration of the detailed resonance information, including nuclear hyperfine interaction, and the local crystal field symmetry. Such detailed identification has not yet been possible in the case of amorphous semiconductors, and thus for the present, the assignment of recombination transitions rests on a comparison of both PL and ODMR data. We note that ODMR signals are *not* expected for either singlet exciton recombination or to free-bound processes. The case of singlet exciton recombination cannot be directly detected by ODMR, as a singlet level cannot split in an external magnetic field, and free-bound recombination, as we shall see, proceeds via a spin-allowed singlet process, eliminating any recombination spin-dependence.

In this paper, we shall describe the technique of ODMR and its application to the determination of recombination processes in semiconductors. Next, we shall describe the ODMR signals which have been obtained in the pnictide and chalcogenide glasses and the related crystals, and we shall show how these results have led to an evaluation of the various recombination paths in these materials.

II. THE ODMR TECHNIQUE

The experiments described in this paper were carried out with a variety of ODMR systems, and details of these can be found in Cavenett.[5] However, the variable temperature ODMR system derived from a commercial ESR spectrometer has been invaluable for the investigation of amorphous semiconductors, and, therefore, is worth a brief description.

This spectrometer, shown schematically in Fig. 1, has been constructed from a modified Bruker 200D X-band ESR system. The microwave bridge was locked to an external reference microwave cavity, and power was directed to a microwave p.i.n. switch, amplified with a travelling wave tube (TWT), and then matched to a standard Bruker rectangular microwave cavity. The end-plate of this cavity was modified to allow laser beam access for luminescence excitation, and to optimize the solid angle for luminescence collection with the lens system shown in Fig. 1. The ODMR signals (observed as change in PL intensity) are detected using a North Coast Ge detector, EO-817, and a Brookdeal 9503 lock-in. The ESR spectrometer console was modified to allow monitoring of both the reference and the sample cavity. The microwave power from the bridge was set from the console, and the output from the TWT was adjusted by a separate attenuator. The maximum rf power used was 3 Watts. Either microwave modulation or magnetic field modulation at audio-frequencies could be used; the standard 100 kHz ESR field coils could be used for enhancing the ODMR signals as described by Davies.[6]

Using an Oxford Instruments ESR 10 gas flow system, ODMR signals have been investigated from room temperature[7] down to the minimum

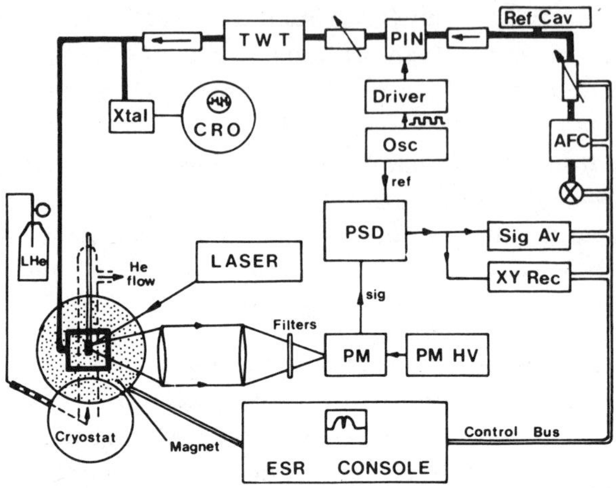

Fig. 1. Variable Temperature ODMR system, which uses a modified Bruker 200D ESR spectrometer. Temperatures from 2.5-300K can be obtained by overpumping the vapour or liquid through an Oxford Instruments ESR10 flow cryostat.

PIN = p.i.n. diode microwave switch
XTAL = microwave detector diode
PM = photodetector

temperature of 2.5K. However care must be exercised with the laser power levels used to maintain this lowest temperature. The possibility of changing samples during the experiment is an important advantage of this system, as is the fact that standard ESR samples can be incorporated in the microwave cavity for g-value determination without affecting the optically detected results.

III. PAIR PROCESSES IN CRYSTALLINE AND AMORPHOUS SEMICONDUCTORS

A. FREE-BOUND RECOMBINATION

Free-Bound recombination processes are well known in crystalline semiconductors, particularly in materials with low defect concentrations. However, in amorphous semiconductors, free-bound recombination is expected to be of negligible importance, because of the high density of localized states, due both to intrinsic defects, and to the localized tail states[2] of the conduction and valence bands. In any case, ODMR from free-bound recombination will be small or non-existent, as can be seen by considering the schematic diagram of Fig. 2(a). A free electron is shown recombining with one of several holes trapped at shallow acceptor centres. The infinite spatial extension of the free-electron wavefunction is illustrated by the shaded part of the diagram, which represents the spatial probability density $\psi_e^2(r)$. Due to this infinite extent, a free carrier interacts simultaneously with several trapped holes, each of which has an identical overlap integral with the electron wavefunction. Thus the recombination process always proceeds via a spin-allowed singlet radiative path as illustrated in Fig. 2(a), and magnetic resonance, while flipping either the electron or the hole spins, cannot influence the net recombination rate, so that negligible ODMR signals are expected.

B. BOUND-BOUND RECOMBINATION

In this case, both recombining carriers are spatially localized, as illustrated in the schematic diagram of Fig. 2(b), and the recombination rate is described by

$$\tau(r) = \tau_o \exp(r/R_o) \tag{1}$$

where r is the pair separation, τ_o is of the order of 10 ns, and R_o corresponds to the Bohr radius of the more extended carrier. As has been shown by several authors,[8] the exponential term in Eq. (1) ensures that recombination is predominantly nearest-neighbour at moderate excitation densities, irrespective of the spin-orientation of the nearest-neighbour pair. Such pair processes are well known in crystalline semiconductors, where the binding centres are more usually conventional donor and acceptor centres, so that the recombination process may be written

$$D^o + A^o \rightarrow D^+ + A^- + \hbar\omega_{pair} \tag{2}$$

The spatial separation of the charged D^+ and A^- centres in the final state of the recombination transition gives rise to the well-known "pair Coulomb term" for such donor-acceptor emission processes. Close pairs recombine quickly, from Eq. (1), and also have a maximum Coulomb interaction, causing

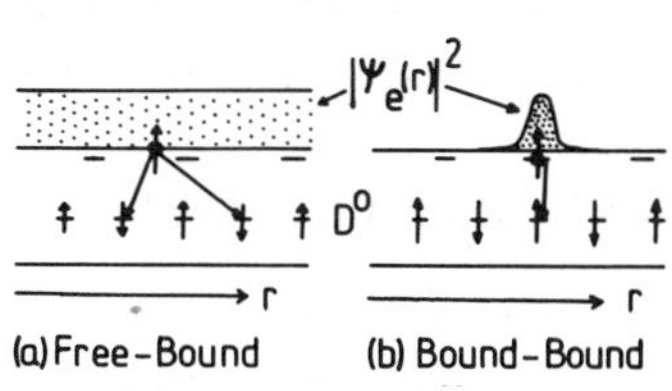

Fig. 2. Schematic diagram of (a) free-bound and (b) bound-bound electron-hole recombination.

the fast close pair recombination to occur at higher energy than the slow distant pair processes, giving a time-shift of the luminescence peak after an excitation pulse. In the case of more complicated binding centres, however, the Coulomb interaction may well be absent, and a simple example of this is the well-known bound-bound recombination process involving the (Cd–O) isoelectronic centre in GaP.[9] Thus radiative recombination can occur between an electron trapped at a (Cd–O) centre, and a hole trapped at an acceptor centre, such as a Zn (A^o_{Zn}). The resulting emission process is:

$$(Cd{-}O)^- + A^o_{Zn} \rightarrow (Cd{-}O)^o + A^-_{Zn} + \hbar\omega_{pair} \tag{3}$$

and the emission energy is clearly independent of the separation between the (Cd–O) and Zn centres. Finally, a Coulomb shift of opposite sign to the conventional DA model is expected if both centres are charged before recombination and neutral afterwards:

$$C^- + C^+ \rightarrow C^o + C^o + \hbar\omega_{pair} \tag{4}$$

where C represents an unspecified localized electron or hole centre.

However, even though the characteristic Coulomb shift of the time resolved emission spectrum may be absent in a particular case, it is nevertheless possible to determine the pair nature of such an emission process from a consideration of the ODMR signals. Fig. 3(a) shows the schematic energy level diagram for such a weakly-coupled spin pair in a magnetic field, B. Strong ODMR signals are observed for such a pair system, particularly if the recombination time τ is much shorter than the spin relaxation time T_1, in which case the system is "unthermalized". Microwave resonance of the electron or the hole, shown by the arrows on Fig. 3(a), transfers spins from the long-lived triplet to the highly radiative singlet pair states, resulting in an increase in the emission intensity. Resonant ODMR signals are thus observed at magnetic field positions determined by the electron and hole centre (i.e. donor and acceptor, in crystals) g-values. Similar ODMR signals are expected wherever bound-bound pair recombination is of importance, although in amorphous semiconductors both electron and hole resonances are expected to overlap, since both g_e and g_h (electron and hole g-values) are close to 2.0 for these materials.[4]

IV. TRIPLET EXCITON RECOMBINATION IN SEMICONDUCTORS

Triplet exciton recombination is well established in crystalline semiconductors, where triplet exciton can often be unambiguously identified from an examination of the Zeeman or stress splitting of sharp luminescence

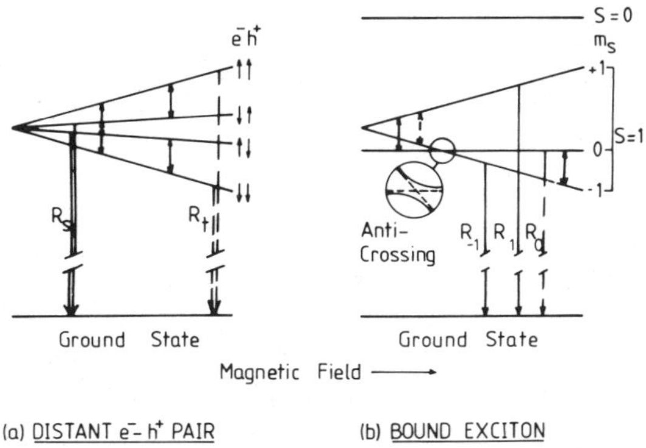

Fig. 3. Energy levels of (a) a distant e^--h^+ pair, and (b) a bound J = 0,1 exciton, in an external magnetic field. In (b) the external field is applied along the principal exciton axis determined by crystal-field symmetry and/or anisotropic exchange. Allowed transitions are shown by continuous arrows; forbidden transitions by dashed arrows.

lines. Triplet excitons can be formed either from the combination of a $j = 1/2$ electron state with a $j = 3/2$ hole state (giving rise to $J = 1$ and $J = 2$ exciton states), or from the combination of a $j = 1/2$ electron and a $j = 1/2$ hole state (giving rise to $J = 0$ and $J = 1$ exciton states). ODMR signals have not yet been observed corresponding to the $J = 1,2$ exciton systems such as N in GaP. This is not, however, a serious problem, as in amorphous semiconductors it is invariably observed that the hole states observed by ESR have almost isotropic g-value, close to $g = 2$, and a spin-doublet nature ($j = 1/2$), implying that the latter type of $J = 0,1$ exciton is more likely to be observed. Fortunately, the $J = 0,1$ exciton system can indeed give rise to strong, well-characterized ODMR spectra as outlined in the Introduction, and Fig. (3b) gives a schematic diagram of such an exciton in an external magnetic field, showing how the ODMR signals can arise. The level diagram of Fig. 3(b) is actually shown for the case of an exciton localized at a site of axial symmetry (cf. GaS). where the magnetic field is applied along the principal axis. Radiative recombination is electric-dipole-forbidden from the $m_s = 0$ substate, causing the build-up of excess population on the $m_s = 0$ level since the recombination time τ, is shorter than the spin relaxation time, T_1. Microwave resonance transfers spins from the non-emitting $m_s = 0$ substate to the outer $m_s = \pm 1$ levels, causing a resonant increase in emission intensity, as shown in Fig. 3(b). The triplet energy levels can be described by an effective-spin Hamiltonian

$$\mathbf{H} = \mu_B \underline{B} \cdot \underline{g} \cdot \underline{S} + D\left(S_z^2 - 1/3\,S(S+1)\right) + E\left(S_x^2 - S_y^2\right) \tag{5}$$

where the D and E terms describe axial and orthorhombic field components respectively, which may be due to the local crystal field and/or anisotropic exchange. In a layered material such a GaS, it is observed that the major crystal field axis is uniquely determined by the axial crystal structure and thus the ODMR spectra obtained correspond to a single exciton orientation relative to the external magnetic field. The situation is more complicated in the case of, for example, a zincblende-structured semiconductor such as GaP, since here there are three equivalent directions for exciton centres with major axis along a [100] direction, four equivalent directions for a [111] system, or six equivalent directions for an exciton with a [110] major axis, causing the simultaneous observation of several groups of resonances from different exciton orientations. In an amorphous semiconductor, a powder-averaged spectrum is obtained, made up of individual exciton resonances corresponding to all possible orientations of the major axis to the external magnetic field. To consider the effect of this situation, consider the case of an axial triplet exciton state with $g_{\parallel} = 2.0$, $g_{\perp} = 2.5$ and $D = 0.1\ cm^{-1}$. The angular dependence of the resonance for a single exciton state as a function of magnetic field orientation with respect to the exciton major axis is illustrated in Fig. 4(a). The signals have a maximum separation of $2D/g\mu_B$ for $B // z$ ($\theta = 0°$), and collapse for $\theta = 55°$. Averaging the spectra so as to include the probability distribution of the various exciton orientations results in the spectrum of Fig. 4(b) where a small intrinsic resonance linewidth has been assumed. In practice, nuclear hyperfine interaction is strong in materials such as a–P and the amorphous arsenic chalcogenides, causing intrinsic linewidths of at least 50 mT. Recalculating the powder spectrum to take this intrinsic linewidth into consideration results in the expected ODMR spectrum of Fig. 4(c). Finally, in an amorphous material, the "$\Delta m_s = 2$" transition between the $m_s = \pm 1$ sublevels can be of appreciable intensity (see the dotted line on Fig. 3(b)) and this gives rise to the dashed low-field ODMR transition shown on Fig. 4(c).

It is clear that the ODMR spectrum of a triplet exciton in an amorphous material is characterized by a broad spectrum centered near $g = 2$, corresponding to the "$\Delta m_s = 1$" transitions, together with a relatively narrow low-field peak generally near $g = 4$, corresponding to the "$\Delta m_s = 2$" microwave transition. By contrast, the ODMR spectra due to pair emission processes in amorphous semiconductors consist of superimposed narrow resonance lines near $g = 2$, caused by the electron and hole resonances as described in Section III above. We shall now consider the ODMR information obtained for amorphous semiconducting glasses in the light of this previous discussion.

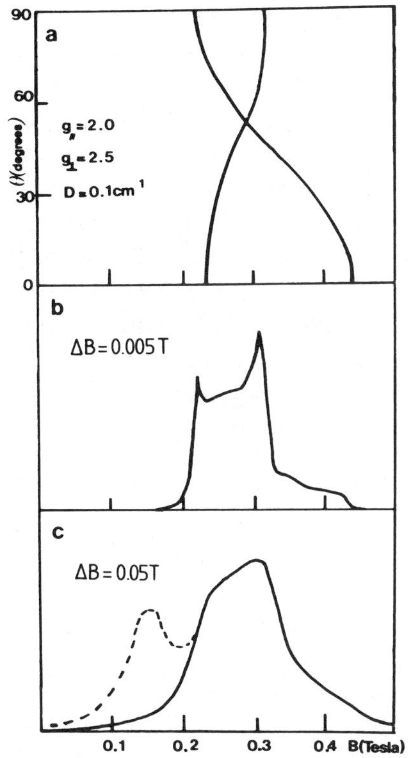

Fig. 4. (a) Magnetic Resonance field variation as a function of the angle (θ) between external magnetic field B and the principal exciton axis for a triplet system. (b) Powder spectrum for the triplet system of (a) obtained for a narrow resonance linewidth. (c) Recalculated powder spectrum with a larger intrinsic linewidth (solid line) and showing the effect of including the "$\Delta M_s = 2$" low-field transition (dashed line).

V. RECOMBINATION IN AMORPHOUS PHOSPHORUS (*a-P*)

The luminescence spectrum of a–P is a single, asymmetric broad band, with a long tail to low emission energies.[10] The PL spectrum peaks near 1.4eV for above-gap excitation, and the optical gap[11] for bulk a–P is close to 2eV. A typical cw luminescence spectrum for a bulk a–P sample is shown in Fig. 5(a), taken with a Ge detector, at 2K, with the data corrected for the response of the detection system. Time-resolved luminescence and PL decay studies[11,12,13,14a] have shown the presence of four principal luminescence components which can be distinguished by their emission spectra and their decay characteristics. First, there is an extremely fast (less than 10 ns) decay component peaking near 1.4eV. The next portion of the decay is dominated by a near-exponential component peaking near 1.1-1.2eV, with lifetime close to 150 ns. Finally, there are two relatively slow emission components; a non-exponential component related to the 1.4eV emission region, and with mean lifetime close to 3-4 ms, and a reasonably exponential decay in the low energy region, with lifetime close to 4 ms. The fast nanosecond component does not concern us here, as firstly, it does not give rise to substantial ODMR signals, and secondly, it is involved in a negligible proportion of recombination events. The other three decay components are, however, important, and the low-energy (LE) band associated with the 150 ns component is illustrated in Fig. 5(a), together with the high-energy (HE) band associated with the non-exponential process, as the dashed lines superimposed on the cw PL spectrum.

First, let us consider the ODMR spectra illustrated in Fig. 6. The full spectrum is seen to consist of a relatively narrow line near g = 4, a very

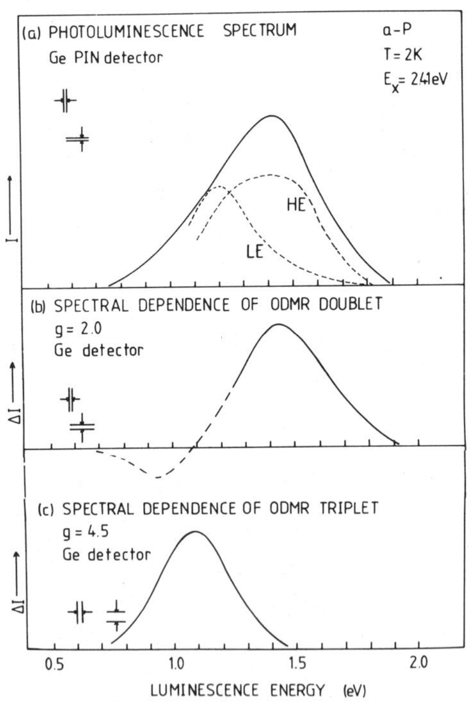

Fig. 5. (a) Photoluminescence spectrum of a–P at 2K, showing the HE and LE TRS components obtained by Fasol and Davis (1981) (dashed lines) for comparison. (b) Spectral dependence of the g = 2.0 ODMR signal from Fig. 6(a). (c) Spectral dependence of the triplet ODMR signal in Fig. 6(b).

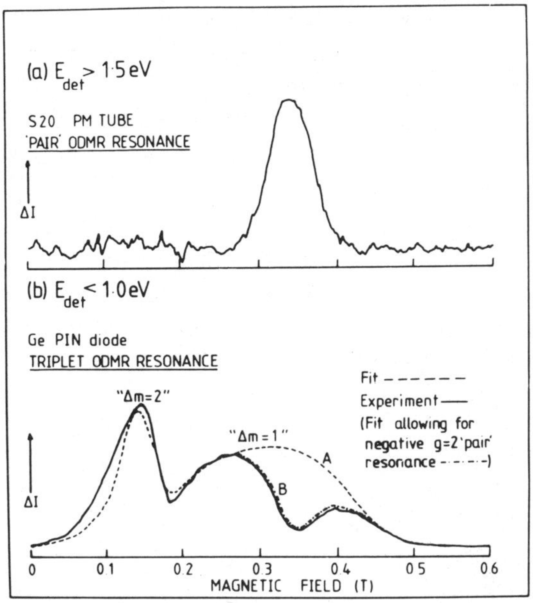

Fig. 6. ODMR spectra in a–P. (a) The pair spectrum obtained from the HE emission. (b) The triplet spectrum (solid line) associated with the LE emission. The dashed "A" curve shows the calculated powder spectrum using $g_{\parallel} = g_{\perp} = 2.13$, $D = 0.13\ \text{cm}^{-1}$ and $E = 0.02\ \text{cm}^{-1}$ as described in text. The chain line "B" shows the effect of including a negative contribution from the pair ODMR signal of Fig. 6(a).

broad underlying component extending from zero-field to near 0.6 Tesla, and a relatively narrow spectrum close to $g = 2$. This complex spectrum can be further simplified by using suitable optical bandpass filters to isolate various portions of the PL spectrum. Thus Fig. 6(a) shows the ODMR signal obtained from the highest energy part of the PL only: the signal consists of a symmetric resonance profile near $g = 2$, as expected for a pair emission process. The signal linewidth (~60 mT) compares well with the ESR and LESR signal linewidths observed[15] for intrinsic centres in a–P and is principally due to nuclear hyperfine interaction with the P nucleus. Fig. 6(b) shows the ODMR spectrum obtained from the LE region of the PL spectrum only. The $g = 4$ signal is seen to be associated with this emission region, as is the broad underlying resonance stretching from 0 - 0.6 Tesla. A negative dip in the ODMR spectrum is observed near $g = 2$, and detailed waveform measurements[16] have associated this negative ODMR signal with a competitive pair process involving $g = 2$ centres. The broad spectrum, together with the $g = 4$ line, are seen to form the typical spectrum expected for a triplet exciton powder spectrum, as may be seen by comparing the experimental data with the calculated spectrum of Fig. 4(c) above. Carrying out a numerical fit to the spin Hamiltonian of Eq. (3.4) yields the best fit shown in Fig. 6(b) as the dashed line, where the g-value has been assumed isotropic, and the g, D and E values, together with the relative intensity of the "$\Delta m_s = 2$" transition, have been taken as the variable parameters. Such a fit, while providing an example of the qualitative agreement between the experimental data and a simple triplet exciton model, cannot be expected to provide accurate estimates of the ESR parameters, due to the necessary simplifying assumptions (such as that of isotropic g-value) required to make the fitting procedure tractable. Bearing in mind this qualification, the fit shown was obtained for the following parameters: $g_{\parallel} = g_{\perp} = 2.13$, $|D| = 0.13\ cm^{-1}$, $|E| = 0.02\ cm^{-1}$.

Measurement of the spectral dependence of the Pair and Triplet ODMR signals can be achieved either point-by-point by measuring the ODMR signals at various emission energies using a monochromator to isolate the emission region, or by setting the magnetic field to the peak of a given ODMR signal and scanning the spectrometer throughout the entire emission band. Both of these methods give identical results for a–P and the spectral dependence of the pair resonance is shown in Fig. 5(b), together with that of the triplet resonance in Fig. 5(c). Comparison with the positions of the HE and LE spectra (shown in Fig. 5(a)) obtained from the TRS data of Fasol and Davis[12] show that the triplet resonance is associated with the LE luminescence region, and the pair ODMR signal is associated with the HE emission. The former assignment is certainly consistent with the near-exponential 4 ms emission component observed in the LE region: exponential decay is characteristic of free-bound, excitonic or internal transition recombination in semiconductors. Of these, free-bound recombination is very unlikely, for reasons discussed above, and excitonic and internal transitions are not only difficult to distinguish, but may in some cases be essentially equivalent. Thus bound exciton recombination at a (Cd–O) centre in GaP may be equivalently described as an internal molecular transition of the (Cd–O) centre, and the precise term used is largely a matter of individual preference. As for the assignment of the pair ODMR signal to the HE region, this is also consistent with the PL decay results mentioned above, which find a non-exponential decay in this emission region. Such non-exponential decay is typical of pair recombination due to the wide distribution of possible pair separations, r, which in turn gives a quasi-continuous PL lifetime distribution from Eq. (1).

Further information may be deduced from studying the transient ODMR effects, such as the ODMR signal waveform, and lifetime-resolved ODMR. The former technique has been described in detail by Dunstan and Davies.[8] Briefly, one monitors the transient effect of the application and removal of microwave resonance: the characteristic waveform observed yields information on the optical recombination and spin relaxation times. The waveforms of both the triplet and pair ODMR signals in a–P are shown in Fig. 7(a) and 7(b) respectively: the waveforms are characteristic of unthermalized radiative centres,[7] and yield a mean recombination time of 3 - 4 ms for both the triplet and pair processes. This lifetime clearly identifies the triplet ODMR signal with the slow, near-exponential, 4 ms decay component in the LE region, rather than with the fast 150 ns component. This result suggests that the LE region is composed of two excitonic emission processes, probably corresponding to the singlet (J=0) and triplet (J=1) excitonic states formed from the j = 1/2 electron and j = 1/2 hole states. The singlet emission is spin-allowed, hence the short 150 ns lifetime, while the triplet emission is spin-forbidden causing the characteristic slow millisecond recombination time. Confirmation of this assignment comes from preliminary lifetime-resolved ODMR signals, obtained using the double modulation technique of frequency response spectroscopy, described by Depinna and Cavenett[14c] and Dunstan, Depinna and Cavenett.[17] The results of this preliminary investigation are summarized in Fig. 8 for a–P, and show that the triplet ODMR signal is associated with PL of lifetimes longer than 10^{-5}s, while the pair ODMR signal is negative on the fast emission components, and positive on the slow components, suggesting that the pair emission competes with the singlet and triplet exciton emissions.

It is possible to arrive at a tentative recombination model which can account satisfactorily for the PL, TRS and ODMR results. This model has been described elsewhere[14,14a] and makes use of the IVAP concept established for Group V and Group VI amorphous semiconductors.[18,19] The IVAP, or "intimate valence-alternation pair" consists of a nearest-neighbor isoelectronic

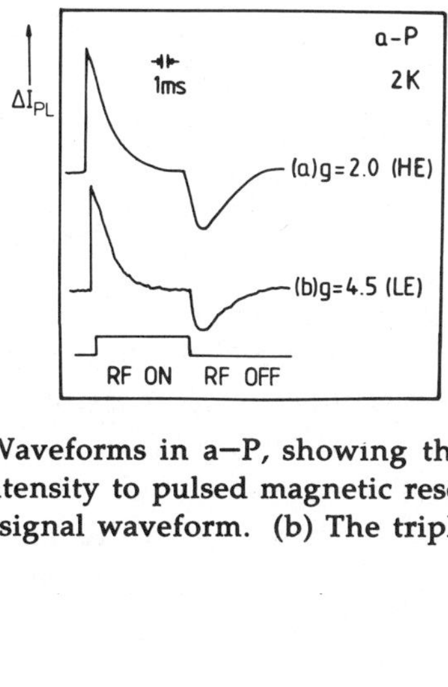

Fig. 7. ODMR Waveforms in a–P, showing the transient response of the emission intensity to pulsed magnetic resonance. (a) The pair (Fig. 6(a)) ODMR signal waveform. (b) The triplet (Fig. 6(b)) ODMR signal waveform.

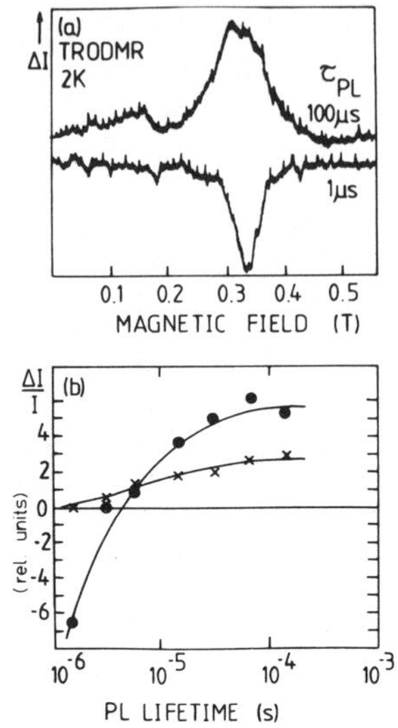

Fig. 8. Preliminary lifetime-resolved ODMR signals in a–P, using the technique of quadrature frequency response spectroscopy. (a) Lifetime-resolved ODMR spectra, for emission components with $\tau = 100\ \mu s$ and $\tau = 1\ \mu s$ respectively. (b) Plot of normalized resonance intensity ΔI/I against emission lifetimes for the triplet (X) and (•) ODMR signals.

pair of intrinsic bonding defects, usually described as (D^+, D^-), which is in many ways analogous to the well-known isoelectronic exciton binding centres in crystalline semiconductors such as (Cd–O) and (Zn–O) in GaP. The distant pair emission process can be described as

$$(D^+, D^o) + (D^o, D^-) \rightarrow 2(D^+, D^-) + \hbar\omega_{pair} \tag{6}$$

and the excitonic process can be described as

$$(D^o, D^o) \rightarrow (D^+, D^-) + \hbar\omega_{exciton} \tag{7}$$

The advantage of this model is that, as well as providing a natural candidate for exciton and single-carrier binding centres, it is capable of explaining the apparent linking of pair and exciton emissions (inferred from the observation of the negative pair ODMR signal on the LE emission), and also from the fact that the pair emission occurs at higher energy than the exciton spectra. The linking of the pair and exciton emissions on this model follows from the fact that sequential formation of an exciton in the above scheme requires the successive capture of both an electron and a hole at the same IVAP. At the intermediate stage, when only a single carrier has been captured, pair resonance increase the single-carrier recombination rate and thus inhibits exciton formation. Secondly, when the IVAP is in its ground state (D^+, D^-), there is a strong Coulomb attraction between the nearest-neighbour members of this molecular centre. Capturing either one or two carriers removes this net Coulomb term, thus increasing the energy of the centre. Examination of Eqs. (6) and (7) shows that the pair emission involves two IVAP's, compared to only one such centre for the excitonic process, and thus the pair emission is expected to lie higher in energy than the exciton emissions by an amount equal to the nearest-neighbour Coulomb term. The precise Coulomb interaction is difficult to calculate, and requires an exact knowledge of the configuration of the centre. However, the analogous (Cd–O) centre in GaP has a nearest-neighbour Coulomb term of 0.5eV,[9] which is certainly in reasonable agreement with the separation of the pair and exciton emission bands in a–P (0.3eV).

It should be noted that an objection has been raised to this model by Fasol and Davis,[12] who point out that no Coulomb emission peak shift is observed in the distant pair HE emission process from the TRS data. However, we have shown elsewhere[16] that a detailed consideration of the HE TRS results shows that calculated Coulomb shift for the HE emission region is less than the experimental resolution of the TRS data, and thus there is no actual conflict with the TRS evidence.

Confirmation of the triplet nature of the slow LE emission component comes from the observation of anomalous changes in the luminescence intensity as a function of applied magnetic field, in the absence of microwave excitation.[14] In the region of the "crossing" of the $m_s = 1$ and the $m_s = 0$ Zeeman triplet sub-levels (see Fig. 3(b)) an enhancement of the emission

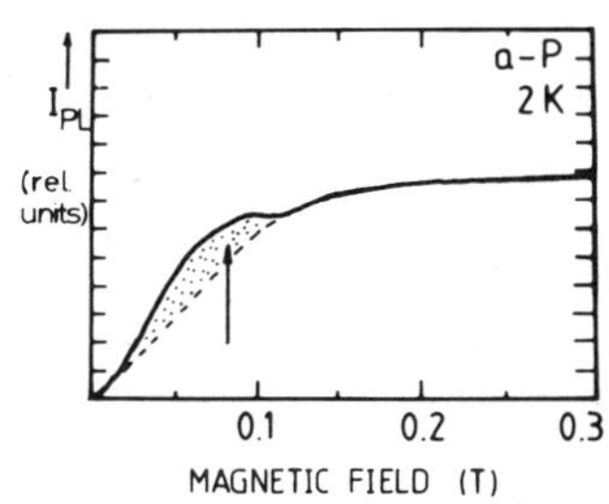

Fig. 9. Typical level anti-crossing signal in a–P, showing luminescence intensity as a function of magnetic field. The arrow shows the approximate anti-crossing position.

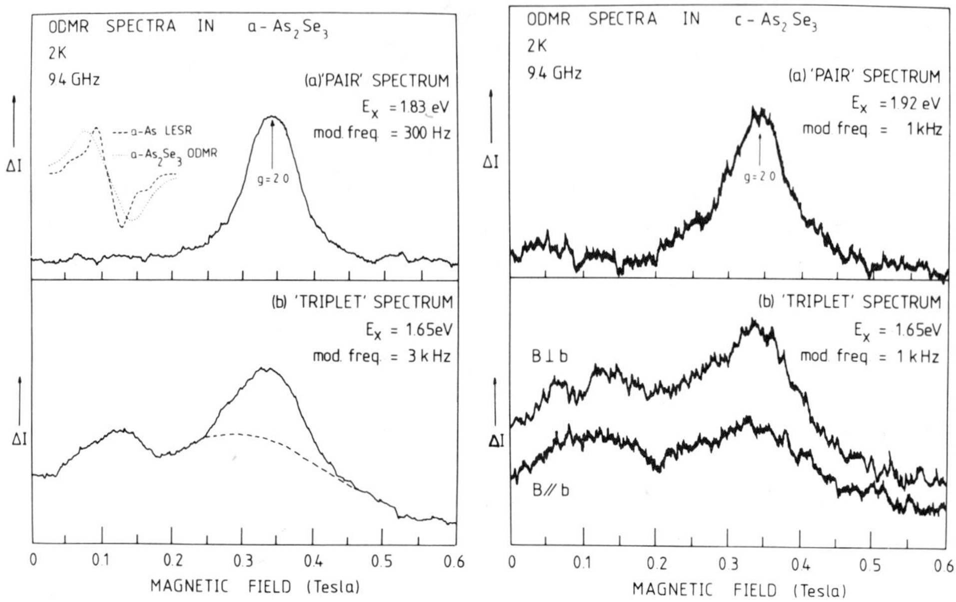

Fig. 10. ODMR spectra in (a) amorphous and (b) crystalline As_2Se_3, showing pair (upper) and triplet (lower) ODMR.

intensity is usually observed for an unthermalized triplet exciton emission process. Such "level anti-crossing" signals are well-known in crystalline studies e.g. GaSe,[20] GaS,[21] and have indeed been observed in a–P. A typical level anti-crossing signal is shown for bulk a–P in Fig. 9.

VI. RECOMBINATION IN THE ARSENIC CHALCOGENIDE GLASSES: a-As_2Se_3, a-As_2S_3

The above detailed discussion of the ODMR results in a–P largely carries over to describing the situation for c– and a–As_2Se_3 and c– and a–As_2S_3. The ODMR signals obtained in a– and c–As_2Se_3 are summarized in Fig. 10(a) and Fig. 10(b) respectively. Exciting the crystal above the band-gap gives rise to a broad emission band with a single lifetime.[22] This is probably due to free-bound recombination[14c] and hence should not give an ODMR signal. As expected, no ODMR signal is observed under these conditions. For the glass, above-gap excitation results in a broad emission band with a non-exponential decay,[22] which is characteristic of distant pair recombination, as described previously, and which also causes strong ODMR signals near $g = 2$ (Fig. 10(a)). Exciting below the band-gap still gives pair resonances in the case of the glass, but now also gives ODMR signals in the crystal which have identical lineshapes to that of the glass, within experimental error

(Fig. 10(b)). Finally, exciting well below the band-gap (1.5-1.6eV) for either the crystal or the glass causes substantial excitonic emission, as inferred from the characteristic triplet ODMR spectrum clearly visible in Figs. 10(a) and 10(b).

Very similar results have also been observed[14d] in the case of a$-As_2S_3$, and the ODMR spectra are summarized in Fig. 11. Exciting well above the band-gap causes both pair and exciton emissions in a$-As_2S_3$, as inferred from the triplet spectrum evident in Fig. 11(a), with the negative $g = 2$ line due to competitive pair recombination superimposed. Exciting near the band-edge (Fig. 11(b)) gives rise to predominantly triplet emission in the glass. As for c$-As_2S_3$, no ODMR signals are observed for above-gap excitation, in resemblance to the results for c$-As_2Se_3$. Exciting well below the gap, however, causes the observation of pair ODMR spectra in both the crystal and the glass phase of As_2S_3. The glass ODMR spectrum is shown in Fig. 11(c) and the crystal spectrum is shown in Fig. 12. The crystal spectrum is clearly observed to have two superimposed components; the narrow

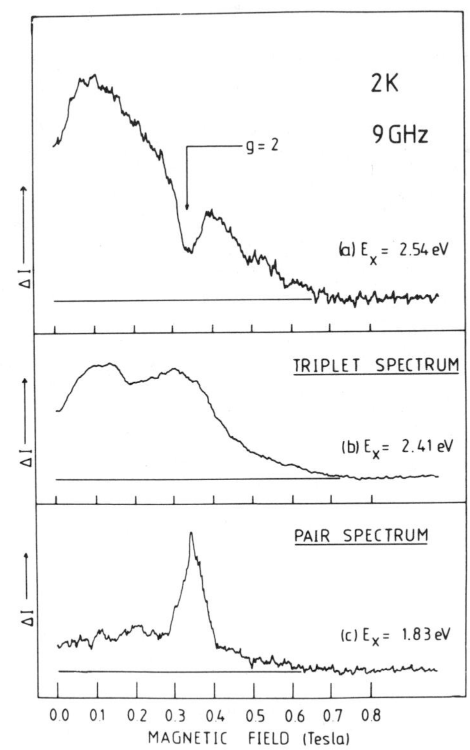

Fig. 11. ODMR spectra in a$-As_2S_3$ for various excitation energies E_x.

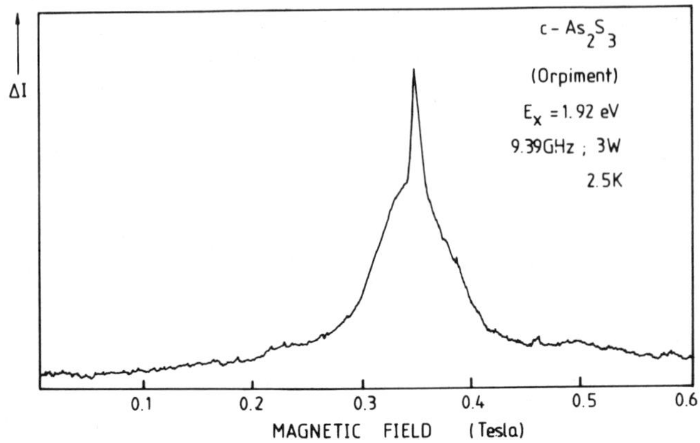

Fig. 12. Pair ODMR spectrum in c–As_2S_3: two components are clearly observed; a broad line near g = 2.06 and a narrow line (~4 mT) near g = 2.00.

component is probably due to a S-related centre, while the broader component is probably an As-related centre, which derives its linewidth from hyperfine interaction with the As nucleus.

A significant difference between the results for a–As_2Se_3 and a–As_2S_3, compared with those of a–P, lies in the spectral dependence measurements. While in a–P it was possible to relate the pair and triplet ODMR signals to distinct emission regions, and the IVAP recombination model was suggested by the observation that the pair emission lies about 0.3eV above the exciton spectrum, in a–As_2Se_3 and a–As_2S_3 preliminary measurements indicate that both the exciton and pair emissions overlap much more strongly than for a–P, and any peak shift is correspondingly smaller. As discussed elsewhere,[23] this observation suggests that in the arsenic chalcogenides the pair recombination mechanism may always involve at least one isolated defect or tail-state centre, e.g.

$$e^- + (D^+, D^o) \rightarrow (D^+, D^-) + \hbar\omega_{pair}$$

$$(D^o, D^o) \rightarrow (D^+, D^-) + \hbar\omega_{exciton} \tag{8}$$

in which case identical Coulomb terms contribute to the pair and exciton emission processes, unlike the situation of a–P.

VII. SUMMARY

We have seen that the ODMR results obtained thus far in the amorphous semiconducting glasses have yielded the first comprehensive information regarding the presence of both exciton and pair mechanisms of recombination in these materials. The results are qualitatively similar in all the amorphous semiconducting glasses examined so far and the high sensitivity of the technique will clearly be of considerable value, enabling, for example, a magnetic resonance study of the related intrinsic defects in the crystalline and glass phases of these materials for the first time. A continuation of this work should be aimed first at extending the technique to other related materials such as a–Se and a–As, and secondly towards a more detailed theoretical analysis of the optically detected spectra, which should provide a rich source of information concerning the detailed bonding structure of recombination centres in amorphous semiconductors.

ACKNOWLEDGEMENTS

We are grateful for the generous financial support of SERC, The Royal Society and the Ministry of Defense which has made this work possible. We are also indebted to Dr. G. Fasol and Professor E. A. Davis for the a–P samples, and to Dr. T. M. Searle and Professor I. G. Austin for the chalcogenide samples used in these experiments.

REFERENCES

1. R. A. Street, Adv. Phys. **25**, 347 (1976); Adv. Phys. **30**, 593 (1981).
2. N. F. Mott and E. A. Davis, "Electronic Processes in Non-Crystalline Materials," Oxford, 2nd ed. (1979).
3. T. M. Searle, T. S. Nashashibi, I. G. Austin, R. Devonshire, and G. Lockwood, Phil. Mag. **B, 39**, 389 (1979).
4. S. G. Bishop, U. Strom, and P. C. Taylor, Phys. Rev. **B, 15**, 2278 (1977).
5. B. C. Cavenett, Adv. Phys. **30**, 475 (1981).
6. J. J. Davies, J. Phys. **C, 11**, 1907 (1978).
7. S. Depinna, B. C. Cavenett, T. M. Searle, and I. G. Austin, Solid State Commun., **43**, 79 (1982a).
8. D. J. Dunstan and J. J. Davies, J. Phys. **C, 12**, 2927 (1979).
9. P. J. Dean, Prog. Sol. St. Chem., **8**, 1 (1973), ed. J. O. McCaldin and G. Somorjai.
10. P. D. Kirby and E. A. Davis, Proc. 14th Int. Conf. on the Physics of Semiconductors, ed. B. L. H. Wilson, p. 1309 (1978).
11. G. Fasol, thesis, University of Cambridge (1982).
12. G. Fasol and E. A. Davis, J. de. Physique, **25-26**, Colloque C4-571 (1981).
13. G. Fasol, A. D. Yoffe and E. A. Davis, J. Phys. C., **15** 5851 (1982).
14. S. Depinna and B. C. Cavenett, Solid State Comm., **40**, 813 (1981); a, Solid State Comm., **43**, 25 (1982); b, Phil. Mag. **B, 46**, 71 (1982); c, J. Phys. **C, 15**, L489 (1982); d, Phys. Rev. Lett., **48** 556 (1982).
15. B. V. Shanabrook, S. G. Bishop, and P. C. Taylor, J. de Physique **25-26**, Colloque C4-865 (1981).
16. S. Depinna, thesis, University of Hull (1982).
17. D. J. Dunstan, S. Depinna, B. C. Cavenett, J. Phys. **C, 14**, L425 (1982).

18. M. A. Kastner, D. Alder, and H. Fritszche, Phys. Rev. Lett., **37**, 1504 (1976).
19. M. A. Kastner and H. Fritszche, Phil. Mag. **B, 37**, 199 (1978).
20. K. Morigaki, P. Dawson, and B. C. Cavenett, Sol. St. Comm. **28**, 829 (1978).
21. P. Dawson, K. Morigaki, and B. C. Cavenett, Proc. 14th Int. Conf. on the Physics of Semiconductors, ed. B. L. H. Wilson, p. 1023 (11979); P. Dawson, N. Killoran, and B. C. Cavenett, Solid State Commun. **32**, 1163 (1979).
22. G. S. Higashi and M. A. Kastner, 1982, private communication, (to be published).
23. S. Depinna, B. C. Cavenett, and W. E. Lamb, Phil. Mag. **B, 47** 99 (1983); B. C. Cavenett, S. Depinna, and W. E. Lamb, Proc. Int. Conf. on the Physics of Semiconductors, Montepellier, (1982), to be published.

My co-author of this article, Dr. S. Depinna, aged 27, was tragically killed in a car accident in November 1983 after having just been appointed as lecturer in Physics at the University of Leicester, England. He had published 24 papers covering spectroscopy of defects and excitons in a wide range of crystalline and amorphous semiconductors and this note pays tribute to him as a friend, a colleague and an outstanding scientist.

B.C.C.

DISCUSSION

Optically Detected Magnetic Resonance

It was clear from Professor Cavenett's talk that ODMR (optically-detected magnetic resonance) is one of the most important and direct probes of defects in glasses, and that the data are rich in detail. The discussion dealt mainly with the detailed interpretation of the ODMR data.

To summarize the most important aspects of the data, both doublet (pair) and triplet (exciton) processes can be resolved in the ODMR spectra of a-P, a$-As_2S_3$, a$-As_2Se_3$ and c$-As_2Se_3$. These two processes can also be separated in the optical emission spectra, with the exciton emission occurring at *lower* energy than the pair emission. To explain these results, Cavenett said the basic defect was the IVAP - a nearest-neighbor pair of oppositely charged defects. The exciton emission arises from an electron and a hole trapped on a single IVAP and the pair emission from the recombination of an electron on one IVAP with a hole on the other. The discussion started with the question of why it was assumed that the basic defects are pairs as opposed to single defects. The answer was that this is required in order to give the exciton emission a lower frequency than the pair emission. It was argued on fundamental grounds however that the exciton emission should always be higher in energy than the pair emission. One of the difficulties appears to be that Cavenett's exciton was really an electron-hole pair trapped close together so that exchange was sufficiently large to produce a large singlet-triplet splitting.

J. Orenstein, Chairman
M. L. Theye, Discussion Leader

COMMENTS

The Application of ODMR to Studies of Recombination in a-Si:H

Various studies of optically-detected magnetic resonance (ODMR) in a-Si:H all agree that the observed signals are due to distant pairs. Both quenching and enhancing ODMR signals are observed, corresponding respectively to resonant decreases and increases in luminescence intensity. From a consideration of the ODMR response to transient microwave excitation, we show that the enhancing signal in sputtered a-Si:H has the characteristic response of a radiative distant pair mechanism, while the quenching signal is due to a competitive nonradiative channel. This result is confirmed by recent lifetime-resolved ODMR data.

While differences in the detailed interpretation of ODMR spectra in a-Si:H have been noted,[1–3] it is nevertheless clear that all studies propose an interpretation of the ODMR spectra in terms of recombination at distant, weakly-coupled spin pairs. However, qualitatively different models have been proposed to describe the quenching ODMR signal near g = 2.0050-2.0055. Thus while the Hull and Tokyo groups[1,3] consider an interpretation of this resonance based on the increase of a nonradiative channel, the Xerox group has suggested an alternative explanation[2], based on radiative geminate pairs. In this discussion, we show how the techniques of ODMR transient waveform analysis and lifetime-resolved ODMR can resolve this problem.

First, consider a weakly-coupled unthermalized spin pair. The rate equations describing this system may be written[4]

$$\dot{n}_s = 2P_sN_0-(R_s+M+2P_s)n_s+(M-2P_s)n_T$$

$$\dot{n}_T = 2P_TN_0-(R_T+M+2P_T)n_T+(M-2P_T)n_s \tag{1}$$

where P_s, P_T are the singlet and triplet excitation rates, R_s, R_T are the singlet and triplet recombination rates, M is the resonant microwave transition rate, and N_0 is the total number of pairs (excited + ground state). For low excitation density, these rate equations can be readily solved algebraically. First consider a radiative geminate pair model. Here $P_T = 0$, and we have

$$I_{PL} \propto n_s = \frac{2P_sN_0}{R_s+R_T} - \frac{P_sN_0}{R_s} \exp\left[-\left(\frac{R_s+R_T}{2}\right)t\right]\left[\frac{R_s-R_T}{R_s+R_T} - \exp(-2Mt)\right] \quad (2)$$

immediately after the switch-on of a microwave pulse. This gives a rapid decrease in emission intensity ($\sim\mu s$) with time constant 2M, followed by a recovery with time constant $-\left(\frac{R_s+R_T}{2}\right)$. A transient "spike" observed on microwave switch-on is, in fact, a general characteristic of radiative unthermalized pair systems[4].

On the other hand, for a nonradiative transition, ODMR affects the rate at which electrons and holes tunnel to nonradiative centers. The tunneling rate thus depends on the total number of vacant or available nonradiative states, i.e. N^{NR} (vacant) $= N_0^{NR} - (n_s+n_T)^{NR}$ for nonradiative pairs. Solving (1) for $P_s = P_T$ gives

$$(n_s+n_T) = \frac{8PN_0}{R_s+R_T} + \frac{2PN_0(R_s-R_T)^2}{R_sR_T(R_s+R_T)} \exp\left\{-\left(\frac{R_s+R_T}{2}\right)t\right] \quad (3)$$

Unlike a radiative pair, a nonradiative pair has a monotonic ODMR response to transient microwave excitation, with time constant $-\left(\frac{R_s+R_T}{2}\right)$. The response is in qualitative agreement with a nonradiative pair waveform but does not compare well with the "radiative geminate" waveform mentioned above.

A similar deduction can be made from a consideration of the lifetime-resolved ODMR data reported elsewhere.[5] Here we have observed that the quenching ODMR signal in sputtered a-Si:H is linked to the longest-lived emission components ($\tau \gtrsim 300\mu s$), while the enhancing ODMR spectra predominate in the short lifetime regime ($\tau \lesssim 300\mu s$).

Were the quenching resonance due to geminate radiative pairs, it should be linked to the *fastest* emission components, with $\tau << T_1$, due to the assumption of singlet spin memory required by the geminate model. The experimental observation that the quenching ODMR is in fact linked to the *slowest* emission components supports the identification of this resonance with nonradiative pairs, as the slowest emission components come from the most distant radiative pairs, for which nonradiative tunneling is particularly probable.

In summary, independent observations of the transient ODMR waveform and lifetime-resolved ODMR show that the quenching ODMR resonance from sputtered a-Si:H is due to a nonradiative pair process. It is interesting to note that the characteristic transient "spike" of a radiative, unthermalized distant pair process is indeed observed in the enhancing ODMR case.

Acknowledgments

We wish to thank Professor I. G. Austin, Dr. T. M. Searle, Dr. M. J. Thompson, and Dr. J. Allison for the samples used in these experiments. S. Depinna is grateful to S.E.R.C. for a research assistantship.

S. P. Depinna
B. C. Cavenett
Department of Physics
Hull University
Hull, England

REFERENCES

1. K. Morigaki, J. Phys. Soc. Japan *50,* 2279 (1981).
2. D. K. Biegelsen, J. C. Knights, R. A. Street, C. Tsang, and R. M. White, Phil. Mag. *B37,* 477 (1978).
3. S. P. Depinna, B. C. Cavenett, I. G. Austin, and T. M. Searle, Solid State Comm. *41,* 263 (1982).
4. D. J. Dunstan and J. J. Davies, J. Phys. C, *12,* 2927 (1979).
5. S. P. Depinna and B. C. Cavenett, J. Phys. C, *15,* L489 (1982).

ANDERSON LOCALISATION

D. Weaire

Physics Department
University College
Dublin, Ireland

The history of the theory of Anderson localisation is reviewed, from its beginning in 1958 to the recent development of scaling theory.

I. INTRODUCTION

Are the energy eigenstates of the Schrodinger equation localised or extended when the potential is not periodic? It is surprising that this question did not arise explicitly in the earliest days of modern solid state theory. Although Landauer and Helland[1] and others touched upon it in the 1950's it was not until 1958 that it was clearly posed by Anderson[2], who also gave a tentative answer - that the states can be either extended or exponentially localised, depending on the amount of disorder which is incorporated in the potential. By "extended" we mean that a state has the same amplitude everywhere, apart from local fluctuations. An exponentially localised state has the asymptotic form $\exp(-\alpha r)$, apart from local fluctuations.

In later years Mott[3] clarified, extended and applied the concept of Anderson localisation. It has been most commonly invoked by Mott and others in the description of electronic transport in disordered solids, particularly amorphous semiconductors, but it certainly has a wider significance - see, for example, the article of Gibbs in this volume.

The original paper of Anderson laid the foundations of the subject but did not offer any exact solutions. Today there is still a remarkable lack of exact results, even of a limited nature, except in the special case of one dimension, for which progress was quite rapid in the 1960's, with the conclusion that all states are localised for any degree of disorder.[4,5] Two and three dimensions proved more intractable and were attacked in the 1970's by a variety of analytical and numerical methods.[6] Most of these were full of uncertainties, but gradually the location of the Anderson transition (from extended to localised states) was identified for some simple Hamiltonians.

In 1979, Abrahams *et al*[7] outlined a new framework for the study of Anderson localisation in terms of a scaling theory. While originally somewhat speculative, it has received growing support from formal developments and numerical tests. The implications were quite startling, particularly in relation to the two-dimensional case, since it was claimed that all states were localised for any degree of disorder! The previously identified Anderson transition was seen to have only an approximate significance, as a transition from "weak" to "strong" localisation, in this case. There are also consequences for three dimensions, as explained in the following sections.

With increasing confidence in the scaling theory, it may be that the original problem posed by Anderson has been solved, at least in broad terms. He was however, careful to point out that it constituted "only the irreducible minimum from which a theory of this kind of transport, if it exists, must start".

In the following section, the history of the subject is examined in greater detail.

II. A HISTORY OF THE MAIN DEVELOPMENTS OF THE THEORY.

As we have already noted, the most successful of the early work was on the one-dimensional case. It seems to have been regarded as rather academic and its relevance to conduction in thin wires was pointed out by Thouless only at a much later date.[8] Mott and Twose[4] gave a proof that all states were exponentially localised for infinitesimal disorder, quite different in style from the original approach of Anderson (which was directed at three dimensions). Localisation was indeed seen in several direct numerical calculations of eigenstates, such as that shown in Figure 1.

The one-dimensional case has since been treated by many different methods (see, e.g. the review of Ishii[9]) and has recently been given a surprising new twist in the form of incommensurate potentials.[10]

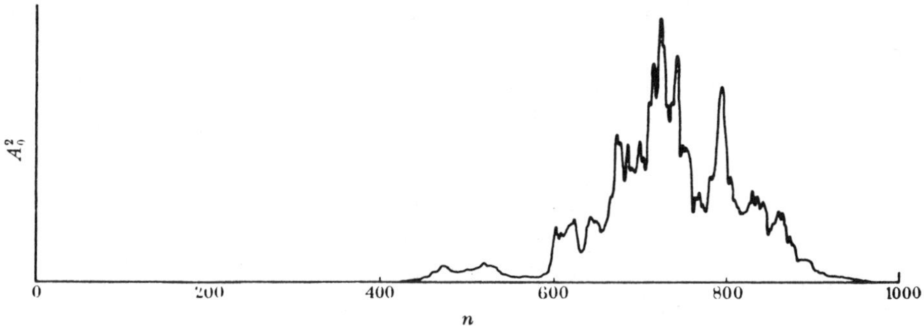

Fig. 1. Envelope of a localised eigenfunction of a one-dimensional Hamiltonian, calculated by Borland.[5]

Herbert and Jones[11] and Thouless[12] derived a useful relation between the localisation length, which is the inverse of the constant in the asymptotic form $|\psi| \sim \exp - \alpha r$, and the density of states n(E). This means that in one special case, that of the Lloyd model[13], for which the density of states is known exactly, we also have in one dimension an exact formula for the localisation length. However, even this is open to question - Gogolin[14] claims, in a recent review, that the correct formula for the localisation length differs by a factor of two!

As for two and three dimensions, the picture that was developed by Mott and others in the 1960's was as follows.

States could be either extended or exponentially localised, with a sharp transition between the two regimes, if either energy or strength of disorder is varied. The transition was not supposed to have any effect upon the density of states n(E). The Lloyd model[13] (Lorentzian distribution of diagonal matrix elements), for which n(E) could be specified exactly, provided support for this assertion, since n(E) was always smooth. More support was recently given by a proof due to Wegner,[15] that there can be no vanishing or divergence of n(E) under fairly general conditions.

The energy E_c at which the Anderson transition takes place within a band has been dubbed the "mobility edge". In any theory of activated conductivity it will play the role which the band edge plays in a crystalline semiconductor. Mott surmised that the mobility should rise *abruptly* at the mobility edge. As we shall see, this idea is currently regarded as incorrect even in three dimensions. It was always controversial but arguments from both sides were somewhat intuitive. Mott's general approach was based on the assumption that the primary effect of disorder is to randomise the phase of the electronic wave function, while others argued from a more classical point of view, stressing the continuous decrease to zero of the conductivity in a percolation transition, which is the obvious classical analogue of the Anderson transition.

If, as Mott emphasised, the identification of E_c and the understanding of the variation of the mobility close to it were essential ingredients of transport theory for, *inter alia,* amorphous semiconductors, it was desirable that Anderson's original work be improved upon. Anderson's treatment was based on remormalised Green's function expansions whose convergence/divergence indicated the localised/extended character of states, and the obvious next step was to refine the various rough approximations that he used in examining their convergence. This was undertaken by Cohen and collaborators (for a review see Economou and Licciardello[16]). Extensive results were produced in two and three dimensions for the same simple type of model Hamiltonian used by Anderson,

$$H = \sum_i \epsilon_i |i\rangle\langle i| + V \sum_{\substack{ij \\ \text{neighbours}}} |i\rangle\langle j| \tag{1}$$

Here ϵ_i is a random variable usually chosen to be uniformly distributed over a range W, so that the strength of disorder may be represented by W/V or W/ZV (Z = number of nearest neighbours).

The results were most conveniently presented as "mobility edge trajectories" which give the position of the mobility edge as a function of W, that is, the strength of disorder. For example, see Figure 2.

Complementary to this work was that of Edwards and others[17,18] who used a path integral formulation to estimate the critical behaviour of the localisation length, finding

$$\begin{aligned} \nu &= 2/3 \text{ in 2d} \\ &= 3/4 \text{ in 3d} \end{aligned} \tag{2}$$

for the critical index defined by

$$\alpha \sim (E - E_c)^{\nu} \tag{3}$$

Despite all this analytical development there did not emerge any key exact results for d $\geqslant$ 2 analogous to the Onsager solution of the 2d Ising model in theory of phase transitions, which might be used as benchmarks for the evaluation of less exact theory. In the absence of these (or directly related experiments) it was necessary to pursue numerical methods to provide points of comparison.

One of the earliest of numerical calculations was that of Yoshino and Okazaki[19], whose results are shown in Figure 3. This explicit representation of localised states in two dimensions gave the subject a reality which, for many, it had previously lacked.

Yoshino and Okazaki extracted the critical index $\nu = 2.0$ from their results (*cf* Eq. 3). Their method was inherently limited to two dimensions so more flexible schemes were necessary. Thouless and Licciardello[20] exploited the sensitivity of eigenenergies to changes of boundary conditions, Weaire and Srivastava[21,22] used the equation-of-motion method and Stein and Krey[23] used the recursion method. Some detailed results from the equation-of-motion method are shown in Figure 2. The broad agreement between different methods, typified by Figure 2 was satisfactory, but there was already a hint in some of the work of Thouless and Licciardello that all was not well in the case of two dimensions. Also, when attempts were made to calculate the conductivity, a wide variety of estimates resulted.[6] The method of Thouless and Licciardello contained the seminal idea for the next development of the theory which was indeed to change matters, namely, the scaling theory of Abrahams *et al.*[7]

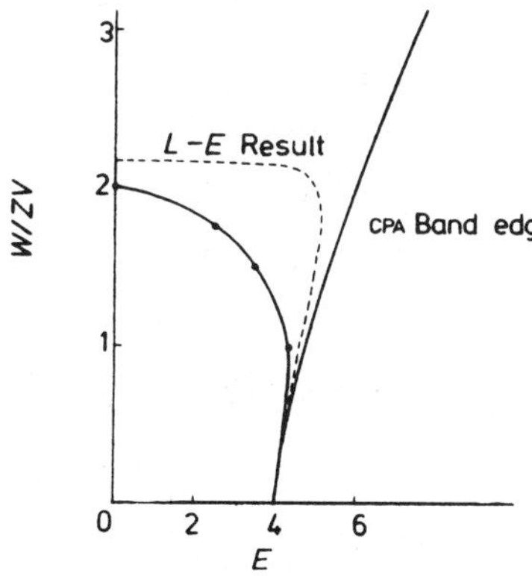

Fig. 2. Mobility edge trajectory (critical W versus E) as calculated by Weaire and Srivastava[21] and compared with the theory of Licciardello and Economou,[16] for the Hamiltonian (1) defined on the diamond cubic structure.

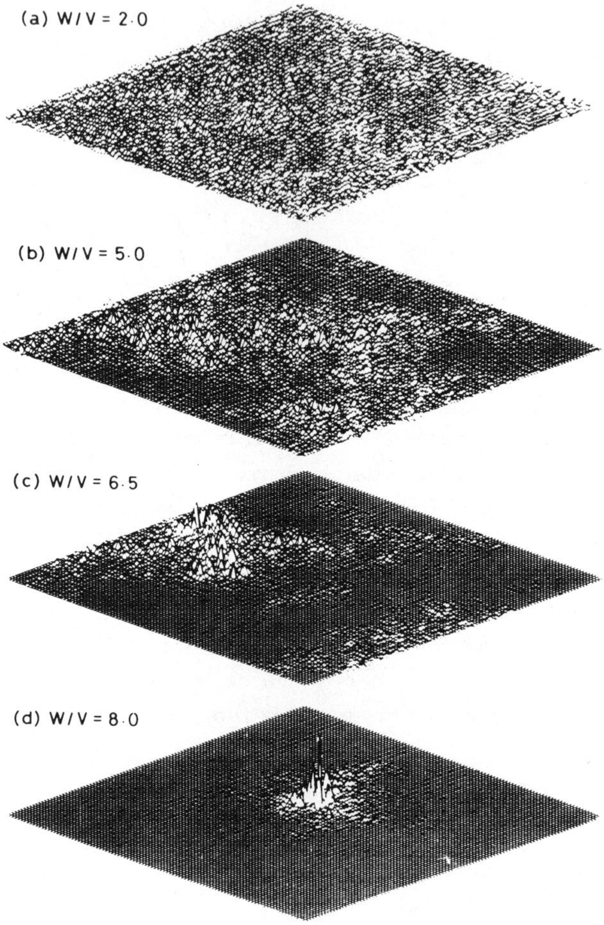

Fig.3. Map of the probability density of eigenfunctions of the Hamiltonian (1) for the two-dimensional square lattice, calculated by Yoshino and Okazaki.[19]

Abrahams *et al*[7] based their novel description of the problem on the assumption that

$$\frac{d\ln g}{d\ln L} = \beta(g) \,, \tag{4}$$

where g is the dimensionless conductivity related to conductance σ by $\sigma = \frac{e^2}{\hbar} L^{d-2} g$, and L is the length of each side of a sample in d dimensions.

The motivation for this owed much to the earlier work of Thouless[20], which related g to the effect of boundary conditions upon eigenenergies and hence to the strength of coupling between blocks of side L, when combined to make blocks of side 2L. It follows from (14) that scaling trajectories may be drawn as in Figure 4, representing the variation of g with L for a wide class of Hamiltonians including (1), provided L is large enough. The limiting behaviour for low g is determined from the assumption of exponential localisation in that limit. In the other limit the usual scaling of microscopic conductivity is used. In fact, recourse to perturbation expansions and assumptions about "reasonable" forms for the scaling trajectories is necessary to generate Figure 4. Given this, the conclusions are

(1) All states are exponentially localised in one dimension (as in the conventional wisdom).

(2) All states are exponentially localised in two dimensions but for weak disorder this will not manifest itself until the sample reaches the size corresponding to the "elbow" in the curve shown.

(3) In three dimensions, there is an Anderson transition but the conductivity goes to zero smoothly. There is a connection between the corresponding critical index ($\sigma \sim (E-E_c)^s$) and that defined for the localisation length on the

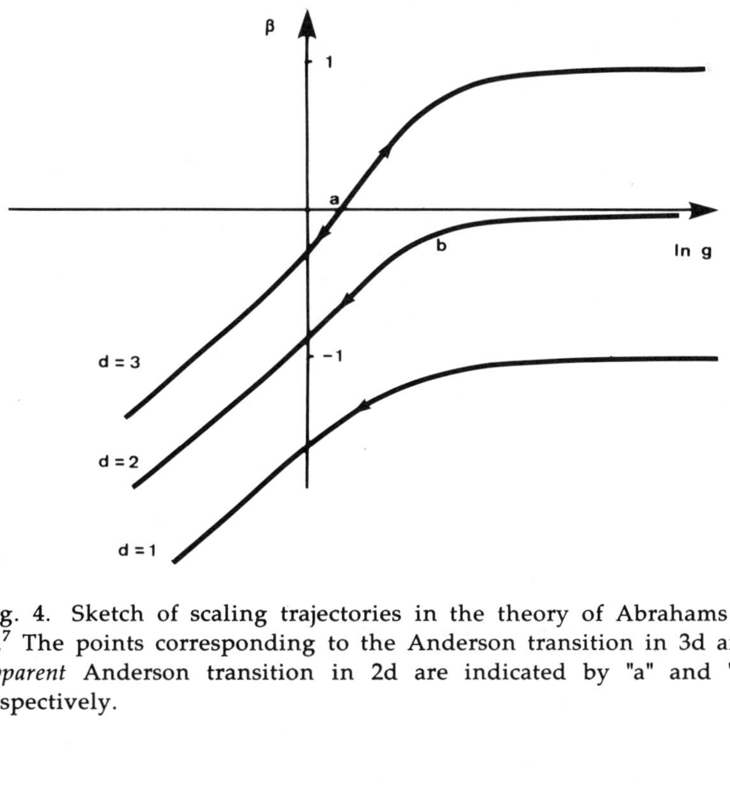

Fig. 4. Sketch of scaling trajectories in the theory of Abrahams *et al.*[7] The points corresponding to the Anderson transition in 3d and *apparent* Anderson transition in 2d are indicated by "a" and "b" respectively.

other side of the transition, according to $s = \nu$. (This was consistent with results of more formal scaling theory due to Wegner.[24])

Finally, it should be mentioned that an approximate analytical theory due to Götze[25] was found to be reasonably consistent with the above conclusions and indeed could be put into close correspondence with the scaling theory itself.[26]

Scaling theory is often presented as a "gedanken" numerical calculation so it is natural to put it to the test by directly computing the quantities involved. In the case of the scaling theory of Abrahams *et al*, its numerical counterpart is the work by Lee[27] which found (eventual) agreement with its predictions.

A different numerical scheme inspired by scaling ideas has emerged more recently[28,29] and again confirmed the same general features, and in addition provided the estimate (see Fig. 5)

$$s = \nu = 1.2 \pm 0.3 \tag{5}$$

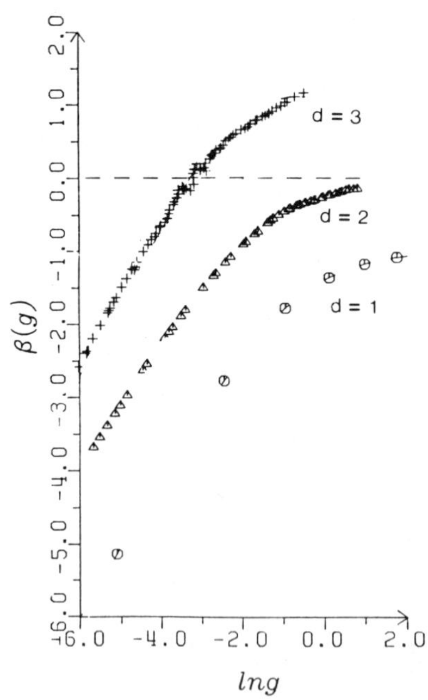

Fig. 5. Scaling trajectories calculated by MacKinnon *et al.*[28,29]

A third scheme, that of decimation[30], has run into difficulties of interpretation.

The absence of localised states in two dimensions appears at first to be in stark contradiction to the earlier numerical work which we have described. However it must be remembered[31] that states are only weakly localised below what may look like an Anderson transition. Whether the earlier analytical work which resulted in estimates for critical indices in two dimensions has any approximate validity is still not clear - it may be so.

Finally, it should be mentioned that, once discovered, weak localisation in two dimensions has been illuminated from other points of view. In particular, Hodges[32] relates it to a property of random walks, namely that random walks in two dimensions always return to the origin eventually (Polya's Theorem).

III. AMORPHOUS SEMICONDUCTORS

Because of the above developments, interest has been concentrated on two-dimensional systems of late and is currently being further stimulated by the remarkable recent results for the quantised Hall effect.[33]

Our understanding of activated conduction in amorphous semiconductors remains rather limited. In the case of hydrogenated amorphous Si, there has been a tendency to continue to fit experiments with the kind of theory advocated by Mott before the recent revision on localisation theory. This entails the use of

$$\sigma = \sigma_o \exp\left(-(E_c - E_F)/kT\right) \tag{6}$$

where σ_o is the minimum metallic conductivity.

This has been pursued at a purely empirical level - it is extremely difficult even to make a sensible estimate of E_c in this case, although attempts have certainly been made.[34] However the formula itself will probably have to be revised once the variation of mobility around the mobility edge is better understood.

ACKNOWLEDGMENTS

I would like to thank the ICTP, Trieste for its hospitality in the period during which this review was written. Research support by NBST is acknowledged.

REFERENCES

1. R. Landauer and J. Helland, J. Chem. Phys. *22*, 1655 (1954).
2. P. W. Anderson, Phys. Rev. **109**, 1492 (1958).
3. N. F. Mott and E. A. Davis, Electronic Processes in Non-Crystalline Solids (Oxford University Press, London 1971).
4. N. F. Mott and W. D. Twose, Adv. Phys. **10**, 107 (11961).
5. R. E. Borland, Proc. Roy. Soc. **A274**, 529 (1963).
6. D. Weaire and B. Kramer, J. Non-cryst. Solids. **35/36**, 9 (1980).
7. E. Abrahams, P. W. Anderson, D. C. Licciardello and T. V. Ramakrishnan, Phys. Rev. Letters **42**, 673 (1979).
8. D. J. Thouless, Phys. Rev. Letters **39**, 1167 (1977).
9. K. Ishii, Suppl. Prog. Theor. Physics **53**, 77 (1973).
10. S. Aubry, Annals of the Irael Physical Society, ed. C. G. Kuper (Adam Hilger, Bristol, 1979) Vol. 3, p 133.
11. D. C. Herbert and R. Jones, J. Phys. **C4**, 1145 (1971).
12. D. J. Thouless, J. Phys. **C5**, 77 (1972).
13. P. Lloyd, J. Phys. **C2**, 1717 (1969).
14. A. A. Gogolin, Physics Reports **86**, 2 (1982).
15. F. Wegner, Z. Phys. **B44**, 9 (1981).
16. D. C. Licciardello and E. N. Economou, Phys. Rev. **B11**, 3697 (1975).
17. R. A. Abram and S. F. Edwards, J. Phys. **C5**, 1196 (1972).
18. N. F. Mott, Commun. Phys. **1**, 203 (1976).
19. S. Yoshino and M. Okazaki, J. Phys. Soc. Japan **43**, 415, (1979).
20. D. C. Licciardello and D. J. Thouless, J. Phys. C **11**, 925 (1978).
21. D. Weaire and V. Srivastava, J. Phys. C **10**, 4309 (1977).
22. D. Weaire and V. Srivastava, in "Amorphous and Liquid Semiconductors", ed. by W. Spear (CICL, University of Edinburgh 1977) p. 286.
23. J. Stein and U. Krey, Solid. State. Comm. **27**, 1405 (1978).
24. F. Wegner, Z. Phys. **B35**, 207 (1979).
25. W. Götze, P. Prelovsek and P. Wölfle, Solid State Comm. **30**, 369 (1979).
26. D. Vollhardt and P. Wölfle, Phys. Rev. Letters **48**, 699 (1982).
27. P. A. Lee and D. S. Fisher, Phys. Rev. Letters **47**, 882 (1981).
28. A. MacKinnon and B. Kramer, Phys. Rev. Letters **47**, 1546 (1981).
29. A. MacKinnon, to be published.
30. H. Aoki, Solid State Commun. **31**, 999 (1973).
31. D. Weaire and C. J. Lambert, J. de Physique **42**, C4 (1981).
32. C. H. Hodges, J. Phys. C **14**, L247 (1981).
33. K. von Klitzing, G. Dorda and M. Pepper, Phys. Rev. Letters **45**, 494 (1980).
34. J. H. Davies, J. Non-cryst Solids **35/36**, 67 (1980).

DISCUSSION

Anderson Localization

In response to the query, "Why can't you look for localization phenomena in phonons rather than electrons?", the problem is that experiments must either be done at "high" temperatures, so that there are thermal phonons, or in excited states. In both cases, anharmonicity leads to phonon-phonon scattering and hence a finite lifetime. This would mask any effects of localization due to a static random field of impurities.

Following questions regarding the accuracy of numerical calculations near the localization transition, the scaling curve as presently calculated by Mackinnon does not look continuous through the critical point at $\beta(g_c) = 0$. It is very difficult to make numerical calculations here, and Weaire suggested that improved methods would lead to agreement with analytical theory near the critical point; agreement away from this region is already good.

There is an apparently identical computer calculation by Pichard and Sarma, which gives differing results. The difference is not fully understood, but is probably connected with an unreliable extrapolation to infinite size.

The one-parameter scaling theory as described in Weaire's paper assumes noninteracting particles, which makes verification by experiment difficult. Currently, the best results available for three-dimensional materials are those of Paalanen, Rosenbaum, and Thomas[1] on crystalline Si:P (nominally uncompensated). They find that the critical behavior of the electrical conductivity σ (extrapolated to zero temperature) as a function of the impurity concentration n, is

$$\sigma \sim (n-n_c)^{\alpha}$$

where the index α has the value 0.50 ± 0.05. For comparison, classical percolation theory predicts $\alpha \approx 1.6$, and scaling theory predicts α close to unity, possibly $\alpha \approx 1.2$. It was pointed out that the experiment did not vary n directly but merely stress, so that some caution in relating Δn to changes in stress and strain should be exercised in extracting the critical index, α.

It is clear that a good treatment of electron-electron interaction as well as disorder is needed to account for these experiments.

J. H. Davies, Chairman
D. L. Huber, Discussion Leader

REFERENCE

1. M. A. Paalanen, T. F. Rosenbaum, G. A. Thomas, and R. N. Bhatt, Phys. Rev. Lett., *48*, 1284 (1982).

RESONANT ENERGY TRANSFER IN RUBY AND ANDERSON LOCALIZATION

H. M. Gibbs

Optical Sciences Center
University of Arizona
Tucson, Arizona 85721

S. Chu, S. L. McCall and A. Passner

Bell Telephone Laboratories
600 Mountain Avenue
Murray Hill, New Jersey 07974

Direct measurements of energy transfer among single chromium ions in ruby show that the dominant interaction is dipole-dipole and the transfer time is long (approximately milliseconds for 0.25 wt.% Cr concentration). The direct methods utilize transient-grating, transient-electric-field, and fluorescence-line-narrowing techniques. Indirect measurements of ion-ion transfer which utilize third- and fourth-nearest-neighbor traps were first thought to signify fast (submicrosecond) transfer. However, simulations of single-ion and trap decays which neglect ion-ion transfer but include ion-trap transfer and a small amount of radiation trapping explain the data. Since the ion-ion transfer is dipole-dipole, it cannot give rise to an Anderson transition between localized and extended states. This explains recent failures to observe mobility edges in ruby with Cr concentrations up to 1%.

I. INTRODUCTION

Anderson[1] has shown that three-dimensional energy transfer mediated by an interaction potential dropping off faster than the inverse third power of the separation between sites can undergo an Anderson transition. That is, at low concentrations of sites, energy is not transferred between sites but remains localized on the sites excited. However, above a critical concentration the excitation is a coherent superposition of true eigenstates which are extended throughout the crystal. Because of alternative explanations for mobility edges in more complicated systems such as semiconductors, it would be desirable to demonstrate an Anderson transition in a system known to satisfy the assumptions of the theory. For this reason the report[2] of a mobility edge in ruby led Anderson to refer to it as the only direct evidence for an Anderson transition in all of physics.[3] However, even in the case of ruby there were other experiments[4,5] which indicated that the energy transfer might be slow and dominated by ion-trap rather than ion-ion transfer. It was suggested that the observed mobility edges might arise from resonance transfer between a single ion and excited states of the fourth-nearest-neighbor trap.[6]

Because of the importance of a doubt-free demonstration of an Anderson transition and the questions surrounding the Koo et al.[2] ruby experiment, that experiment was repeated.[7] A dye laser permitted rapid tuning across the entire inhomogeneous line compared with the thermal tuning on one side only in the original experiment. No convincing evidence for a mobility edge was found in the new experiment.[7]

The fluorescence ratio data did not rule out fast resonant energy transfer in a convincing manner. Consequently, the efforts then turned to searching for direct ways of determining the nature of the ion-ion transfer interaction. These direct methods have shown that the dominant interaction is dipole-dipole and the transfer takes about a millisecond. This rules out resonant energy transfer in ruby as a candidate for an Anderson transition.

The usual discussion of ruby and Anderson localization begins by reviewing all of the history of why the transfer was believed to occur by an exchange interaction and the evidence for a mobility edge. In this talk we will take a different approach. In Section II are described the direct measurements of energy transfer in ruby which show that it is dipole-dipole and slow. It is also calculated that dipole-dipole transfer correctly accounts for the amount of transfer observed. This disposes of ruby as a candidate for an Anderson transition and establishes the nature of the transfer. In Sections III and IV we back up and clean up--succinctly because Section II settles the central issues. In Section III we describe indirect measurements of ion-ion transfer which involve trap fluorescence. We discuss why these results first suggested fast ion-ion transfer by exchange and how they are actually consistent with the slow dipole-dipole interaction established in Section II. In Section IV we describe the beautiful idea of using ruby to see a mobility edge assuming an exchange interaction, the original experiment which reported evidence for an edge, and the repeat experiment which did not. In Section V we draw the conclusion that the overwhelming evidence at present is that ruby does not satisfy the conditions for an Anderson transition, and we must seek another system for the ideal demonstration.

II. DIRECT MEASUREMENTS OF NONRADIATIVE RESONANT ION-ION ENERGY TRANSFER IN RUBY

A. GRATING EXPERIMENTS VIA DEGENERATE FOUR-WAVE MIXING

Three independent experiments have placed an upper limit of 100 to 300 Å on the energy migration distance in ruby (Cr^{3+}-doped Al_2O_3) during the fluorescence lifetime for concentrations between 0.05 and 1.55 at.%.[8–10] (Concentrations are given in wt.% defined as the percentage weight of Cr in the sample and at.% defined as the percentage of Cr atoms in the total number of Cr and Al atoms. Coincidentally, for the range of interest here, namely $\lesssim 1\%$, these are essentially numerically equal for our purposes.) The typical experiment involves two counterpropagating pump beams and a probe beam at a small angle; the phase conjugate wave generated in this degenerate four-wave mixing scheme comes out opposite to the probe. The conjugate is created by the scattering of one pump beam by the grating created by the interference fringes of the probe beam and the other pump beam. The grating formed by the probe and the nearly copropagating pump beam has a large spacing ($\simeq 10\ \mu m$). The other grating produced by nearly counterpropagating beams, has a much smaller spacing ($\simeq 0.2\ \mu m$). Physically the gratings arise from the modulation of the index of refraction resulting from the difference in polarizability between optically excited ions and ground-state ions.

When spatial migration of the excitation is absent, the lifetime of both gratings is determined solely by the excited-state lifetime and is independent of grating spacing. Spatial excitation transfer over distances comparable to the grating spacing would make the grating lifetime depend on its spacing. Eichler et al.[8] and Liao et al.[10] monitored the temporal decay of the grating after pulsed excitation. Hamilton et al.[9] determined the spacing dependence by observing the state of polarization of the conjugate wave. The low-temperature[10] and room-temperature[8,9] experiments all conclude that there is no spatial energy diffusion within the experimental resolutions. Liao et al.[10] deduce a transfer rate of $<1.7 \times 10^3\ \mathrm{sec}^{-1}$ for a 0.25-at.% sample assuming an average transfer rate, microscopic inhomogeneity, a 50-MHz homogeneous width, and 30-GHz (1-cm^{-1}) inhomogeneous width. These experiments do not determine the nature of the inhomogeneity or rule out rapid diffusion within 100-Å regions or domains. They are consistent with the slow transfer found by electric-field (II C) and fluorescence-line-narrowing (II B) experiments.

B. FLUORESCENCE LINE NARROWING

The basic idea of fluorescence line narrowing[11] (FLN) is that when an inhomogeneous absorption is probed with a high-resolution laser pulse, only those ions in resonance with the laser (the donors) will fluoresce immediately after the excitation. If an interaction mechanism exists for transferring excitation from ion to ion, then at later times emission is observed from a different set of ions (the acceptors). If the acceptor energy level is in exact resonance with the donor, no changes in the emission line shape occur in time. Resonant energy transfer in ruby can be detected, however, because the ground state $\left({}^4A_2\right)$ has two levels separated by $\Delta = 0.38\ \mathrm{cm}^{-1}$. If an ion in the lower level absorbs at ν_L, the fluorescence contains components at ν_L and $\nu_L - \Delta$. If an upper-level ion absorbs at ν_L, the fluorescence is at ν_L and $\nu_L + \Delta$. In a sample in which the inhomogeneous broadening exceeds Δ, absorption can occur from both levels simultaneously and fluorescence occurs at three frequencies, namely, ν_L and $\nu_L \pm \Delta$. Now if fluorescence is absorbed by the lower level, emission at $\nu_L - 2\Delta$ can occur; this is *radiative* resonant transfer. Likewise a lower-level ion excited by $h\nu_L$ can undergo resonant *nonradiative* energy transfer in which it drops to the upper level of the ground state and transfers energy $h(\nu - \Delta)$ to a lower-level ion. The latter can then fluoresce at $\nu_L - \Delta$ and $\nu_L - 2\Delta$.

Selzer and Yen[12] studied radiative transfer in ruby, both phonon-assisted nonresonant (exponential temperature dependence) and resonant (temperature independent) transfer. They found that fluorescence at $\nu_L - 2\Delta$ (or $\nu_L + 2\Delta$) arises slowly. They point out that if rapid ion-ion transfer occurs within a homogeneous width then $\nu_L \pm 2\Delta$ fluorescence should also appear rapidly if the inhomogeneous broadening is microscopic. Accepting rapid ion-ion transfer as highly probable, they surmised that their samples were macroscopically inhomogeneously broadened. Macroscopic broadening would permit rapid single-single transfer, but a $\nu_L - \Delta$ transfer could not occur because ions with different frequencies in the inhomogeneous line would not be close in space. Radiative transfer could occur since fluorescence connects ions far apart. Since other experiments show that ion-ion transfer is in fact slow, Selzer and Yen's observations are consistent with microscopic inhomogeneity. In fact, their study of nonradiative phonon-assisted transfer[13–14] points directly to microscopic inhomogeneity--they never see a gradual spreading of the narrow components until they fill the inhomogeneous line. Instead the nonradiative transfer leads to a background which always has the full inhomogeneous width, pointing toward microscopic inhomogeneous broadening. These FLN results of Selzer et al. for crystals of 0.03 to 1 at.% are consistent with similar measurements of Chu et al.[15] In an attempt to reconcile fast transfer with their observations, Selzer et al. introduce several classes of ions.[14] We believe that this is unnecessary for the major features of their data now that slow transfer has been established.[15e]

Selzer et al. focused their attention on transfer via radiative and nonresonant nonradiative interactions. The emphasis in Chu et al. is upon resonant nonradiative transfer because of its importance in the Anderson transition question. Selzer et al. were kind enough to send us some of their FLN data taken in the 20 μs following 5-ns excitation of 0.25 and 0.9 at.% samples at 5 K. Already a peak at $\nu_L+2\Delta$ is seen which must arise from resonant nonradiative transfer since radiative transfer occurs on a millisecond time scale. In the 0.9 at.% sample a nonradiative nonresonant background is already present and broad. Chu et al. have studied the nonradiative resonant transfer by FLN techniques. The most thorough study has been with the time-resolved electric-field technique discussed in the next section. However, it is important to note that the growth of the fourth peak is slow (approximately milliseconds) even in zero electric field, disproving the hypothesis[16] that the electric field separates the ions into two classes, A and B, which transfer rapidly within each class but not between classes.

C. TRANSIENT ELECTRIC FIELDS

Cr^{3+} ions in ruby are located at inequivalent sites (labeled A and B) such that R_1 transitions shift in opposite directions in an electric field.[17] The experiment of Chu et al.[15] can be understood with the aid of Fig. 1. In (a) the A and B transitions are separated by 0.5 to 1.0 cm^{-1} with 2.5 kV and a small subset of A ions is resonantly excited by the output of a CW dye laser with a bandwidth less than 0.1 cm^{-1}. The FLN three-peaked spectrum with a growing fourth peak can be seen from those A ions excited, but the B ions are unexcited and do not fluoresce. In (b), the electric field is reduced to zero for a variable time to allow A-to-B transfer. In (c), the field is reapplied;

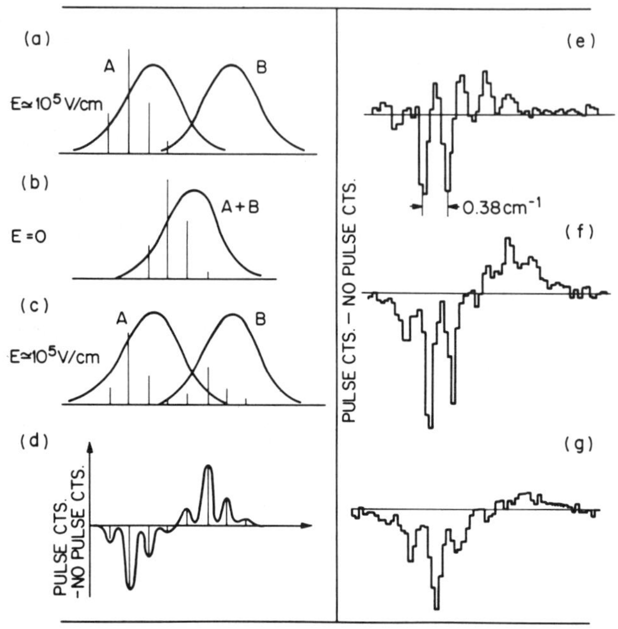

Fig. 1. (a)-(c) Schematic of the electric-field sequence and a transfer of energy from A to B sites. (d) Expected appearance of fluorescence when a pulse minus no-pulse subtraction is made (e)-(g) Fluorescence spectra for 0.25 wt.%, 0.69 wt.%, and 1.2 wt.% samples, respectively. The field-off time is 800 μs. Ref. 15(c).

some B ions now exhibit the FLN spectrum and the A spectrum is reduced. In (d), the expected appearance of the A "flop-out" and B "flop-in" fluorescence spectrum monitored for 8 ms is displayed. The excitation light was amplitude-stabilized and limited to a few milliwatts focused to approximately 0.5 mm^2 for a few milliseconds. Fluorescence was analyzed with a Fabry-Perot interferometer (4 cm^{-1} free spectral range and <0.15 cm^{-1} laser-plus-interferometer linewidth) and a single-photon-counting RCA photomultiplier tube. The ruby samples, 80- to 100 μm thick, were sandwiched between two glass plates with transparent electrodes and immersed in liquid He at $\simeq$1.7 K.

Data were taken with the electric field alternately pulsed off and not pulsed off; a pulse minus no-pulse subtraction eliminated features in the data not associated with the pulsing of the electric field. Typical interferometer scans taken for a fixed field-off time of 800 μs are shown in Figs. 1(e) to (g), revealing resonant transfer of excitation from A to B ions. At 0.25 wt.%, the

B spectral features are sharp, and the flop-in counts equal the flop-out counts, i.e., resonant transfer dominates. However, for the higher concentrations, the B spectrum is spectrally diffuse, and the flop-in signal is less than the flop-out signal. At 1.2 wt.%, the spectrally diffuse fluorescence dominates over the sharp-line emission; most of the fluorescence is not even in the R_1 line.

When the interferometer was tuned to one of the A or B fluorescence lines, and the off-time was varied, the data of Fig. 2 were obtained. The solid lines are least-squares fits of the data to

$$\alpha + \beta t + \gamma t^{3/N} \tag{1}$$

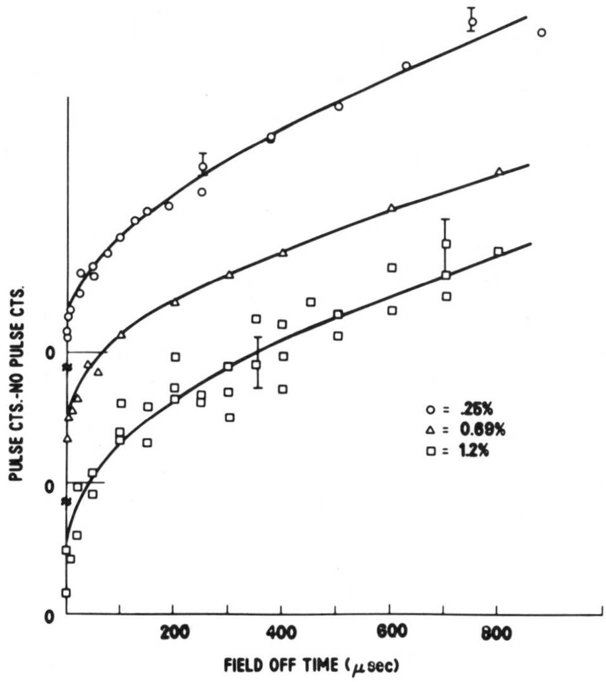

Fig. 2. Time dependence of the energy transfer with the interferometer tuned to a particular flop-in emission line (0.25 wt.% and 0.69 wt.%) or to a flop-out line (1.2 wt.%). The solid lines are nonlinear least-squares fits (see text). The fitted values of N are 5.6 ± 1.6 (5.4 ± 1.4 for a second run), 10.9 ± 4.5, and 7.4 ± 2.7 for 0.25 wt.%, 0.69 wt.%, and 1.2 wt.% samples, respectively. The fitted (partially trapped) R_1 lifetime of 4.1 ms of 0.25 wt.% ruby agrees well with the observed value of 4.0 ms. Ref. 15(c).

where t is the field-off time, α represents a background, β represents weak long-range resonant radiative transfer, and $\gamma t^{3/N}$ accounts for a nonradiative resonant transfer rate that varies as r^{-N} (see Section IID). The 0.69 wt.% and 1.2 wt.% data do not rule out N = 6, 8, or 10, but the 0.25 wt.% data (for which resonant transfer is dominant) point to N = 6, i.e., a dipole-dipole coupling.

The fraction that transfers from A to B during the first 100 μs is roughly 3.5% and 6.5% for the 0.25 wt.% and 0.69 wt.% concentrations, respectively. Thus, about 7% and 13% of the Cr^{3+} ions make an A→B or A→A resonant transfer during the first 100 μs. Using N = 6 to extrapolate for 0.25 wt.%, one expects 7% (3.7 ms/100 μs)$^{3/6}$ = 43% to transfer during the 3.7-ms lifetime.

Because of the random distribution of Cr ions in the Al_2O_3 lattice, there is a wide range of ion-ion coupling strengths. One might worry that the electric field was pulsed too slowly, so that an appreciable fraction of the ions adiabatically followed the field and returned to the A state. This was shown not to be the case by an electric-field-reversal experiment and FLN studies.[15]

In Section II B, FLN experiments pointed to microscopic inhomogeneous broadening. Spectral diffusion in the presence of an electric field shows A→B diffusion, showing that there is no gross segregation of A and B ions.

A variation of the pulsed-field experiment has given a rough measure of the homogeneous width. Suppose the field is cut off long enough for near equilibrium to be established between the A ions excited and the resonant B ions. Now if the field is increased by δV, where δV is larger than the homogeneous width measured in volts, the A ions should now be resonant with a new set of unexcited B ions. This was found to be the case as shown

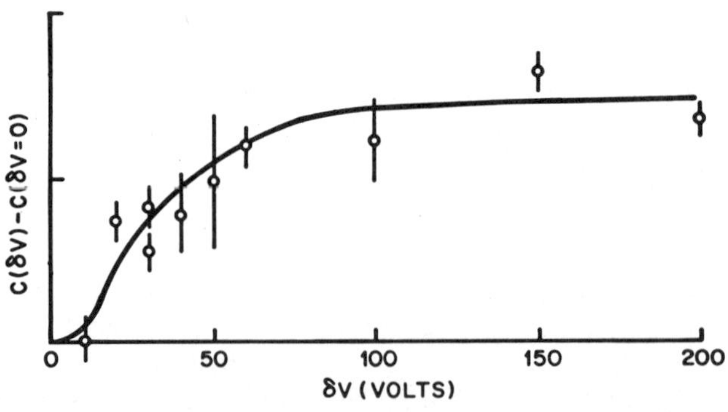

Fig. 3. Additional A → B transfer seen as a function of a second step in voltage δV for a 0.25 wt.% sample of ruby. Ref. 15e.

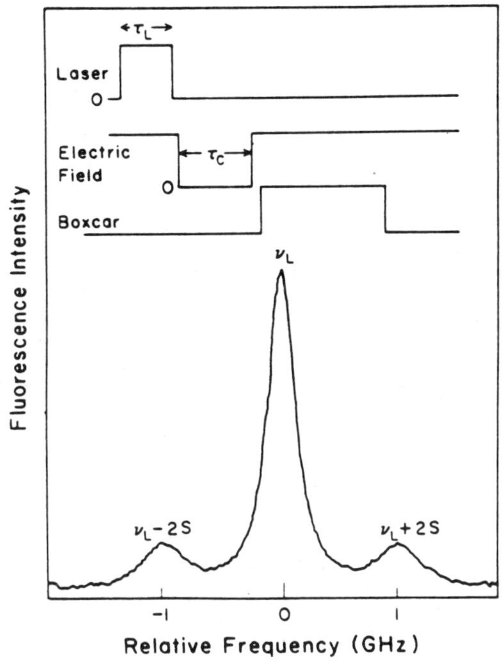

Fig. 4. Fluorescence line narrowing signal showing transfer lines at $\nu_L \pm 2S$ and directly excited line at the laser frequency ν_L in 0.35-wt.% Cr_2O_3 ruby at 5 K. Inset shows measurement time sequence where the laser pulse length is $\tau_L = 0.8$ ms, followed by a Stark pulse of length τ_c, and finally the boxcar gate of 2 ms length. Ref. 18b.

in Fig. 3; from the δV required to increase the A flop-out signal, the homogeneous width was found to be $\simeq$130 MHz for 0.25 wt.%.

Learning[18] of the pulsed-field technique in a discussion with the first author, Szabo and Jessop conducted very similar but independent pulsed-field measurements, first[18a] reporting no nonradiative resonant transfer and then later[18b] finding weak transfer in agreement with Chu et al. They use a cw dye laser with 1-MHz linewidth to resonantly excite at frequency ν_L both A and B ions split by a total Stark splitting of 2S. When the electric field is removed, the excited A ions have resonance frequencies ν_L+S. If transfer from A ions to B ions occurs in zero field then when the field is reapplied the new B ions will fluoresce at ν_L+2S. Likewise B-to-A transfer gives rise to fluorescence at ν_L−2S. An example of fluorescence at $\nu_L \pm 2S$ is shown in Fig. 4, and the dependence of the transfer upon the field-off contact time τ_c is given in Fig. 5. Jessop and Szabo interpret Fig. 5 as a fast nonradiative transfer involving a few percent of the ions in the first few hundred

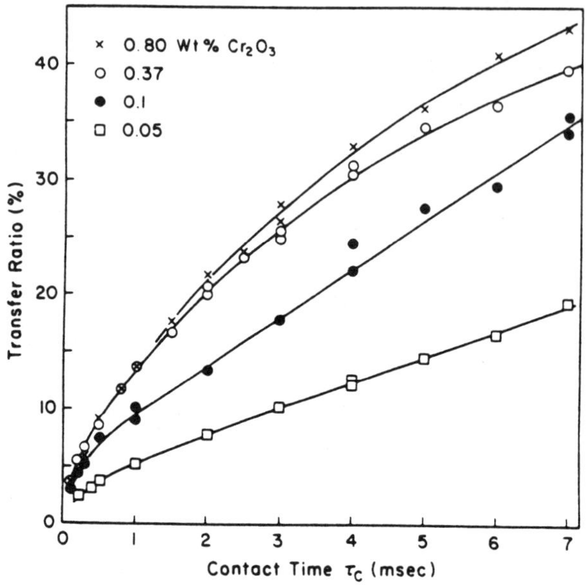

Fig. 5. Experimental variation of the transfer ratio (sum of the two transfer peaks divided by central peak) vs contact time for various Cr concentrations. Ref. 18b.

microseconds followed by a slower radiative process. They use a simple model to remove radiative transfer from the transfer curves and conclude that only 6% to 16% of the ions resonantly transfer energy assuming that the inequivalent A and B sites are equally occupied by impurity ions. From the measured ion-ion transfer rate they conclude that the interaction could be dipole-dipole and that the matrix elements are not too much larger than the dipole-dipole matrix element calculated by Imbusch.[19]

D. CALCULATION OF DIPOLAR-DIPOLAR TRANSFER

So far in this section it has been established that resonant nonradiative transfer in ruby is slow and most likely dipole-dipole. Evidence for the dipolar nature came from fitting the transfer signal to the function $\alpha+\beta t+\gamma t^{3/N}$ and determining that $N = 6$ (dipole-dipole) gave the best fit. In this section the magnitude of dipole-dipole transfer is calculated and found to be in good agreement with that observed.

It is the range dependence of the transfer which leads to good agreement between the measured rate and the calculated dipole-dipole rate. Past analyses of ruby transfer have used an average transfer rate, with no explicit radial or angular dependence. We treat the radial dependence in a continuum approximation. The calculation could be refined by beginning the radial integration at a minimum radius to exclude ions so close that both ions' energy levels are shifted too much or by summing over a discrete lattice. One does not expect that such refinements would make much difference for low concentrations.

Assume that the density of single ions is low enough that the fraction of ions which transfer is small. The A ions are excited and then brought into contact with B ions for a time t. What is the number of excited B ions as a function of t? Let an A ion be at $\vec{r} = 0$ and excited at $t = 0$. Let the nearest B ion be at $\vec{r}$. The dipole-dipole interaction Hamiltonian is

$$W = \frac{p_A p_B}{\kappa r^3}\left[\hat{p}_A \cdot \hat{p}_B^* - 3\left(\hat{p}_A \cdot \hat{r}\right)\left(\hat{p}_B^* \cdot \hat{r}\right)\right] \tag{2}$$

where $\kappa \simeq (1.76)^2$ is the dielectric constant and $p_A\hat{p}_A$ and $p_B\hat{p}_B$ are the ground-excited state dipole moment matrix elements. The transition probability per unit time for transfer is:

$$\Gamma = \frac{2\pi}{\hbar}|W|^2 \int_{-\infty}^{+\infty} g_A(E) g_B(E) dE \tag{3}$$

where $g_A(E)$ and $g_B(E)$ are the normalized energy distribution functions.[19] There are several possible transitions as shown in Fig. 6:

$$\vec{p}_1 = p_1(\hat{x} + i\hat{y})/\sqrt{2} \qquad p_1^2 = 0.390\, p_T^2 \tag{4a}$$

$$\vec{p}_2 = p_2\hat{z} \qquad p_2^2 = 0.0569\, p_T^2 \tag{4b}$$

$$\vec{p}_3 = p_3(\hat{x} + i\hat{y})/\sqrt{2} \qquad p_3^2 = 0.553\, p_T^2 \tag{4c}$$

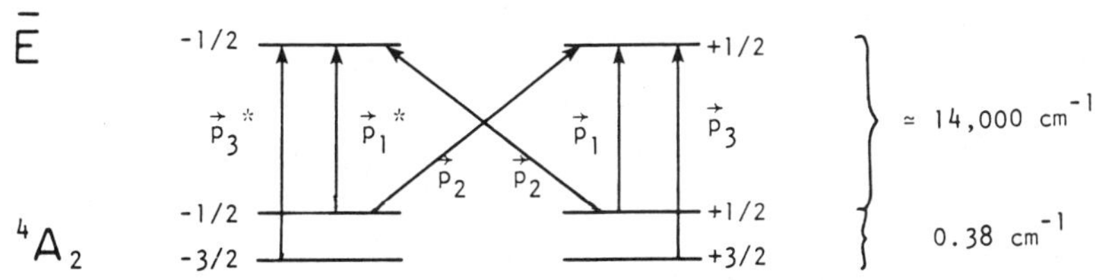

Fig. 6. Transition dipole moments of the R_1 line of ruby.

where the ratios are taken from Nelson and Sturge.[20] One might think of setting p_T^2 equal to $p^2 = 1.63 \times 10^{-40}$ cgs from $1/\tau_{R_1} = (3.7\ \text{ms})^{-1} = 4\kappa^{1/2}\omega^3 p^2/3\hbar c^3$ but p^2 also contains ≅20% contributions from one-phonon assisted decay which should be excluded from p_T^2. From the measurements of Nelson and Sturge one finds $p_T^2 = 1.2 \times 10^{-40}$ cgs.

Consider first the case that A is excited and B is in the 1/2 ground state. The possible states of the system can then be written as:

TABLE I

Probability	A ion in state	&	B ion in state
a	$\bar{E}(+1/2)$		$^4A_2(+1/2)$
b	$\bar{E}(+1/2)$		$^4A_2(-1/2)$
c	$\bar{E}(-1/2)$		$^4A_2(+1/2)$
d	$\bar{E}(-1/2)$		$^4A_2(-1/2)$
u	$^4A_2(+1/2)$		$\bar{E}(+1/2)$
v	$^4A_2(-1/2)$		$\bar{E}(+1/2)$
w	$^4A_2(+1/2)$		$\bar{E}(-1/2)$
x	$^4A_2(-1/2)$		$\bar{E}(-1/2)$

The transition matrix table is then

	u	v	w	x
a	F	E	E	H
b	E	H	G	E
c	E	G	H	E
d	H	E	E	F

where $E = \Gamma_{21}$, $F = \Gamma_{11}$, $G = \Gamma_{1^*1}$, $H = \Gamma_{22}$ and Γ is given by Eq. (3) and the subscripts refer to the transitions labelled as p_1 and p_2 in Fig. 6. The set of eight coupled differential rate equations can then be written as

$$\begin{aligned} \dot{a} &= Fu + Ev + Ew + Hx - (F + 2E + H)a \\ &0 \\ \dot{u} &= Fa + Eb + Ec + Hd - (F + 2E + H)u \end{aligned} \tag{5}$$

etc. Or Eqs. (5) can be greatly simplified by writing them as

$$\frac{d}{dt}\begin{pmatrix}\alpha \\ \beta\end{pmatrix} = -2\begin{pmatrix}R & E \\ E & T\end{pmatrix}\begin{pmatrix}\alpha \\ \beta\end{pmatrix} \tag{6}$$

where

$$\alpha \equiv a + d - u - x\,, \tag{7a}$$

$$\beta \equiv b + c - v - w\,, \tag{7b}$$

$$R \equiv F + E + H\,, \tag{7c}$$

$$T \equiv G + E + H\,. \tag{7d}$$

Eq. (6) has solutions of the form $e^{-\Lambda t}$ where the eigenvalues are

$$\Lambda_\pm = R + T \pm ((R-T)^2 + 4E^2)^{1/2} \tag{8}$$

and the eigenvectors are

$$\begin{pmatrix}\alpha_\pm \\ \\ \beta_\pm\end{pmatrix} = \begin{pmatrix}R - T \pm ((R-T)^2 + 4E^2)^{1/2} \\ \\ 2E\end{pmatrix}. \tag{9}$$

One can write

$$\begin{pmatrix}\alpha \\ \\ \beta\end{pmatrix} = C_+\begin{pmatrix}\alpha_+ \\ \\ \beta_+\end{pmatrix} + C_-\begin{pmatrix}\alpha_- \\ \\ \beta_-\end{pmatrix}. \tag{10}$$

If at $t = 0$, $\alpha = 1$ and $\beta = 0$ i.e. the A ion is excited and the B ion is not,

$C_+^\alpha = \beta_-/\det$ and $C_-^\alpha = -\beta_+/\det$, where $\det = \alpha_+\beta_- - \alpha_-\beta_+$. If at $t = 0$, $\beta = 1$ then $C_+^\beta = -\alpha_-/\det$ and $C_-^\beta = \alpha_+/\det$. Half of the time the C^α coefficients are in effect and the other half the C^β are, so the sum is needed.

Let ρ be the density of Cr^{3+} ions. Assuming an equal number of A and B ions, there are $\rho d^3r/8$ ions in each of the four ground states $^4A_2(\pm 1/2)$ and $^4A_2(\pm 3/2)$. Since $\alpha + \beta = a + b + c + d - (u + v + w + x)$,

$$\frac{\rho}{8}\, d^3r(\alpha+\beta) = \rho\, \frac{d^3r}{8} \left\{ \left[C_+^\alpha\alpha_+ + C_+^\alpha\beta_+\right] \exp(-\Lambda_+ t) + \left[C_-^\alpha\alpha_- + C_-^\alpha\beta_-\right] \exp(-\Lambda_- t) \right\} \tag{11}$$

is the number of A ions excited minus the number of B ions excited in d^3r for initial states $|\alpha,\gamma\rangle$ or $|\beta,\delta\rangle$. The total number of A and B ions excited is $\rho d^3r/8$ in d^3r. The signal is proportional to the B ions excited by transfer, i.e., to $(\rho d^3r/16)[1-(\alpha+\beta)]$. The number of A ions excited by transfers from B ions can be calculated similarly for $\beta = 1$ at $t = 0$. The sum is the total signal from d^3r, i.e., the probability $dS_{1/2}$ that a transfer has occurred from an excited A to a B in $^4A_2(\pm 1/2)$ or from an excited B to an A in $^4A_2(\pm 1/2)$:

$$dS_{1/2} = \frac{\rho d^3r}{16} \left\{ 2 - \left(C_+^\alpha + C_+^\beta\right)\left(\alpha_+ + \beta_+\right) \exp(-\Lambda_+ t) - \left(C_-^\alpha + C_-^\beta\right)\left(\alpha_- + \beta_-\right) \exp(-\Lambda_- t) \right\} \tag{12}$$

$$= \frac{\rho d^3r}{16} \left\{ Q_+(1 - \exp(-\Lambda_+ t)) + Q_-(1-(\exp(-\Lambda_- t)) \right\} \tag{13}$$

where

$$Q_\pm \equiv 1 \pm \frac{2E}{((R-T)^2+4E^2)^{1/2}} . \tag{14}$$

To obtain the signal, one needs to integrate over $d^3r \equiv r^2 dr d\Omega$. Note that the entire r dependence in Eq. (13) is contained in the r^{-6} dependence in $\Lambda_\pm$ from Eqs. (2), (3), and (8). Let

$$\Lambda_\pm \equiv \lambda_\pm / r^6 , \tag{15}$$

so the integrals of interest are

$$\int_0^\infty r^2 dr \left[1 - \exp\left(-\lambda_\pm t/r^6\right)\right] = \frac{\left(\lambda_\pm t\right)^{1/2}}{6} \int_0^\infty \frac{du}{u\sqrt{u}}(1-e^{-u}) = \frac{\sqrt{\pi}}{3}\sqrt{\lambda_\pm t} \quad (16)$$

showing the $t^{3/N}$ dependence of the transfer for the dipole-dipole case (N = 6). (Radiative resonant transfer is independent of radius, i.e., $1 - \exp(-\Gamma_I t)$ which reduces to $\Gamma_I t$ for small t giving the linear term in Eq. (1).) Then the signal is

$$S_{1/2} = \frac{\rho}{48}\sqrt{\pi}\sqrt{t}\int [Q_+\sqrt{\lambda_+} + Q_-\sqrt{\lambda_-}]d\Omega \ . \quad (17)$$

The angular factors can be evaluated using Eqs. (2) and (4) to give

$$E = Pp_1^2p_2^2 \left|\frac{3}{\sqrt{2}} \sin\theta \cos\theta \right|^2 = Pp_T^4\, \epsilon(\theta) \quad (18a)$$

$$F = Pp_1^4 \left| \frac{3}{2} \cos^2\theta - \frac{1}{2} \right|^2 \quad (18b)$$

$$G = Pp_1^4 \left| \frac{3}{2} \sin^2\theta \right|^2 \quad (18c)$$

$$H = Pp_2^4 \left|1 - 3\cos^2\theta\right|^2 \quad (18d)$$

where

$$P \equiv \frac{2\pi}{\hbar\kappa^2 r^6} \int_{-\infty}^{\infty} g_A(E)g_B(E)dE \quad (19)$$

and $p_T^2 = p_1^2 + p_2^2 + p_3^2$. Then

$$R = Pp_T^4\rho(\theta) \quad (20a)$$

$$T = Pp_T^4\tau(\theta) \quad (20b)$$

where

$$\rho(\theta) \equiv \left\{\frac{p_1^4}{p_T^4} \left| \frac{3}{2}\cos^2\theta - \frac{1}{2} \right|^2 + \frac{p_2^4}{p_T^4} \left| 1-3\cos^2\theta\right|^2\right.$$

$$+ \frac{p_1^2 p_2^2}{p_T^4} \left| \frac{3}{\sqrt{2}} \sin\theta \cos\theta \right|^2 \Bigg\} \tag{21}$$

and

$$\tau(\theta) \equiv \Bigg\{ \frac{p_1^4}{p_T^4} \left| \frac{3}{2} \sin^2\theta \right|^2 + \frac{p_2^4}{p_T^4} \left| 1-3\cos^2\theta \right|^2$$

$$+ \frac{p_1^2 p_2^2}{p_T^4} \left| \frac{3}{\sqrt{2}} \sin\theta \cos\theta \right|^2 \Bigg\} \tag{22}$$

so that

$$\Lambda_\pm = \frac{\lambda_\pm}{r^6} = P p_T^4 [\rho + \tau \pm ((\rho-\tau)^2 + 4E^2)^{1/2}] \tag{23}$$

and

$$S_{1/2} = \frac{\rho}{24} \left(\frac{2}{\hbar} \right)^{1/2} \frac{\pi^2 p_T^2}{\kappa} \sqrt{t} \left(\int_{-\infty}^{\infty} g_A(E) g_B(E) dE \right)^{1/2} \int_0^\pi T_{1/2}(\theta) \sin\theta d\theta \tag{24}$$

where

$$T_{1/2}(\theta) \equiv \left(1 + \frac{2\epsilon}{((\rho-\tau)^2 + 4\epsilon^2)^{1/2}} \right) (\rho + \tau + [(\rho-\tau)^2 + 4\epsilon^2]^{1/2})^{1/2}$$

$$+ \left(1 - \frac{2\epsilon}{((\rho-\tau)^2 + 4\epsilon^2)^{1/2}} \right) \{\rho + \tau - [(\rho-\tau)^2 + 4\epsilon^2]^{1/2}\}^{1/2} . \tag{25}$$

One finds numerically that

$$\int_0^\pi T_{1/2}(\theta) \sin\theta d\theta = 1.897 . \tag{26}$$

The signal for B in the 3/2 ground state is the same as above with E and H equal to zero and p_1 replaced by p_3, so that

$$S_{3/2} = \frac{\rho}{24} \left(\frac{2}{\hbar} \right)^{1/2} \frac{\pi^2 p_T^2}{\kappa} \left(\int_{-\infty}^{\infty} g_A(E) g_B(E) dE \right)^{1/2} \sqrt{t} \int_0^\pi T_{3/2}(\theta) \sin\theta \, d\theta , \tag{27}$$

where

$$T_{3/2}(\theta) \equiv (2\rho_{3/2})^{1/2} + (2\tau_{3/2})^{1/2} , \tag{28a}$$

$$\rho_{3/2} = \frac{p_3^4}{p_T^4} \left| \frac{3}{2} \cos^2\theta - \frac{1}{2} \right|^2 , \tag{28b}$$

$$\tau_{3/2} = \frac{p_3^4}{p_T^4} \left| \frac{3}{2} \sin^2\theta \right|^2 , \tag{28c}$$

and

$$\int_0^\pi T_{3/2}(\theta) \sin\theta \, d\theta = 2.165 \tag{29}$$

is found numerically. The total signal is then

$$S = S_{1/2} + S_{3/2} + 2.363 \frac{\rho p_T^2}{\sqrt{\hbar\kappa}} \sqrt{t} \left[\int_{-\infty}^{\infty} g_A(E) \, g_B(E) dE \right]^{1/2} . \tag{30}$$

To evaluate the transfer, we use $\kappa = (1.76)^2$, $\rho(0.25\%) = 1.16{\times}10^{20}$ cm^{-3}, $p_T^2 = 1.2{\times}10^{-40}$ esu. The integral

$$I \equiv \int_{-\infty}^{\infty} g_A(E) g_B(E) dE \tag{31}$$

depends, of course, upon the details of the lineshapes of the interacting ions. If $g_A(E)$ and $g_B(E)$ are normalized Gaussians with HWHM widths $E_{1/2}$ (i.e., $E_{1/2} = \pi\hbar\Delta\nu_H$ where $\Delta\nu_H$ is the FWHM homogeneous linewidth) and separation ϵ, then

$$I(\epsilon) = \int \frac{\ln 2}{\pi E_{1/2}^2} \exp\left\{ - \left(\ln 2 / E_{1/2}^2 \right) [(E+\epsilon/2)^2 + (E-\epsilon/2)^2] \right\} dE$$

$$= \frac{1}{E_{1/2}} \left(\frac{\ln 2}{2\pi} \right)^{1/2} \exp\left(-\epsilon^2 \ln 2 / 2E_{1/2}^2 \right) . \tag{32}$$

If there were no inhomogeneous broadening ($\epsilon = 0$), $S(0.25\%) = 0.30$ for $t = 1$ μs and $\Delta\nu_H = 130$ MHz, i.e., the characteristic transfer time would be a few microseconds. Of course, in the electric field switching experiment the inhomogeneous width (FWHM) was $\Delta\nu_I \cong 9$ GHz. The integral in Eq. (30)

should be replaced by

$$\mathrm{I} = \int \sqrt{\mathrm{I}(\epsilon)}\, g_{\mathrm{I}}(\epsilon)\, d\epsilon \tag{33}$$

where $g_I(\epsilon)$ is the normalized inhomogeneous distribution function. Since $\Delta\nu_H << \Delta\nu_I$, $g_I(\epsilon)$ is approximately constant where $I(\epsilon)$ is nonnegligible, i.e.,

$$\mathrm{I} \cong g_{\mathrm{I}}(0) \int \sqrt{\mathrm{I}(\epsilon)}\, d\epsilon = \left(\frac{\ln 2}{2\pi}\right)^{1/4} \frac{2}{\sqrt{\pi\hbar\Delta\nu_H}} \frac{\Delta\nu_H}{\Delta\nu_I} \tag{34}$$

at the peak of the inhomogeneous line. Then Eq. (30) yields S = 8.3% for t = 100 μs, compared with the 3.5% measured. The discrepancy may result from uncertainties in the widths and distribution functions of the homogeneous and inhomogeneous broadening. For example, the homogeneous widths measured by hole burning (Ref. 21) and electric-field increment (Ref. 15) experiments are much larger ($\cong$100 MHz) than those measured by photon echos (Ref. 22). The hyperfine interaction between the Cr^{3+} ions and the Al nuclei complicates the problem so that the homogeneous distribution is neither Gaussian nor Lorentzian. At any rate the calculated transfer exceeds that observed, so we *conclude that the dipole-dipole transfer mechanism with no adjustable parameters quantitatively accounts for the absolute magnitude of the observed nonradiative resonant transfer as well as its square-root dependence upon time.*

E. CONCLUSIONS

The direct measurements of ion-ion nonradiative resonant energy transfer in ruby for Cr^{3+} concentrations less than 1% show that it is dipole-dipole and slow (about half of the single ions transfer their excitation in the 3.7-ms R_1 lifetime for 0.25 wt.%). Electric field experiments show that nonradiative resonant transfer between inequivalent A and B ions is slow and dipole-dipole and that the inhomogeneous broadening is microscopic. Fluorescence-line-narrowing experiments show that nonradiative resonant transfer between ions in zero electric field is also slow. The growth of nonradiative nonresonant transfer occurs over the entire inhomogeneous line showing again the microscopic character of the inhomogeneous broadening. Finally the calculated transfer via a dipole-dipole interaction is in good agreement with that observed. In summary, direct measurements and calculations are consistent with slow dipole-dipole resonant nonradiative transfer.

III. INDIRECT MEASUREMENTS OF NONRADIATIVE RESONANT ION-ION ENERGY TRANSFER IN RUBY

In Section II it was emphasized that the direct observations of nonradiative resonant ion-ion transfer show that it is slow and dipole-dipole. The measurements which seem to conflict with this conclusion all involve observations of fluorescence from traps, i.e., pairs of Cr^{3+} ions close enough together to substantially modify each other's energy levels. For example, the fourth-nearest-neighbor pair (N_2) has an emission line at 7009 Å at 77 K corresponding to the 6934 Å line of an isolated ion. From the ratio of N_2 to R_1 fluorescence and a comparison of their decay curves, several workers[2,14,19,23,24] have concluded that ion-ion transfer must be much faster than found in Section II. Others have concluded that it is slow.[4,5,15] In this section it will be shown that the trap fluorescence data are consistent with the conclusions of Section II.

A. TRAP FLUORESCENCE DECAYS

Imbusch[19] noted that for intermediate concentrations (≃0.2%) and pulsed excitation the N_2 decay rate approached that of the R_1 line (≃10 ms for his thick, strongly radiatively-trapped samples) after 5 to 10 ms. Selzer et al.[14] found similar decay curves using powdered samples to largely eliminate the trapping. Chu et al.[15e] obtained similar curves in thin platelet samples with weak trapping. The rapid ion-ion transfer hypothesis was a logical explanation for the observations, since a rate equation analysis shows that such a hypothesis leads to nearly equal long-time decay rates for the R_1 and N_2 lines. But Section II rules out rapid ion-ion transfer. This section will show that the shape of the N_2 decay can be explained assuming dipole-dipole ion-trap transfer and no ion-ion transfer.

The results of Section II D for ion-ion transfer can be used for comparing the N_2 and R_1 decay curves. With no electric field, Eq. (3) can be written as

$$\Gamma_{II} = \frac{2\pi}{\hbar} |M|^2 \int_{-\infty}^{\infty} g_I(E) g_I'(E) dE \tag{35}$$

for ion-ion transfer and as

$$\Gamma_{IT} = \frac{2\pi}{\hbar} |M|^2| \int_{-\infty}^{+\infty} g_I(E) g_T(E) dE \tag{36}$$

for ion-trap transfer. In Eq. (36) the integral may be dominated by the phonon-assisted tail of the ion and absorption lines in $g_T(E)$.

Let I(t) be the probability that an ion at r=0, which was excited at t=0, is still excited at t. Let T(t) be the probability that a trap at r is excited at time t. Then:

$$\frac{dI}{dt} = -\Gamma_I I - \Gamma_{IT} I , \tag{37}$$

and

$$\frac{dT}{dt} = -\Gamma_T T + \Gamma_{IT} I , \tag{38}$$

where Γ_I and Γ_T are the radiative decay rates of an ion and trap, respectively. Note that ion-ion transfer is being neglected. The ion fluorescence signal F_I is then proportional to

$$I(t) = I(0)e^{-(\Gamma_I+\Gamma_{IT})t} \tag{39}$$

wheres the trap fluorescence signal F_T is

$$F_T(t) = \int_0^\infty d^3r\ T(r,t) . \tag{40}$$

Then

$$\frac{dF_T}{dt} = \int_0^\infty d^3r\ \frac{dT}{dt} = \int_0^\infty d^3r[-\Gamma_T T + \Gamma_{IT} I]$$

$$= -\Gamma_T F_T + I(0) \int_0^\infty d^3r\ e^{-\Gamma_{IT}t}\ \Gamma_{IT} e^{-\Gamma_I t} . \tag{41}$$

Assuming that the ion-trap transfer is also dipole-dipole,

$$\Gamma_{IT} = \Gamma_0/r^6 . \tag{42}$$

Let $v^2 \equiv \Gamma_0 t/r^6$, then

$$\frac{dF_T}{dt} = -\Gamma_T F_T + I(0)\ \frac{4\pi}{3}\ \sqrt{\Gamma_0/t} \int_0^\infty e^{-v^2}\ dv\ e^{-\Gamma_I t}$$

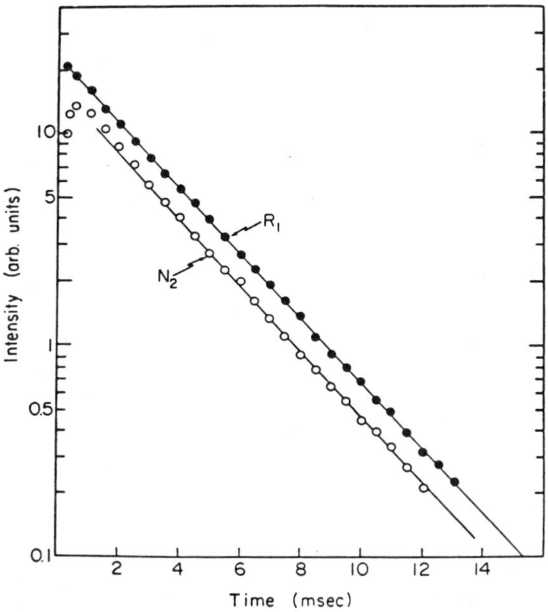

Fig. 7. Semilog plot of broadband (1-nm resolution) R_1 and N_2 lifetimes for a 0.51 at.% powdered sample at 5 K. The straight lines represent a decay time of 2.8 ms. Laser excitation is slightly on the high-energy side of line center. Ref. 14.

$$= -\Gamma_T F_T + \frac{2\pi}{3}\, I(0)\, \sqrt{\Gamma_0 \pi / t}\; e^{-\Gamma_I t} \, . \tag{43}$$

Multiplying by $e^{\Gamma_T t}$:

$$\frac{d\left(e^{\Gamma_T t} F_T\right)}{dt} = \frac{2\pi}{3}\, I(0)\, \sqrt{\Gamma_0 \pi}\; \frac{e^{(\Gamma_T - \Gamma_I)t}}{\sqrt{t}} \tag{44}$$

with $s^2 \equiv (\Gamma_T - \Gamma_I)t$

$$F_T = \frac{4\pi}{3}\, I(0)\, \frac{\sqrt{\Gamma_0 \pi}}{\sqrt{\Gamma_T - \Gamma_I}}\; e^{-\Gamma_T t} \int_0^{\sqrt{(\Gamma_T - \Gamma_I)t}} e^{s^2}\, ds \, . \tag{45}$$

A curve of the form of Eq. (45) has been least-squares best fit to the data of Selzer et al.[14] shown in Fig. 7 and repeated in Fig. 8. The best-fit curve is the

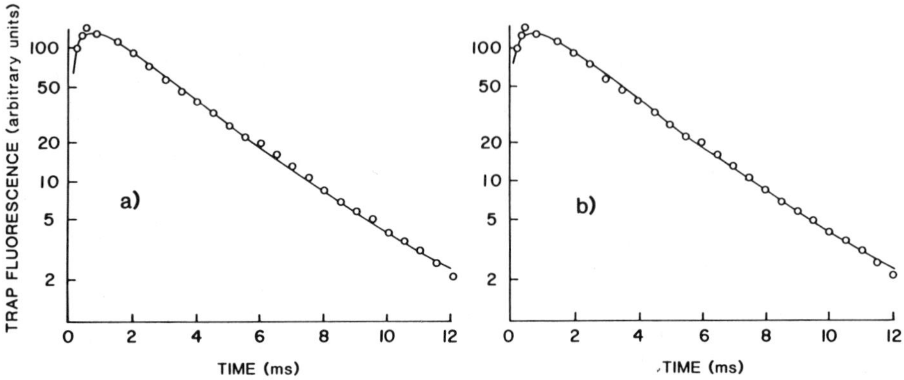

Fig. 8. Trap fluorescence decay data and theoretical fits. The circles are observed points from Fig. 7 (Selzer et al., Ref. 14). (a) Solid curve is a best fit of Eq. (45), i.e., with no radiation trapping. The best-fit parameters are $\Gamma_I = (3.72\ \text{ms})^{-1}$ and $\Gamma_T = (1.08\ \text{ms})^{-1}$. (b) Solid curve is a best fit with radiation trapping included; best fit parameters are $\Gamma_I = (3.34\ \text{ms})^{-1}$, $\Gamma_T = (1.08\ \text{ms})^{-1}$, and 5.4% radiation trapping. Ref. 15e.

solid line in Fig. 8a (the best-fit values were $\Gamma_I = (3.72\ \text{ms})^{-1}$ and $\Gamma_T = (1.08\ \text{ms})^{-1}$). Note that the N_2 decay data are described well by this simple ion-trap dipole-dipole transfer model. The fitted lifetime for the trap is perhaps 50% longer than the accepted[14] value. Perhaps the inclusion[25] of single-single transfer would improve the fit.

For long times, i.e., $\sqrt{(\Gamma_T - \Gamma_I)t} >> 1$,

$$\int_0^x e^{s^2} ds \simeq \int_0^\delta d\delta\ e^{x^2 - 2x\delta} \simeq \frac{e^{x^2}}{2x}, \tag{46}$$

so that

$$F_T \propto \frac{\exp(-\Gamma_I t)}{\sqrt{t}}. \tag{47}$$

That is, the trap fluorescence approximately follows the ion decay for large times. This conclusion also follows when radiation trapping is included. This corresponds to the fact that if one waits long enough the only ions feeding the traps are far away and weakly coupled. Thus the shape of the trap fluorescence decay curves of Imbusch,[19] Selzer et al.,[14] and Chu et al.[15] can be understood with the dipole-dipole model. Note that the model agrees with the data because of the inclusion of the range dependence of the transfer (in a continuum approximation). The need to do this was mentioned by Imbusch[19] and many[26,14,10] since. Our calculation follows Auerbach, Robinson and Zwanzig,[25] who found that with no donor transfer the acceptor fluorescence rises faster and peaks earlier.

Radiation trapping can be included by dividing the ions into ions R not near a trap and ions I near traps, so that Eqs. (37) and (38) become

$$\frac{dR}{dt} = -\Gamma_I R + \sigma R \tag{48}$$

$$\frac{dI}{dt} = -\Gamma_I I + \sigma R - \Gamma_{IT} I \tag{49}$$

$$\frac{dT}{dt} = -\Gamma_T T + \Gamma_{IT} I \tag{50}$$

where $\Gamma_I = (3.7 \text{ ms})^{-1}$ and $\Gamma_T = (0.7 \text{ ms})^{-1}$ are the isolated ion and trap radiative decay rates, σ is trapping rate, and $\Gamma_{IT} = \Gamma_0/r^6$. Then $R = R(0) \exp(-\Gamma_\rho t)$ where $\Gamma_\rho = \Gamma_I - \sigma$ is the observed R_1 decay rate. Then[15e]

$$F_T = \int_0^\infty d^3r\, T(r,t)$$

$$= \text{const.} \left\{ \exp(-\Gamma_T t) \int_0^{\sqrt{(\Gamma_T-\Gamma_I)t}} \exp(s^2)\, ds \left[1 - \frac{\sigma}{\Gamma_T - \Gamma_\rho} \right] \right.$$

$$\left. + \frac{\sqrt{\sigma(\Gamma_T-\Gamma_I)}}{\Gamma_T - \Gamma_\rho} \exp(-\Gamma_\rho t) \int_0^{\sqrt{\sigma t}} \exp(-s^2)\, ds \right\} . \tag{51}$$

A best-fit of Eq. (51) to the Selzer and Yen data yields $\Gamma_I = (3.34 \text{ ms})^{-1}$, $\Gamma_T = (1.08 \text{ ms})^{-1}$, and 5.4% radiation trapping; see Fig. 8b. If one forces $\Gamma_I = (3.7 \text{ ms})^{-1}$ and $\Gamma_T = (0.9 \text{ ms})^{-1}$ the best-fit trapping is 4.8% and the least-squares fit is equally good.

Experimentally the magnitude of the N_2 fluorescence is comparable to that of R_1 for 1% concentration; it is difficult to determine if our model agrees because of insufficient information on $g_I(E)$ in the wings and $g_T(E)$.

Imbusch[19] estimated that dipole-dipole transfer was too weak making reasonable assumptions about $g_I(E)$ and $g_T(E)$. He did not know that there is an energy level of the fourth nearest neighbor trap which is almost coincident with the $\bar{E}$ state of a single ion.[27] Although we have insufficient information to confirm that this resonance enhancement of the single-trap dipole-dipole transfer accounts for the anomalously large fourth-nearest-neighbor fluorescence we regard it as likely. Furthermore the dipole-dipole transfer accounts well for the fluorescence from the third nearest neighbor which has no coincident resonance.[15e] Also the fluorescence-line-narrowing experiments showed that at 1% concentration the transfer is phonon-assisted; it occurs over the entire inhomogeneous line. Just why such is the case is an interesting question, but the large N_2 fluorescence at 1% should no longer be used as evidence for fast resonant ion-ion transfer. The large phonon-assisted transfer probably also accounts for the reduction in R_1 lifetime at high concentrations.

B. OTHER TRAP FLUORESCENCE EXPERIMENTS

Gerlovin[4] repeated Imbusch's decay curve measurements using thin polycrystalline samples to avoid radiation trappings and using broadband flashlamp excitation. He found departures from a single exponential R_1 decay and concluded from his analysis that hardly any energy migration among single ions occurs for 0.4 to 2 at.%. Fast-transfer advocates question whether Gerlovin's results were affected by direct excitation of traps by the broadband pump. That objection was avoided by a number of experiments using narrowband lasers on the R_1 line.[2,7,14,18]

Heber and Murmann[5] have performed heat-pulse experiments in which they estimate $\Gamma_{II} = (250\ \mu s)^{-1}$ and $\Gamma_{IT} = (4\ \mu s)^{-1}$ for 1.35% ruby at 5.3 K and concluded that both processes are phonon assisted. These conclusions are consistent with fluorescence line narrowing results.[15] The possibility that the technique artificially selects ions which are close to traps making $\Gamma_{IT} >> \Gamma_{II}$ has prevented their results from receiving more weight.

Even though they may interpret their data differently, we claim that the data of Imbusch,[19] Gerlovin,[4] Heber and Murmann,[5] Selzer et al.,[14] Chu et al.,[15] and Jessop and Szabo,[18] as well as the transient grating experiments, are all consistent with slow dipole-dipole transfer between single ions. Monteil and Duval[23] have conducted energy transfer studies using 20-Hz modulation of cw light exciting the $^4A_2 \rightarrow {}^4T_2$ band and measured the phase difference between R_1 and N_2 fluorescence. They conclude[23c] that the transfer is mainly an exchange interaction in the range of 0.1 to 1.4% concentration, but they do not rule out dipole-dipole.[16] Recently they have performed an experiment to test the hypothesis that transfer may be slow between A and B ions but fast among the A's and among the B's. They applied a uniaxial stress to separate the ions into two sublattices 1 and 2 different from the electric field sublattices A and B. They find that the effect of the electric field (0.21 cm^{-1} splitting) on the N_2/R_1 ratio is ~10% for 0.1% concentration, much less than the effect (factor of 2 reduction) on the ratio for a stress causing a 0.1 cm^{-1} splitting. They conclude[16] that "there exists fast resonant transfer in each sublattice differentiated by the electric field and that the Anderson transition is possible in ruby." They do not comment on the fact that fast A↔A transfer would cause a fourth peak to grow rapidly in FLN measurements in zero electric field; it does not. They do not discuss whether the use of 580-nm light for excitation may have produced large electric fields[28] so that the applied external field could have no further effect. All that they showed is that the N_2/R_1 ratio decreased with stress and not with electric fields under their experimental conditions. They did not show that A↔A transfer is fast; FLN data show that is is slow (approximately milliseconds).

C. CONCLUSIONS

There is no trap fluorescence experiment which is inconsistent with the conclusion in Section II E that the ion-ion resonant nonradiative transfer is dipole-dipole and slow. Inclusion of the range dependence of the ion-trap transfer interaction is the key factor in explaining the time dependence of the trap fluorescence.

IV. SEARCHES FOR AN ANDERSON TRANSITION

A. RUBY

The logic for using ruby for a test of Anderson's idea was quite straightforward. Anderson suggested that there is a critical concentration, ρ_{crit}, below which there is no diffusion and the excitation is spatially localized. The requirement on the interaction $M_{ij}(r_{ij})$ giving rise to the transfer of excitation from an ion in the $E(^2E)$ state at site i to an ion in the $^4A_2$ ground state at site j is that M_{ij} fall off faster than r_{ij}^{-3}. Imbusch[19] had concluded from the magnitude and shape of the trap fluorescence decay that the single-trap transfer could not be dipole-dipole or exchange and proposed quadrupole-quadrupole. Birgeneau[26] calculated that anisotropic exchange would dominate over quadrupole-quadrupole. Then Lyo,[29] using the exchange interaction, estimated ρ_{crit} to be roughly 0.3%. Of course, if M_{ij} is dipole-dipole as shown in Sections II and III, the logic breaks down at the beginning. We believe that is the case; otherwise the suggestion was a good one and might find application in other systems.

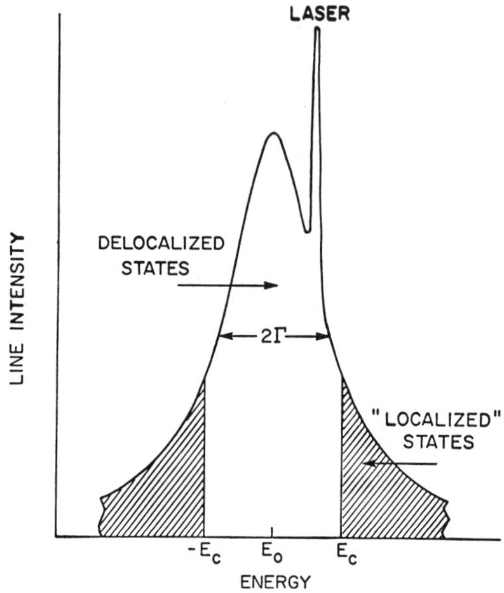

Fig. 9. Schematic division of inhomogeneously broadened line, above the critical concentration for an Anderson transition, into regions of "localized" and delocalized states. Laser light is used to excite these states selectively. Ref. 2.

The idea of using ruby to look for an Anderson transition hinges on the use of the inhomogeneous broadening to vary the concentration; see Fig. 9. Ions within a homogeneous width are assumed to be able to undergo resonant nonradiative transfer. Since the homogeneous width is about 100 times narrower than the inhomogeneous width, the density of Cr^{3+} ions can be chosen so that the central part of the inhomogeneous line is above ρ_{crit}. As a laser is tuned across the R_1 inhomogeneous linewidth, one looks for a sudden drop in N_2 fluorescence, assuming that the ion-ion transfer will experience a mobility edge as ρ drops below ρ_{crit}. (Of course, we now believe that the ion-ion transfer is dipole-dipole and the traps are excited mostly by ion-trap transfer (which may be dipole-dipole) with little ion-ion transfer; see Section II D.)

Koo, Walker, Geschwind[2] reported such an experiment and Fig. 10 summarizes their observed dependence of the ratio of trap to single fluorescence upon laser detuning. The "breaks" were interpreted as mobility edges of an Anderson transition. The breaks do not behave according to the simple picture given above since the ratio of concentration to linewidth is close to 0.2 for all three samples so the break should occur at the same normalized distance from line center in Fig. 10. The experiment was hampered by the fact that the ruby laser could be tuned only to the low energy side of line center. The tuning was slow, requiring magnetic field and temperature changes.

Chu, Gibbs, and Passner[7] repeated the Koo et al. experiment with a cw dye laser which could be quickly tuned across the entire R_1 line. In an attempt to reproduce the Koo et al. results they used a wide variety of concentrations and sizes of powders and platelets. They found a frequency dependence to the N_2/R_1 ratio which was concentration independent, having the roughly universal shape shown in Fig. 11. They concluded that their N_2/R_1 data provided no evidence for a sharp mobility edge.

Although Selzer et al.[12–14] interpreted their fluorescence-line-narrowing data as strongly supporting fast resonant transfer between single ions (a conclusion shown to be unjustified in Sections II D and III A), they report no sudden change in ion-ion transfer as a function of laser tuning. Similarly Chu et al.[15] report no mobility edge in their electric-field and FLN studies.

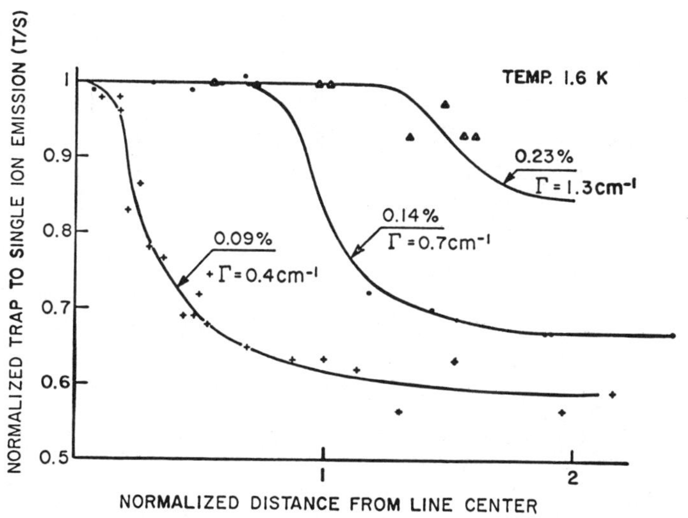

Fig. 10. Normalized trap (N_2 line) to single-ion emission (R_1 line) in ruby as a function of laser excitation in different regions of the low-energy side of the R_1 line. The observed breaks suggest mobility edges separating delocalized states in the central region from the localized states beyond the break. Ref. 2.

Jessop and Szabo[30] looked for a change in homogeneous linewidth or coherence time measured by photon echoes and found no mobility edges in samples up to 0.1 wt.%. Since their inhomogeneous linewidths were only 2 to 5 GHz compared with 24 to 78 GHz for Koo et al., they effectively went up to 1% concentration in the Koo et al. experiment and saw no edge. Monteil and Duval[16] recently proposed that ruby is still a good candidate for an Anderson transition using one of the electric-field sublattices, but FLN experiments show that resonant non-radiative transfer is slow within those sublattices also (see Section III B).

The experiments and conclusions discussed so far have been for Cr concentrations of about 1% or lower. For an ion density $\rho = 2.77\ \text{Å}/c^{1/3}$ where c is the concentration, the radius R for which the probability of finding no Cr ions is 0.5 is given by $\exp[-4\pi R^3\rho/3] = 0.5$. Then, assuming $\Delta\nu_I/\Delta\nu_H = 100$ and $c = 1\%$, $R = 33\ \text{Å}$, so that the conclusion that the interaction is electric dipole-dipole was made for shorter interaction distances. Clearly a shorter range interaction could dominate the transfer for sufficiently close ions. If two ions are brought closer than about 6 Å they form a pair or trap with an absorption line outside the R_1 inhomogeneous line. When several ions cluster together they must be treated as a single quantum mechanical system. If the cluster is isolated from other clusters it might constitute an Anderson localized state. A mobility edge might occur at sufficiently high density when the clusters merge into one cluster throughout the crystal. This transition would deserve to be called an Anderson transition rather than percolation because a quantum mechanical description is required. Coherences requiring a density matrix approach would be as important as diagonal (rate equation) terms. Of course, quantum mechanical

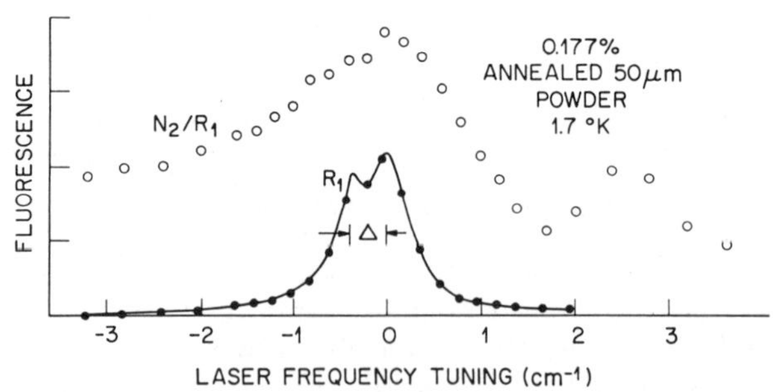

Fig. 11. "Universal" curve of N_2/R_1 ratio versus laser frequency. Note that the frequency region to the red of line center bears some resemblance to the Koo et al. data. Ref. 7.

clusters could also become delocalized by incoherent transfer, i.e., by percolation.

Huber and Ching[31] have suggested that such an Anderson transition might occur in ruby for 10% concentration. From our measurements at concentrations around 1% we point out several discouraging facts. At 1% concentration excited ions transfer almost exclusively to traps. The small amount of ion-ion transfer is almost exclusively nonresonant phonon-assisted transfer over the entire inhomogeneous line. The R_1 decay curves have no measurable long-time component decaying with the isolated-ion radiative time of 3.7 ms; i.e., there are no isolated single ions, and the delocalization is complete. Upon increasing the concentration from 0.25% to 1.2% we find that the peak R_1 absorption coefficient remains almost unchanged, i.e., the inhomogeneous width is being increased. If the homogeneous width remains the same the distance to the nearest resonant ion also remains the same. The distance from an excited single to the nearest trap clearly decreases. In fact at 1% concentration the radius R for which there is a 50% probability that there is not another resonant single is about 32.7 Å assuming $\Delta\nu_I/\Delta\nu_H = 100$. R for the fourth-nearest-neighbor trap is only 8.37 Å. Thus the R^{-6} factor in the transition rate is over 3000 times larger for the trap. This almost certainly more than compensates for the smaller overlap integral for ion-trap than for ion-ion transfer and explains the dominance of ion-trap transfer. At higher temperatures, suggested by Huber and Ching to suppress the delocalizing effects of the dipolar interaction, phonon-assisted transfer will be even faster.

In summary, although the data reported by Koo et al. have not been explained, none of the other experiments has revealed a mobility edge in ruby. Those experiments both repeated the Koo et al. experiment and searched for an edge by observing ion-ion transfer directly without using traps as detectors. Consistent with the slow dipole-dipole character discussed in Section II, ruby does not exhibit an Anderson transition for concentrations up to 1%.

B. OTHER OPTICAL SYSTEMS

1. $LaAlO_3:Cr^{3+}$

Siebold and Heber[32] have studied energy transfer in $LaAlO_3:Cr^{3+}$ in which the Cr^{3+} occupies the Al^{3+} sites. The spectra of the single ions and pairs are very similar to those in ruby, but $LaAlO_3:Cr^{3+}$ has a number of advantages. The fluorescence lifetime of the single ions is 81 ms, much longer than the 3.7 ms in ruby; energy transfer can be observed over much longer times. The ratio of the fluorescence lifetimes of the donors and acceptors is 48 compared with 5 in ruby, giving much better time resolution to transfer transients.

Reabsorption is practically negligible. Only the first-nearest-neighbor pairs accept excitation energy from the single ions. Apparently unaware of the electric-field transfer studies in ruby, they found negligible energy diffusion in $LaAlO_3:Cr^{3+}$ also. From the absence of energy diffusion, the weak temperature dependence, and the small donor-acceptor energy mismatch, they conclude that the energy transfer from the single ions to the pairs is dominated by a nonradiative, one-phonon assisted, direct transfer process.

2. $CaWO_4:Sm^{3+}$

Hsu and Powell[33] have studied energy transfer among Sm^{3+} ions in $CaWO_4$ crystals at a fixed 2 at.% concentration of Sm. The dependence of the fluorescence on time and temperature shows the interaction to be electric quadrupole-quadrupole and the interaction strength to be much less than the inhomogeneous linewidth of the transition. They report energy localization at low temperatures and diffusion at high temperatures when homogeneous broadening of the transitions exceeds the inhomogeneous broadening. Coherent transfer should be temperature independent until the homogeneous width exceeds the interaction width, and then the transfer becomes incoherent. Hsu and Powell may have observed a phonon-assisted ion-ion or ion-trap delocalization.

3. MIXED ORGANIC CRYSTALS

There have been several studies of energy transfer in isotopically substituted molecular crystals which are relevant to a search for a clear Anderson transition.[34] The triplet excitations in mixed organic crystals are particularly attractive because they have tightly bound Frenkel excitons, no dipole-dipole terms, well-known interactions, long lifetimes, random nature of substitutional disorder, ease of spectral resolution, no complications due to backtransfer, no radiative trapping or phonon bottleneck. Smith, Mead, and Zewail[35] studied two isotopic mixed systems: DBN (1,4-dibromonaphthalene) and phenozine solids. Colson, George, Keyes, and Vaida[36] have studied singlet and triplet exciton percolation in benzene isotopic mixed crystals. Kopelman et al. have concentrated on naphthalene.[37]

Klafter and Jortner[38] "conclude that the currently available experimental information on triplet electronic energy transfer in low-temperature isotopic impurity bands is compatible both with the simple kinetic picture as well as with the Anderson localization scheme." Ahlgren and Kopelman[39] believe that the electronic energy transport critical concentrations and their temperature dependence are not accounted for by the Klafter-Jortner models of an Anderson transition. "... the weight of the experimental evidence seems to be against the static Anderson transition models." Francis and Kopelman[34] state that "the occurrence of an Anderson-Mott transition (mobility edge) has not yet been established, either by rigorous theory or by reliable calculations, and experimental verifications have proven elusive. ... In summary, the naphthalene triplet excitation transport experiments, at 1.7 and 4.2 K, are inconsistent with an Anderson-Mott mobility edge model. However, it is obvious that they are consistent with a kinetic model. We claim here that the latter is also true for all the other available triplet excitation experiments, based on data available to date.[37,35] We also claim that the available evidence for *any* Anderson-Mott-like transition in organic crystals[35] is ambiguous and should be tested by the methods mentioned above (use of deep-trap acceptors, variation of their concentration and species, temperature studies with deep-trap acceptors and kinetic studies)." Their percolation critical exponents indicate two-dimensional energy transport for the triplet naphthalene exciton.

4. SEMICONDUCTORS

Cohen and Sturge[40] have studied exciton emission in the mixed semiconductor CdS_xSe_{1-x}. Random fluctuations in the composition produce concomitant fluctuations in the band gap. The exciton states are thus inhomogeneously broadened by an amount which may exceed the energy of the long-wavelength acoustic phonons which can couple to the exciton and give rise to spectral diffusion. "We have shown that fluorescence line narrowing occurs under selective excitation of the low-energy side of the intrinsic exciton absorption line in CdS_xSe_{1-x} below about 10 K. We have argued that this is evidence for exciton localization, on a time scale of the order of nanoseconds, by alloy fluctuations. There is at present no satisfactory theory of such localization and of its spectral consequences, and our arguments are essentially qualitative. The abrupt disappearance of the narrowed line above a certain energy suggests the existence of an effective mobility edge."

Higashi and Kastner[41] question whether or not the excitation-energy dependence of the photoluminescence bandwidth in $a{-}As_2S_3$ should be interpreted as establishing the position of the mobility gap; Bosch[42] thinks not.

V. CONCLUSIONS

In Section II it was concluded that the nonradiative nonresonant ion-ion transfer in ruby is quantitatively accounted for by a slow dipole-dipole interaction between ions within a homogeneous width and that the inhomogeneous broadening is microscopic. Section III concluded that trap fluorescence data can be explained by a slow dipole-dipole ion-trap range-dependent transfer without any ion-ion transfer, so that trap data do not demand fast ion-ion transfer as previously thought. In Section IV it was noted that one does not expect a mobility edge for slow or dipole-dipole transfer; several experiments have found no evidence for one.

Most of the above results were for concentrations around 0.1 to 0.25%. As the concentration is increased above 0.25% the inhomogeneous width increases so that the peak absorption and resonant transfer stay about the same. For concentrations of 1% or more, the phonon-assisted nonradiative transfer dominates over the resonant nonradiative transfer. Thus, one cannot achieve several resonant nonradiative transfers in the R_1 lifetime by increasing the concentration.

In summary, the search for an Anderson transition in ruby has elucidated the mechanism for ion-ion nonradiative resonant transfer and led to the conclusion that the transfer interaction does not satisfy the necessary conditions for such a transition.

Reference is made to other optical studies related to an Anderson transition; the mixed organic crystals appear particularly promising.

REFERENCES

1. P. W. Anderson, Phys. Rev. **109**, 1492 (1958); Comments Solid State Phys. **2**, 193 (1970).
2. J. Koo, L. R. Walker, and S. Geschwind, Phys. Rev. Lett. **35**, 1669 (1975).
3. Talk at International Quantum Electronics Conference, Atlanta, 1978.
4. I. Ya Gerlovin, Fiz. Tverd. Tela. (Leningrad) **16**, 607 (1974) [Sov. Phys. - Solid State **16**, 397 (1974)].
5. J. Heber and H. Murmann, Z. Phys. B **26**, 145 (1977).
6. See, for example, Ref. 14 below.
7. S. Chu, H. M. Gibbs, A. Passner and S. Geschwind, Bull. Am. Phys. Soc. **24**, 894 (1979); S. Chu, H. M. Gibbs, and A. Passner, Phys Rev. B **24**, 7162 (1981).
8. H. J. Eichler, Opt. Acta **24**, 631 (1977); H. J. Eichler, J. Eichler, J. Knof, and C. H. Noak, Phys. Status Solidi **52**, 481 (1979).
9. D. S. Hamilton, D. Heiman, J. Feinberg, and R. W. Hellwarth, Opt. Lett. **4**, 124 (1979).
10. P. F. Liao, D. M. Bloom, L. M. Humphrey, and S. Geschwind, Bull. Am. Phys. Soc. **24**, 586 (1979); P. F. Liao, L. M. Humphrey, D. M. Bloom, and S. Geschwind, Phys. Rev. B **20**, 4145 (1979).
11. A. Szabo, Phys. Rev. Lett. **25**, 924 (1970).
12. P. M. Selzer and W. M. Yen, Opt. Lett. **1**, 90 (1977).
13. P. M. Selzer, D. S. Hamilton, and W. M. Yen, Phys. Rev. Lett. **38**, 858 (1977).
14. P. M. Selzer, D. L. Huber, B. B. Barnett, and W. M. Yen, Phys. Rev. B **17**, 4979 (1978).
15. S. Chu, H. M. Gibbs, S. L. McCall, and A. Passner, (a) Bull. Am. Phys. Soc. **25**, 419 (1980); (b) J. Opt. Soc. Am. **70**, 633 (1980); (c) Phys. Rev. Lett. **45**, 1715 (1980); (d) S. L. McCall, S. Chu, H. M. Gibbs, and A. Passner, Bull. Am. Phys. Soc. **27**, 56 (1981); (e) to be published.
16. A. Monteil and E. Duval, preprint.
17. W. Kaiser, S. Sugano, and D. L. Wood, Phys. Rev. Lett. **38**, 858 (1977); A. Szabo and M. Kroll, Opt. Lett. **2**, 10 (1978).
18. P. E. Jessop and A. Szabo, (a) J. Opt. Soc. Am. **70**, 633 (1980); (b) Phys. Rev. Lett. **45**, 1712 (1980).
19. G. F. Imbusch, Phys. Rev. **153**, 326 (1967).
20. D. F. Nelson and M. D. Sturge, Phys. Rev. **137**, A1117 (1965).
21. A. Szabo, Phys. Rev. Lett. **27**, 323 (1971). P. E. Jessop, T. Muramoto, and A. Szabo, Phys Rev. B **21**, 926 (1980).
22. P. F. Liao and S. R. Hartmann, Opt. Commun. **8**, 310 (1973); A. Szabo and M. Kroll, Opt. Lett. **2**, 10 (1978).
23. A. Monteil and E. Duval, J. Phys. C: Solid State Phys. (a) **12**, L415 (1979); (b) Solid State Phys. **13**, 4565 (1980); (c) J. Luminescence **18/19**, 793 (1979); (d) to be published.
24. A. Monteil, Ph.D. Thesis, Université Claude Bernard-Lyon I, 1982.

25. R. A. Auerbach, G. W. Robinson, and R. W. Zwanzig, J. Chem. Phys. **72**, 3528 (1980).
26. R. J. Birgeneau, J. Chem. Phys. **50**, 4282 (1969).
27. P. Kisliuk, N. C. Chang, P. L. Scott, and M. H. L. Pryce, Phys. Rev. **184**, 367 (1969).
28. P. F. Liao, A. M. Glass, and L. M. Humphrey, Phys. Rev. B **22**, 2276 (1980).
29. S. K. Lyo, Phys. Rev B **3**, 3331 (1971).
30. P. E. Jessop and A. Szabo, Phys. Rev. B **26**, 420 (1982).
31. D. L. Huber and W. Y. Ching, Phys. Rev. B **25**, 6472 (1982).
32. S. Siebold and J. Heber, J. Lumin. **23**, 325 (1981).
33. C. Hsu and R. C. Powell, Phys. Rev. Lett. **35**, 734 (1975). M. J. Treadaway and R. C. Powell, Phys. Rev. B **11**, 862 (1975).
34. A. H. Francis and R. Kopelman, p. 241 in W. M. Yen and P. M. Selzer, eds., *Laser Spectroscopy of Solids* (Springer-Verlag, Berlin, 1981).
35. D. D. Smith, R. D. Mead, and A. H. Zewail, Chem. Phys. Lett. **50**, 358 (1977). See also D. Burland and A. H. Zewail, Adv. Chem. Phys. **40**, 369 (1979).
36. S. D. Colson, S. M. George, T. Keyes, and V. Vaida, J. Chem. Phys. **67**, 4941 (1977).
37. R. Kopelman, E. M. Monberg, F. W. Ochs, P. N. Prasad, J. Chem. Phys. **62**, 292 (1975). E. M. Monberg and R. Kopelman, Chem. Phys. Lett. **58**, 492 and 497 (1978).
38. J. Klafter and J. Jortner, J. Chem. Phys. **73**, 1004 (1980) and references therein. See also C. M. Soukoulis, J. Klafter, and E. N. Economou, Solid State Commun., to be published.
39. D. C. Ahlgren and R. Kopelman, J. Chem. Phys. **73**, 1005 (1980).
40. E. Cohen and M. D. Sturge, Phys. Rev. B **25**, 3828 (1982) and references therein.
41. G. S. Higashi and M. Kastner, Phys. Rev. Lett. **47**, 124 (1981) and **48**, 650 (1982).
42. M. A. Bosch, Phys. Rev. Lett. **48**, 649 (1982).

DISCUSSION

Optical Localization

D. L. Huber pointed out that the failure to detect a mobility edge in ruby was, in retrospect, not surprising. As part of an analysis of the effects of positional disorder on the metal-nonmetal transition in doped semiconductors, N. F. Mott[1] argued that the extended states will have disappeared even in the absence of diagonal disorder (i.e. zero strain broadening in the ruby problem) at the point where $\alpha n^{-1/3} \approx 2.5$. Here α is the range of the exponential coupling, $V_{ij} \sim \exp[-\alpha r_{ij}]$, between centers (ions) and n is the corresponding density. In ruby $\alpha n^{-1/3} \approx 17$ for $n = 0.4$ at.% Cr^{3+}, which is well within the fully localized regime. According to Mott's criterion we would expect an Anderson transition in ruby for a Cr^{3+} concentration in excess of 10 at.%. For further details see D. L. Huber and W. Y. Ching.[2]

J. H. Davies, Chairman
D. L. Huber, Discussion Leader

REFERENCES

1. N. F. Mott, J. Phys. (Paris) *37*, C4-301 (1976).
2. D. L. Huber and W. Y. Ching, Phys. Rev. *B25*, 6472 (1982).

PARTICIPANTS

Richard G. Brewer
IBM Research Lab
San Jose, California 95100, USA

B. Cavenett
Department of Physics
The University
Hull HU6 7RX, England

Alvin J. Cohen
Dept. of Geology
University of Pittsburgh
321 Old Engineering Hall
Pittsburgh, Pennsylvania 15260, USA

J. H. Davies
Cavendish Laboratory
Madingley Road
Cambridge CB3 0HE, England

E. A. Davis
Physics Department
University of Leicester
Leicester, England

Simon Depinna
Department of Physics
The University
Hull HU6 7RX, England

A. M. Elias
Centro de Espectrometria de Massa
Universidade de Lisboa
Avenida Rovisco Pais
1096 Lisboa Cedex, Portugal

S. R. Elliott
Dept. of Physical Chemistry
Lensfield Road
University of Cambridge
Cambridge, England

Paul A. Fleury*
Bell Laboratories
Murray Hill, New Jersey 07974, USA

Hyatt M. Gibbs
Department of Physics
University of Arizona
Tucson, Arizona, USA

Brage Golding*
Bell Laboratories
Murray Hill, New Jersey 07974, USA

M. Gunn
Cavendish Laboratory
Madingley Road
Cambridge CB3 0HE, England

Richard Harley
IBM Research Laboratory
San Jose, California 95100
and Clarendon Laboratory
Oxford, England

John Hegarty
Bell Laboratories
Murray Hill, New Jersey 07974, USA

D. L. Huber
Physics Department
University of Wisconsin
Madison, Wisconsin, USA

J. Jackle
Fachbereich Physik der
Universität Konstanz
D-7750 Konstanz, FRG

J. Joffrin
Université de Paris
Faculté-Sud
Orsay - Essonnes, France

John C. Knights
Xerox Palo Alto Research Center
Palo Alto, California 94304, USA

C. Laermans
Laboratorium voor Vaste Stofen
Hoge Druk-Fysika
Katholieke Universiteit Leuven
Celestijnenlaan 200 D
B-3030 Leuven, Belgium

A. R. Long
Dept. of Natural Philosophy
University of Glasgow
Glasgow G12,800, Scotland

Roger M. Macfarlane
IBM Research Laboratory
San Jose, California 95100, USA

Sir Neville Mott
Cambridge University
Cambridge, England

N. Neuroth
Schott Glaswerke
Hattenbergstrasse 10
Postfach 2480
D-6500 Mainz 1, FRG

Francis V. Pauwels
Air et Chaleur S.A.
Digue du Canal, 112-113
Brussels, Belgium

J. Orenstein
Bell Laboratories
Murray Hill, New Jersey 07974

A. Owen
Dept. of Electrical Engineering
University of Edinburgh
Edinburgh, Scotland

W. A. Phillips*
Cavendish Laboratory
Madingley Road
Cambridge CB3 0HE
England

K. F. Renk
Universität Regersburg
Institut für Physik III
8400 Regensburg, FRG

N. Rivier
Imperial College of
Science and Technology
The Blackett Laboratory
Prince Consort Road
London SW7 2BZ, England

John Ryan
Clarendon Laboratory
South Parks Road
Oxford, England

D. Schoemaker
Departement Natuurkunde
Universitaire Instelling Antwerpen
Universiteitsplein 1
B-2610 Wilrijk, Belgium

S. D. Smith
Department of Physics
Heriot-Watt University
Edinburgh, Scotland

W. J. Stewart
Plessey Research Limited
Caswell Towcester, Northants
England

Roger H. Stolen
Bell Laboratories
Holmdel, New Jersey, USA

P. Craig Taylor
Naval Research Laboratory
Washington, D. C. 20375, USA

M. L. Theye
Laboratoire d'Optique des Solides
Université P. et M. Curie
75230 Paris Cedex 05, France

R. G. Ulbrich
Institut für Physik der
Universität Dortmund
D-4600 Dortmund, FRG

R. Vacher
Université des Sciences
et Techniques du Languedoc
Place Eugene Batallon
34060 Montpellier Cedex, France

H. von Lohneysen
Physikalisches Institut
der Rhein.-Westf. Hochschule
D-5100 Aachen, FRG

Manfred von Schickfus
Max-Planck-Institut FKF
Heisenbergstrasse 1
Stuttgart, FRG

Denis Weaire
University College, Dublin
Dublin, Ireland

A. Wright
Department of Physics
University of Reading
Reading, England

William M. Yen
Physics Department
University of Wisconsin
Madison, Wisconsin 53700, USA

*Conference Organizing Committee.

INDEX